T0134974

Advanced Sciences and Technologies for Security Applications

Series Editor

Anthony J. Masys, Associate Professor, Director of Global Disaster Management, Humanitarian Assistance and Homeland Security, University of South Florida, Tampa, USA

Advisory Editors

Gisela Bichler, California State University, San Bernardino, CA, USA
Thirimachos Bourlai, Statler College of Engineering and Mineral Resources, West Virginia University, Morgantown, WV, USA
Chris Johnson, University of Glasgow, Glasgow, UK
Panagiotis Karampelas, Hellenic Air Force Academy, Attica, Greece
Christian Leuprecht, Royal Military College of Canada, Kingston, ON, Canada
Edward C. Morse, University of California, Berkeley, CA, USA
David Skillicorn, Queen's University, Kingston, ON, Canada
Yoshiki Yamagata, National Institute for Environmental Studies, Tsukuba, Ibaraki, Japan

Indexed by SCOPUS

The series Advanced Sciences and Technologies for Security Applications comprises interdisciplinary research covering the theory, foundations and domain-specific topics pertaining to security. Publications within the series are peer-reviewed monographs and edited works in the areas of:

- biological and chemical threat recognition and detection (e.g., biosensors, aerosols, forensics)
- crisis and disaster management
- terrorism
- cyber security and secure information systems (e.g., encryption, optical and photonic systems)
- traditional and non-traditional security
- energy, food and resource security
- economic security and securitization (including associated infrastructures)
- transnational crime
- human security and health security
- social, political and psychological aspects of security
- recognition and identification (e.g., optical imaging, biometrics, authentication and verification)
- smart surveillance systems
- applications of theoretical frameworks and methodologies (e.g., grounded theory, complexity, network sciences, modelling and simulation)

Together, the high-quality contributions to this series provide a cross-disciplinary overview of forefront research endeavours aiming to make the world a safer place.

The editors encourage prospective authors to correspond with them in advance of submitting a manuscript. Submission of manuscripts should be made to the Editor-in-Chief or one of the Editors.

More information about this series at http://www.springer.com/series/5540

Stefan Rass • Stefan Schauer • Sandra König
Quanyan Zhu

Cyber-Security in Critical Infrastructures

A Game-Theoretic Approach

 Springer

Stefan Rass (ID)
Universitaet Klagenfurt
Klagenfurt, Austria

Stefan Schauer (ID)
Austrian Institute of Technology GmbH
Wien, Austria

Sandra König
Austrian Institute of Technology GmbH
Wien, Austria

Quanyan Zhu
Tandon School of Engineering
New York University
Brooklyn, NY, USA

ISSN 1613-5113   ISSN 2363-9466   (electronic)
Advanced Sciences and Technologies for Security Applications
ISBN 978-3-030-46910-8   ISBN 978-3-030-46908-5   (eBook)
https://doi.org/10.1007/978-3-030-46908-5

This Springer imprint is published by the registered company Springer Nature Switzerland AG
The registered company address is: Gewerbestrasse 11, 6330 Cham, Switzerland

...to our families...

Contents

Part I
Introduction

Chapter 1
Introduction

The man who is a pessimist before 48 knows too much; if he is an optimist after it he knows too little.

M. Twain

Abstract This chapter opens the book by introducing the characteristics and particularities of critical infrastructures. Their existence and interplay forms a vital pillar of contemporary societies, and their protection is a top duty of governments and security research. Recent years have shown a paradigm shift of cyber-attacks from specific individual threat and attack scenarios, to a modern combination of various attack types and strategies to what we call an advanced persistent threat (APT) today. This term describes a diverse class of attacks that all share a set of common characteristics, which presents new challenges to security that demand urgent and continuous action by practitioners, researchers and every stakeholder of a critical infrastructure. The main focus of the book is describing game theory as a tool to establish security against APTs, and to this end, the introduction here starts with the abstract characteristics of an APT, showcasing them with a set of selected real-life documented cases of APTs that ends the chapter.

1.1 What are Critical Infrastructures?

In today's life, society is using and relying on numerous services, which satisfy the basic needs of people and guarantee a smooth flow of the everyday life. Among those services are the supply with basic resources (e.g., electricity, communication, heating, etc.), vital supplies (e.g., water, food, medicine and health care, etc.) as well as industrial goods (e.g., oil, gas, etc.) and general services (e.g., transportation, cash and financial services, etc.). The organizations and companies providing these services are called Critical Infrastructures (CIs) and represent the backbone of today's society. A core characteristic of a CI is that any failure of a CI, either in

S. Rass et al., *Cyber-Security in Critical Infrastructures*, Advanced Sciences and Technologies for Security Applications, https://doi.org/10.1007/978-3-030-46908-5_1

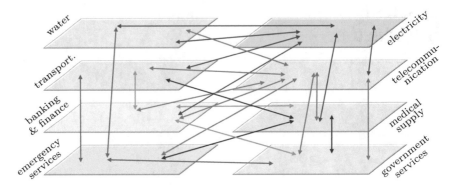

Fig. 1.1 Schematic representation of the complex network of CI

part or as a whole, will have considerable impact not only on the infrastructure itself but also on the social well-being of people [6].

Over the last decades, the interrelations among CIs have increased such that CIs have become more and more dependent on each other. These dependencies emerged due to the fact that national and international supply chains have become more important and many infrastructures depend on the resources other infrastructures provide (cf. Fig. 1.1). The most prominent example is power supply; almost every modern organization relies on a continuous availability of electricity such that CI can provide their main services to full capacity. A shortage in power distribution or even a blackout will either reduce the general output of the depending CIs or even make it impossible for the CIs to provide its service. Figure 1.1 shows a schematic illustration of such dependencies among CIs in multiple domains, which go far beyond the supply with electric power.

Further, CIs also heavily rely on the exchange of information due to the uprising digitalization of those infrastructures. Hence, a smooth operation of the communication infrastructure and a uninterrupted connection to the Internet has become a core requirement. For example, CIs are using Industrial Control Systems (ICSs) and Supervisory Control and Data Acquisition (SCADA) systems to remotely monitor physical systems within their premises and control multiple (even highly important) processes. If communication is interrupted, those systems might shut down immediately or simply cannot be reached any more (cf. real-life examples in Sect. 1.4) and the CI might have to operate the respective systems manually (if this is possible at all).

Due to these facts, the CIs within a nation or even within a region have evolved into a highly complex and sensitive infrastructure network with a variety of interdependencies among them, as illustrated in Fig. 1.1. Thus, incidents within a single CI can have far-reaching consequences, not only causing damages and financial losses within that CI but also affecting multiple other infrastructures as well as society as a whole. For example, a local power outage can cause a cellular antenna tower to shut down and interrupt the mobile communication network in that area. Such indirect consequences are often referred to as *cascading effects* and they impose a major problem. In general, they are hard to identify and thus difficult to assess as well as to prepare for. Therefore, during a risk analysis or the general risk management process of a CI, a particular focus lies on cascading effects. Chapter 2 provides more detailed information on how to model and describe the cascading effects among CIs and how to integrate them into a CI's risk analysis. To better illustrate how cascading effects can manifest themselves in real-life, a few examples of documented incidents over the last two decades follow in Sect. 1.4.

1.2 Security Challenges for Critical Infrastructures

In general, CIs have to face a broad variety of security challenges and threats in their daily business. These are ranging from natural disasters of various kinds over technical faults and human failure up to intentional attacks from the physical and the cyber domain. Depending on the geographical location of a CI, its business domain and the applied technical equipment, these threats can vary. In other words, a CI located at the sea (e.g., a sea port) has to deal with different natural disasters than an infrastructure located in the mountains (e.g., a dam). Further, an infrastructure from the chemical industry has to deal with other technical threats than an airport or a telecommunication provider.

Therefore, CI operators and the respective security officers need to become aware of their individual threat landscape, adapt to the requirements towards risk management and implement a large collection of security measures to be prepared against these threats or counter them. In this context, threats stemming from natural disasters and technical failures are quite static and change slowly (if at all) over time. However, threats from the cyber domain have increased massively, in particular over the last decade, and are considered to be highly dynamic due to the rapidly changing software landscape. In addition to the increasing digitization in CIs, the effects of a cyber attack can be tremendous (as briefly sketched in the following Sects. 1.4.3 and 1.4.4), such that cyber threats have become a major concern for CI operators.

Section 1.2.1 provides a short overview on the security challenges stemming from natural disasters and technical failures. A strong focus is here on the security challenges stemming from cyber threats in Sect. 1.2.2. The information presented in both sections is coming mainly from well-established threat catalogs, i.e., the European Anion Agency for Cybersecurity (ENISA) Threat Landscape [20] and the German Federal Office for Information Security (BSI) IT-Grundschutz Catalog [2].

1.2.1 Natural and Physical Threats

1.2.1.1 Natural Disasters

In the context of the physical domain, *natural disasters* represent threats with a potentially very high impact on an entire CI as well as on its environment. Examples are extreme weather conditions, storms or floods or very high temperatures but also fires or lightning strikes need to be considered. Whether an infrastructure is particularly exposed to some specific natural disaster depends on its geographical location and the environmental conditions.

1.2.1.2 Technical Failures

Critical infrastructures also operate numerous technical equipment, which is required to keep the services provided by the infrastructure running. These physical devices can be subject to different potential *failures and malfunctions*. The reasons for such a failure can be manifold, ranging from aging of material up to flaws in the operating software systems or even human failure during operation and maintenance of those systems. Such failures and malfunctions are usually limited to individual machinery (or parts thereof), data storage or network connections and can be fixed or repaired by the operator of the system (or some sub-contractor) in a certain amount of time.

1.2.1.3 Disruptions and Outages

A special case of a technical failure is a *disruption or outage* of an essential service required by the infrastructure to provide its operation. Any CI is depending on several of such critical services (e.g., power, gas, telecommunication etc.) due to the high interdependencies and interoperability in today's supply networks (cf. Fig. 1.1 above). Hence, any incident in one of those supply networks may lead to far reaching consequences affecting also the CI itself. Although many different reasons can cause an incident in one of these networks (even a cyber attack, for example), for the individual infrastructure as a consumer of the service provided by the network, this can be considered as a technical failure. Such an incident may range from a short-term loss of the service up to a large-scale network outage affecting the functionality of a CI for a longer period. In the context of CIs, the outage of the power supply as well as of communication networks can be seen as major concerns (as briefly mentioned in Sect. 1.1). Additionally, not only technical but also social networks need to be considered, such as, for example, the absence of personnel required to operate the machinery within a CI.

1.2.2 *Cyber Threats*

1.2.2.1 Distributed Denial of Service

With a lot of communication services using the Internet or at least being reachable from the outside world, shutting down such services by generating a high load of traffic is nowadays a common attack vector. This is achieved by creating massive amounts of connection requests targeted at a specific device, which is then no longer able to perform it tasks due to the continuous handling of the requests. Additionally, the current trend towards the Internet of Things (IoT), i.e., connecting small devices to the Internet with low or no security measures installed, makes it much easier for adversaries to take over control of a huge number of devices and thus creating a *Distributed Denial of Service (DDoS)* attack.

1.2.2.2 Malware and Ransomware

In recent years, malicious code and software, i.e., *"malware"*, has been on the rise and becoming a major threat for cyber systems. The variety of malware, its sophistication and the number of systems targeted by it are huge. Hence, it is very difficult to protect systems from being infected by malware. Additionally, a wide range of functions can be implemented by malware such that it can tamper cyber systems on the communication as well as on the operation level. Hence, the proper functioning of a CI can be affected (as described in Sect. 1.4.4), leading to a multitude of different effects. In the context of malware, these effects can range from the retrieval of information by a malicious party up to an injection of fraudulent information or commends into the communication channel. *Ransomware* on the other hand follows a different principle, i.e., it encrypts the entire infected system or crucial data stored on it and demands a ransom (often some amount of Bitcoin) in exchange for the decryption key.

1.2.2.3 Spear Phishing Attacks

Another threat vector that has gained popularity over the last years are *phishing mails*. Such emails appear to the recipient as genuine mails but contain either a link pointing to a malicious website or an attachment (e.g., a Portable Document File (PDF) or Office document) equipped with some malicious code. In both cases, the goal of the attacker is to trick the recipient into clicking on the link or opening the attachment. In this case, a malware or ransomware is loaded onto the recipient's system and is then causing additional damage (see also examples in the following section). In general, such phishing mails are not targeted and sent to a large number of recipients, which makes it easier for an observant user to identify the attempt. However, some phishing attacks are more elaborated and targeted to a certain organization or even to one specific user or a specific user group within an organization. Such a targeted attack is then called *spear phishing attack*. In that case, it is much harder even for a trained user to identify the mail as a phishing attempt. Still, to craft such an attack, a lot more effort is required. In particular, social engineering (see the following paragraph) is applied as an important tool in the preparation phase to tailor the mail and thus the attack to perfectly fit to the recipients.

1.2.2.4 Social Engineering

With all the technical means available for protecting cyber systems today, the main cause of incidents in the cyber domain is human failure. Hence, *social engineering* approaches [12, 19] try to exploit this concept and trick the person operating a system to introduce some malicious software into the system. In this way, an adversary is able to surpass all technological protection mechanism (e.g., firewalls or intrusion protection systems) and get direct access to the internal networks of an organization. Since control and protection mechanisms are usually not as strict in the internal network as they are on the perimeter, this makes social engineering such a highly critical threat.

1.3 Advanced Persistent Threats (APT)

1.3.1 Characteristics

Among today's security challenges for CIs, one of the most complex and severe are *Advanced Persistent Threats (APTs)*. These attacks combine different techniques and attack strategies from the physical and the cyber domain and are very difficult to detect. Although there is no common consensus on how to specifically define an APT due to their variety and diversity, some shared characteristics are commonly accepted as distinguishing features in relation to other attacks. Among these, an APT has at least the following characteristics:

1. *Targeted*: an APT usually aims at a very specific victim and goal. For the victim, this means that the attack is often "hand-crafted" to fit the particularities of the target systems, which in turn means that standard technical security precautions like firewalls, virus scanners or similar are of limited efficacy. With respect to the attacker's goal, we can broadly distinguish two classes of APTs, both of which require distinct modeling and treatment:

   • *Gaining control over a running service*: the adversary's goal is to take over the system for as long as s/he can. However, this does not necessarily mean that the adversary wishes to shut down the targeted system, in particular, if the system should act as a platform from which further attacks are mounted (like a base camp). For example, the APT could aim at limiting bandwidth or blocking certain content in a multimedia streaming network. In this case, the attacker's goal would be to keep the system alive, but at a level that s/he can maintain or cut down at will. An example of a game model to counteract this type of APT is FlipIt, which we discuss in Sect. 9.3.

- *Hitting a vital target*: the purpose of this attack is to shut down the targeted system or cause at least some permanent damage. Here, unlike in the previous case, the time spent in the system is of less importance than the point that the attacker can reach within the system. For example, if the attacker wants to disrupt a nuclear power reactor, remaining in the systems for any period of time, no matter how long or short, that lets the adversary cause a nuclear meltdown or a comparable catastrophe would be the APT goal here. An example game model designed for defense against this second type of APT is Cut-The-Rope, discussed in Sect. 9.4.

2. *Stealthy and Slow*: an additional goal of an APT is often to stay undetected. While this is not surprising and can be considered as a goal for many attacks, the distinguishing fact of an APT is the time scale; an APT is hardly an ad hoc nor quick strike, but rather follows a sophisticated plan of penetration to never cause much attention and therefore always remain under the radar. The rationale is to prepare the attack up to a point where it is hardly – if at all – possible to counteract if the attack is detected.

3. *Unbounded resources*: budget limits are usually hard to assume reliably for an APT. Following contemporary findings, most reported cases of successfully implemented APTs were driven by well organized teams with very high, up to presumably unlimited, resources. In this context, we often see that the achievement of the goal (i.e., the infiltration of a system) weighs much higher than the investment made for this purpose. This is due to the fact that many of these groups are – or are supposed to be – state sponsored [8]. In this sense, the economic reasoning of security to be given if an attack is more expensive than the revenues upon success, this may not apply for an APT.

1.3.2   Life-Cycle

In terms of their chronology and life-cycle as well as by their aforementioned characteristics, APTs are diverse in the actions they take and strategy they follow. However, a sequence of steps can be outlined in accordance with [1, 7, 28], which describes the general modus operandi of an APT. This is called the *kill chain*.

1.3.2.1   Step 1: Reconnaissance

In the initial step of an APT, an attacker chooses the target system or network of his attack. Therefore, the attacker starts to "hunt" for potential targets (e.g., IT systems or pieces of infrastructure with specific vulnerabilities); alternatively, the attack can already be tailored to a specific organization or sector. If the target system is identified, the attacker starts to collect as much information available as possible about it by using open data sources like websites, social media and others. The aim is to gain detailed knowledge about the target's infrastructure, e.g., hardware and software in use, organizational structure, employees and others.

1.3.2.2   Step 2: Initial Compromise

When this information is available, the attacker starts to craft an attack with the aim to compromise a weak systems within the organization. Therefore, technical as well as social skills are used. On the technical side, the attackers are looking for known vulnerabilities in the identified IT systems by searching respective databases, e.g., the National Vulnerability Database (NVD) operated by the National Institute of Standards and Technology (NIST), or crawling the dark web for *zero-day exploits* (i.e., vulnerabilities in software, which are yet unknown, and guidelines how to exploit them). On the social side, the attackers start to identify personnel at the targeted organization (e.g., security officers, network administrators but also normal users) and use social engineering (cf. also Sect. 1.2.2) to obtain detailed information about them. The aim is to find either vulnerable systems within the organization, which are accessible from the outside, or personnel which can easily be infiltrated. Therefore, the attackers create emails containing malicious links or documents (i.e., spear phishing mails, cf. also Sect. 1.2.2). These emails are specifically tailored to the targeted person such that there is a high probability that the person will click on the link or open the document. In this way, a malware is downloaded or an exploit is created, which can then be used by the attacker to gain access to the targeted system (cf. also Sect. 1.2.2).

1.3.2.3   Step 3: Establish Foothold

The malware initially loaded onto the target systems in the previous step usually creates a backdoor for the attacker to install additional software on the system. This software has some remote administrative functionality, which allows the attacker to install additional tools (e.g., command and control software) or establish communication lines (e.g., Virtual Private Network (VPN) tunnels) to the outside world. The attacker's communication with the compromised system is in general stealthy and hard to spot by an administrator, since it usually blends into the normal network traffic. This makes it more difficult for the organization's security officers to detect the attacker's presence in the internal network.

1.3.2.4   Step 4: Escalate Privileges

After having established the presence in the compromised system, the attacker tries to identify additional user accounts with higher privileges (e.g., domain administrators or accounts used for maintenance or service). Most commonly, key loggers or network sniffing tools are used to achieve that; however, passwords from other users can also be exfiltrated from respective databases. In this way, the attacker can escalate the privileges gained by the initial compromise and gain access to additional systems and domains in the organization's network.

1.3.2.5   Step 5: Internal Reconnaissance

Besides looking for user accounts with a higher security level on the initially compromised system, the attacker also scans the network for other important systems, domains or networks. In this way, the attacker obtains an overview on the network structure and the systems running in the organization. This information allows him to identify additional vulnerabilities and load malware to exploit them via the command and control tools installed in Step 2. Since the attacker is now operating within the organization's network, he has additional capabilities to exploit vulnerabilities of systems, which would not be accessible from the outside world (i.e., from the general Internet).

1.3.2.6   Step 6: Move Laterally

If the internal reconnaissance was successful and the attacker was able to identify other vulnerable infrastructure in the network, he will start compromising those other systems by performing Steps 2, 3 and 4 on them. In short, the attacker will exploit the detected vulnerabilities to gain a foothold in the new system, then install command and control software therein to finally escalate privileges on the newly compromised systems. As one reason for moving laterally is to gain access to more important systems in the organization, the risk of being detected by an administrator or security officer can be another reason for an attacker to move to a different system.

1.3.2.7   Step 7: Maintain Presence

With the attacker obtaining control over an increased number of systems in the organization's network, he is able to establish multiple communication channels to the outside world to maintain his connection to the remote command and control tools. This leaves him with additional capabilities to operate in the network even if one of the existing channels is discovered and closed or if his presence is detected and removed from one system. At this point, it is extremely difficult to completely remove the presence of the attacker from the organization's network.

1.3.2.8   Step 8: Complete Mission

When the attacker has reached its final target, i.e., a specific server or (industrial) control system, and gained control over it, the last phase starts. In case of an (industrial) control system, the attacker can gain or already has gained enough privileges on the device due to the malware and exploits installed in the previous steps to perform any desired task (e.g., switch off a transformer substation as happened in the example of the Ukraine described in Sect. 1.4.3). In case the attacker wants to exfiltrate specific data hosted on a server, he can achieve that via the communication channels established in the previous steps. Since an organization usually handles a lot of data going in and out of its network, it is difficult to identify data exfiltration at that point. The attacker has the opportunity to use various tools to obfuscate the source or destination of the data stream, to encrypt the payload or hide the data in a standard data stream.

1.3.2.9   Step 9: Cover Tracks

After completing the mission, the attacker erases all traces of his presence in the system. However, the attacker might also leave some backdoors, tools and communication channels operational to be able to return and compromise the organization again.

1.4   Selected Real-Life Incidents

In the following, we will present a list of a few selected and well documented attacks on CIs. We neither claim this list to be complete nor covering the most severe cases that ever existed, and there might be an undocumented black-count. Nonetheless, the examples in the next sections shall illustrate how past incidents caused failures within a CI and furthermore point out the direct consequences of those incidents as well as on the aforementioned cascading effects. We will loosely distinguish between natural and man-made causes as well as unintentional and intentional causes. The examples shall give an idea how vulnerabilities within CIs could look like and which potential cascading effects a single incident might have.

1.4.1   The Blackout in Italy (2003)

In 2003, Italy experienced a major blackout on Sunday, September 28th, starting shortly after 3:30 in the morning and lasting for several hours [30]. The blackout had far-reaching consequences for various infrastructures all over Italy as well as the Italian population and caused a financial damage of over 1.100 million Euros [26].

The blackout was caused by a series of events resulting in an overload of several main power lines coming from Switzerland, France, Austria and Slovenia. The initial event was a tree flashover at the Swiss 380 kV line "Mettlen-Lavorgo", which caused the tripping of this important supply line [30]. With the failure of the "Mettlen Lavorgo" line, other 380 kV power lines took over the load while several attempts for re-closing the line were executed. As a result of the balancing attempt, one of those lines, the "Sils-Soazza" line, which is the closest to the "Mettlen-Lavorgo" line, also suffered an overload. This overload was acceptable for 15 minutes according to operational standards (and according to expert opinions) for such an emergency. During that period, several countermeasures (i.e., reducing the power consumption for about 300 MW) were implemented in Italy and Switzerland to restore the agreed schedule [30]. However, these measures were not sufficient and about 25 minutes after the outage of the "Mettlen-Lavorgo" line, also the "Sils-Soazza" line tripped.

The loss of two important power lines had a strong impact on the remaining connections to France, Austria and Slovenia, causing them to collapse almost immediately after the outage of the "Sils-Soazza" line [30]. As a result, the Italian power grid was isolated from the European grid. This affected the network in Northern Italy as instability phenomena and overloads happened, resulting in an unsatisfactory low voltage level in the entire Italian network. Although frequency control measures were set in place automatically, turbine tripping, underfrequency relay opening, loss of excitation and other incidents caused the outage of several generating units, which should cushion the effects from the overload [30]. The result was a drop of the frequency below the 47.5 Hz threshold and the blackout of the entire network.

The recovery process started immediately, such that the northern part of the Italian network was back online about 5 hours after the blackout, the central part about 7 hours and main Italy about 13 hours after the blackout happened. As the last part, the island of Sicily was energized about 18 hours after the initial events [30].

1.4.2   The Transportation Gridlock in Switzerland (2005)

In 2005, the entire Swiss railway service was shut down for three hours due to human misjudgment and maintenance work. Over 200.000 commuters were affected and got stuck either on the trains or at the train stations; the shutdown caused a financial damage of over five million Swiss Francs [27].

In the late afternoon of June 22nd 2005, two power lines "Amsteg – Steinen" in the central Swiss Kanton Uri were switched off due to construction work at the tracks at that time [13]. This switch off was planned and checked about a week before the incident and according to existing documentation, the remaining power line was supposed to handle the additional load without problems. However, the two lines could not be switched back on after the construction work was completed and the additional third line could not hold the overload.

It has to be noted at this point that the Swiss railway operates their own power grid, including power plants and the distribution network. Due to the special topology of the Swiss railway power grid, the power line "Amsteg – Steinen" is crucial since it connects three of the main power plants of the Swiss railway operator with the rest of the network. After the third line tripped due to the overload, the Kantons Uri and Tessin were cut off the remaining grid, which caused an overload of the power grid and the shutdown of three power plants in this region [13]. Hence, a blackout occurred in this region and the railway operator started the recovery process to bring them back online.

On the remaining side of the network, the power loss was supposed to be compensated via connections to the power grid of the German railway operator. Due to the resulting heavy load, those connections failed; at this point, the Swiss railway operator was also selling electricity to neighboring countries. Since the operators were mainly occupied with recovering from the first blackout in the Kantons Uri and Tessin, the problems with the connections to the German power grid stayed unobserved. Those issues combined caused also a blackout in the remaining network.

1.4.3 The Attack on the Ukrainian Power Grid (2015)

The Ukrainian power grid fell victim to a major cyber attack in 2015, when hackers managed to gain access to the critical systems of three major power distribution companies (so called "oblenergos") in the Ukraine. They caused a large power outage in most parts of the country, which lasted for about six hours and left approximately 225.000 households without electricity by switching off about 30 substations [14, 18]. The attack is seen as an APT, since it was very well prepared by the adversaries as well as highly sophisticated (cf. also Sect. 1.3 for details on the characteristics of an APT). Further, the 2015 attack is also known the first cyber attack that directly caused power outages.

The attack applied six adversarial techniques over a time period of several months (cf. also Fig. 1.2) to prepare and execute the attack. With these techniques, the attackers followed the main steps of the ICS Cyber Kill Chain [1] (which are also reflected in the individual steps of an APT as described in Sect. 1.3). The attack started with an analysis of open source information about the three power distributors followed by a serious of spear phishing attacks on them. Those attacks were particularly targeted at the power providers by using information from openly available sources (e.g., on Remote Terminal Unit (RTU) and ICS vendors used in these companies) and social media (e.g., on senior personnel working in these companies) [18]. The targets received crafted emails with word documents including an exploit to install the *BlackEnergy 3 malware* [29] on the system. This malware provides several tools to an attacker, starting from simple network scanning up to key logging and remote desktop functionality, which enable a hostile takeover of the infected system and the potential to infiltrate the network the system is connected to.

Fig. 1.2 Illustration of the steps during the hack of the Ukranian power grid [18]

Using the capabilities of the BlackEnergy 3 malware, the adversaries started gain a foothold in the infected infrastructures. This was achieved by collecting user credentials (account name and password) from various users in the systems and, in this way, escalate their privileges in the infected infrastructures, e.g., by identifying administrators and exfiltrating their passwords or by tampering the authentication functionalities of the system. In this way, the adversaries were able to use genuine user accounts and follow standard ways of communication in the networks, which made it more difficult to identify them as malicious parties.

One possibility used by the adversaries to get access to the ICS was to use VPN tunnels from the business network to the ICS network. This was achieved mainly because the VPN connections did not use two-factor authentication, which made it easier for the adversaries to gain access remotely. Additionally, the adversaries were able to control systems in the ICS network due to a remote access functionalities, which were natively built into the ICS systems, and firewall not blocking these commands [14].

In this way, the adversaries finally gained access to the command and control systems in the distribution networks, which allowed them to manipulate their operation and shut them down remotely. In detail, the adversaries developed on the one hand applications to communicate with the Distribution Management System (DMS) environments operated by the organizations and on the other hand malicious firmware for the serial-to-ethernet devices used by the power distributors [18].

Using the stolen credentials, the adversaries then were able to take over the relevant workstations and install the crafted tools on them. This allowed them to take 27 substations offline and upload the firmware to the serial-to-ethernet devices, which locked the operators out of the infected workstations since the altered firmware did not allow remote commands to be issued.

In addition to the above activities, the adversaries started a Denial of Service (DoS) attack on the call service center of the power providers [34]. One effect of this was that the operators in the service center were overwhelmed with fake calls and real customers, who were experiencing the power outage, could not get through. As a consequence, the operators were not aware of the full impact of the situation. A second effect of the DoS attack was that the customers became frustrated because they could not reach the distribution network operators to obtain more information about the blackout situation.

1.4.4 The WannaCry and NotPetya Malware Infections (2017)

In the year 2017, two major malware infections took place all over the globe; the *WannaCry* [31] infection in May and the *NotPetya* [32] infection in June. Both malware infections had a particularly high expansion all over the world, with WannaCry affecting over 200.000 computers globally and NotPetya causing about 10 billion Dollar of damage due to the shutdown of critical systems [11]. In that way, NotPetya had the largest economic effect of any malware infection yet (even more than WannaCry) and has been of major concern for cyber security experts, information security officers and infrastructure operators in 2017 and beyond.

NotPetya originated in the Ukraine, spreading from the premises of a software company named Linkos Group. One of their main products was the tax accounting tool M.E.Doc [4], which almost every company doing business in the Ukraine was using. Due to a lack of security functionality in the update servers of Linkos, a back-door was established by some hackers. This back-door granted them the ability to push a malicious patch of M.E.Doc to the company's customers. This patch contained the NotPetya malware and was automatically distributed to all organizations who were using the accounting software.

The malware itself is applying the *EternalBlue* exploit [21], which takes advantage of a vulnerability in Microsoft's Server Message Block (SMB) protocol and allows an attacker to execute arbitrary code on the target computer. Using this functionality, NotPetya is able to infect computers on the same (local) network and also remote computers that are connected via SMB. Further, it uses the *Mimikatz* malware [10], which allows the attackers to extract Windows passwords from the system's RAM and thus get access to the local system as well as other systems where the local user has access to. By using these two exploits, NotPetya became the fastest spreading malware so far [11].

On an infected system, NotPetya starts to encrypt all files on the hard drive of the system (in contrast to the related malware Petya, which encrypts only the master file table containing all information on the files, directories and metadata located on the hard drive). Additionally, the malware could also completely wipe all files on the system or rewrite them such that they can't be recovered after decryption. This functionality is initialized by forcing a restart of the system, after which the system can't reboot again and a screen is displayed telling that some ransom has to be paid.

Among the multitude of companies hit by the NotPetya Malware, the highest-grossing impact was encountered by the pharmaceutical company Merck (around 870 million Dollars), the European branch of the delivery company FedEx (roughly 400 million US Dollar), the French construction company Saint-Gobain (around 384 million US Dollar) and the Danish shipping and logistics company Maersk (around 300 million US Dollar) [11]. In particular in the case of Maersk, the consequences were highly visible: the infection with the NotPetya malware and the consequent shutdown of a high number of critical systems within Maersk own network caused several processes in the logistic supply chain to break down. For example, operations at the Maersk terminal at the second largest port in India, the Jawaharlal Nehru Port Trust near Mumbai, have been disrupted due to the infection of Maersk's systems such that loading and unloading of vessels was not possible any more [24]. At another of Maersk's terminals in Elizabeth, New Jersey, the registration of trucks picking up or dropping off containers at the port and the general logistics of the containers (i.e., their location etc.) is completely digitalized and supported by the Maersk systems [11]. Hence, the NotPetya infection of Maersk's network caused a shutdown of the registration at the entrance gate to the port, leaving the trucks no way to either enter or leave the port.

1.4.5   The Blackout in Venezuela (2019)

From March 7th to March 14th 2019, a blackout took place in Venezuela and lasted one week. It was the biggest power outage in the history of Venezuela [5]. Although Venezuela has been experiencing power outages on a frequent basis over the last years, the one in March 2019 had wide-spreading impact on several major cities in Venezuela, including the capital Caracas, influencing the social life of the population. Outages of telecommunication and banking as well as shortages of food and water supply, irregularities in health care and also outages of the transportation system were just a few consequences of the blackout. Additionally, the blackout was followed by a serious of smaller outages from March to July with shorter duration and minor impacts each.

Venezuela is in a particular situation when it comes to power supply, since about 70% to 80% of the nation's overall power supply is coming from the hydroelectric power plant at the Gurí Dam [5, 23]. Additionally, most of the bigger cities in Venezuela are powered by the San Geronimo B substation, which is directly connected to the Gurí Dam. A vegetation fire overheated these power lines connecting the substation to the dam and caused the lines to shut down. The resulting power loss caused the turbines at the hydroelectric plant to increase their speed and overload the systems. Although safety systems were in place inside the dam and tried to decrease the power input, the system became uncontrollable and the power generators of the dam had to be disconnected. As a consequence, the frequency in the grid could not be stabilized and started to fluctuate, thus overloading two other power plants at Caruachi and Macagua [17].

As a measure to increase resilience, a number of thermal power plants are put in place in Venezuela, which should take over the load in case of a power shortage or blackout. However, these power plants were not running at the time the incidents happened due to maintenance reasons and a shortage of fuel [16]. Hence, it was particularly difficult to stabilize the frequency of the electrical grid and bring it back online.

The consequences of the blackout were manifold, particularly due to the length of the outage. In detail, almost the entire telecommunication infrastructure of Venezuela was shut down [22], which also affected television (i.e., news broadcasting) and the connection to the Internet. As a second (and related) consequence, banks and the financial system were not able to operate any more, which affected the availability of cash. The use of payment cards and ATMs was not possible and most financial transactions were falling back onto the US Dollar [25]. Without a sufficient power supply, refrigerators were also not working any more, which directly affected the food supply and caused retailers of meat and fish to close down since it was no longer possible to keep the goods fresh [9]. Further, the water distribution system was equally affected by the power outage, leaving large parts of the major cities in Venezuela without water supply for days.

Furthermore, the medical supply and health care suffered from the blackout. Although hospitals were equipped with power generators, they were unable to run at full capacity due to a shortage in the fuel supply. As a result, only the emergency rooms had lighting and standard surgeries could not be carried out properly as the chance of an infection was too high. Consequently, several deaths were reported in different hospitals, which could be directly related to the blackout [15]. From an industry perspective, the power outage heavily influenced the oil production, causing some oil rigs to shut down completely and cutting the overall oil production of Venezuela in half for some time [33].

After seven days, most parts of the country regained a stable power supply; however, 30-day power rationing was necessary after the incident because the grid was not able to run at full capacity. In that time, additional blackouts were encountered at the end of March and beginning of April, which also affected large parts of the country, reducing the operation of telecommunication, traffic and the public sector [3].

References

1. Assante MJ, Lee RM (2015) The Industrial Control System Cyber Kill Chain. SANS White Paper, SANS, Bethesda. https://www.sans.org/reading-room/whitepapers/ICS/industrial-control-system-cyber-kill-chain-36297

2. Bundesamt für Sicherheit in der Informationstechnik (2016) IT-Grundschutz Catalogue. BSI, Bonn. https://www.bsi.bund.de/EN/Topics/ITGrundschutz/itgrundschutz_node.html. English Version
3. CGTN (2019) Maduro announces 30 days of electricity rationing in Venezuela. Egypt independent. https://news.cgtn.com/news/3d3d514f31557a4e33457a6333566d54/index.html
4. Cimpanu C (2017) Petya ransomware outbreak originated in ukraine via tainted accounting software. https://www.bleepingcomputer.com/news/security/petya-ransomware-outbreak-originated-in-ukraine-via-tainted-accounting-software/
5. Dube R, Castro M (2019) Venezuela blackout plunges millions into darkness. Wall Street J. https://www.wsj.com/articles/venezuela-blackout-stretches-across-country-closing-schools-and-businesses-11552053011
6. European Commission (2008) COUNCIL DIRECTIVE 2008/114/EC of 8 December 2008 on the identification and designation of European critical infrastructures and the assessment of the need to improve their protection. Off J Eur Union (L345):75–82. http://eur-lex.europa.eu/legal-content/EN/TXT/PDF/?uri=CELEX:32008L0114&from=EN
7. FireEye (2013) APT1. Exposing one of China's cyber espionage units. Technical report, FireEye Inc., Milpitas. https://www.fireeye.com/content/dam/fireeye-www/services/pdfs/mandiant-apt1-report.pdf
8. FireEye (2019) Cyber threat intelligence reports. https://www.fireeye.com/current-threats/threat-intelligence-reports.html
9. France24 (2019) Race against time in blackout-hit Venezuela to save food stocks. France 24. https://www.france24.com/en/20190311-race-against-time-blackout-hit-venezuela-save-food-stocks
10. Greenberg A (2017) How the Mimikatz hacker tool stole the world's passwords. Wired. https://www.wired.com/story/how-mimikatz-became-go-to-hacker-tool/
11. Greenberg A (2018) WIRED: the untold story of NotPetya, the most devastating cyberattck in history. https://www.wired.com/story/notpetya-cyberattack-ukraine-russia-code-crashed-the-world/
12. Hadnagy C (2011) Social engineering. Wiley, Indianapolis. http://media.obvsg.at/AC08377030-1001
13. Hess H, Lehmann P (2005) Neue Erkenntnisse zum Stromausfall. https://www.gotthardbahn.ch/downloads/stromausfall_medienkonferenz2.pdf
14. ICS-CERT (2016) Cyber-attack against Ukrainian critical infrastructure. https://ics-cert.us-cert.gov/alerts/IR-ALERT-H-16-056-01
15. Jones S (2019) Venezuela blackout: what caused it and what happens next? The Guardian. https://www.theguardian.com/world/2019/mar/13/venezuela-blackout-what-caused-it-and-what-happens-next
16. Kurmanaev A, Herrera I, Krauss C (2019) Venezuela blackout, in 2nd day, threatens food supplies and patient lives. The New York Times. https://www.nytimes.com/2019/03/08/world/americas/venezuela-blackout-power.html
17. La Patilla (2019) El origen de la falla que causó el mega apagón en Venezuela (informe de la UCV). LaPatilla.com. http://www.lapatilla.com/2019/03/13/el-origen-de-la-falla-que-causo-el-mega-apagon-en-venezuela/
18. Lee RM, Assante MJ, Conway T (2016) Analysis of the cyber attack on the Ukrainian power grid. Technical report, E-ISAC, Washington. https://ics.sans.org/media/E-ISAC_SANS_Ukraine_DUC_5.pdf
19. Mann I (2008) Hacking the human. Gower, Aldershot
20. Marinos L, Lourenco M (2019) ENISA threat landscape 2018. Technical report, ENISA, Ispra. https://www.enisa.europa.eu/publications/enisa-threat-landscape-report-2018/at_download/fullReport
21. MS-ISAC (2019) EternalBlue. Security primer SP2019-0101, multi-state information sharing & analysis center (MS-ISAC). https://www.cisecurity.org/wp-content/uploads/2019/01/Security-Primer-EternalBlue.pdf

22. NetBlocks (2019) Second national power outage detected across Venezuela. https://netblocks. org/reports/second-national-power-outage-detected-across-venezuela-dQ8o728n
23. Newman LH (2019) Why it's so hard to restart Venezuela's power grid. Wired. https://www. wired.com/story/venezuela-power-outage-black-start/
24. PTI (2017) New malware hits JNPT operations as APM terminals hacked globally. http://indianexpress.com/article/india/cyber-attack-new-malware-hits-jnpt-ops-as-apm-terminals-hacked-globally-4725102/
25. Rosati A (2019) Venezuela is now awash in U.S. dollars. Bloomberg.com. https:// www.bloomberg.com/news/articles/2019-06-18/once-forbidden-u-s-dollar-is-suddenly-everywhere-in-venezuela
26. Schmidthaler M, Reichl J (2016) Assessing the socio-economic effects of power outages ad hoc. Comput Sci Res Dev 31(3):157–161. https://doi.org/10.1007/s00450-014-0281-9
27. Schweizer Radio und Fernsehen (SRF) (2014) Schweiz – Der Blackout 2005 – ein schwarzer Tag für die SBB. Schweizer Radio und Fernsehen (SRF). https://www.srf.ch/news/schweiz/der-blackout-2005-ein-schwarzer-tag-fuer-die-sbb
28. Secureworks (2019) Advanced persistent threats – learn the ABCs of APT: part A. https:// www.secureworks.com/blog/advanced-persistent-threats-apt-a
29. ThreatStop (2016) Black energy. Security report, ThreatStop, Carlsbad. https://www. threatstop.com/sites/default/files/threatstop_blackenergy.pdf
30. UCTE (2004) Final report of the investigation committee on the 28 September 2003 blackout in Italy. Technical report, Union for the Coordination of Electricity Transmission (UCTE). http://www.rae.gr/old/cases/C13/italy/UCTE_rept.pdf
31. US-CERT (2017) Alert (TA17-132a) indicators associated with WannaCry ransomware. https://www.us-cert.gov/ncas/alerts/TA17-132A
32. US-CERT (2017) Alert (TA17-181a) petya ransomware. https://www.us-cert.gov/ncas/alerts/TA17-181A
33. Zerpa F (2019) Venezuela blackouts cut oil output by half in March. Houston Chronicle. https://www.chron.com/business/energy/article/Venezuela-Blackouts-Cut-Oil-Output-by-Half-13743951.php
34. Zetter K (2016) Everything we know about Ukraine's power plant hack | WIRED. https://www. wired.com/2016/01/everything-we-know-about-ukraines-power-plant-hack/

Chapter 2
Critical Infrastructures

None of us knows what might happen even the next minute, yet still we go forward. Because we trust. Because we have Faith.
P. Coelho

Abstract This chapter refines the introduction of security in critical infrastructures by going into deeper details about how threats and countermeasures differ and are specific for the physical domain, the cyber domain and intermediate areas. Gaining an understanding of these differences is crucial for the design of effective countermeasures against the diverse nature of today's advanced persistent threats (APTs). As even local incidents may have far-reaching consequences beyond the logical or physical boundaries of a critical infrastructure, we devote parts of the chapter to a discussion and overview of simulation methods that help to model and estimate possible effects of security incidents across interwoven infrastructures. Such simulation models form an invaluable source of information and data for the subsequent construction of game-theoretic security models discussed in the rest of the book.

2.1 Examples and Definitions of Critical Infrastructures

2.1.1 What Makes an Infrastructure "Critical"?

Critical Infrastructures (CIs) are essential for security and welfare of a society as they provide crucial products and services. Goods such as drinking water or electricity are part of our everyday life and are typically assumed to be available 24 hours a day. However, recent incidents in connection with the security of cyber-physical systems within CIs such as the cases already mentioned in Chap. 1 or the more complex examples of Stuxnet [52] and WannaCry [92], raised awareness of the vulnerability of CIs and the strong dependence on computer networks. As also described to some degree in the previous chapter, such incidents demonstrate the strong interdependencies among CIs within a region, a country or also all over the world. While a rough classification of these interdependencies is possible [83], the effects of these relations are more challenging to assess. The analysis of an entire network of CIs is complex, where the biggest challenge is understanding the dynamics inside the network.

A formal definition of a critical infrastructure is difficult to find due to the high variability among the different CIs and the changes in their interdependencies. Infrastructures are influenced by emerging technologies, digitalization, growing interconnections and interdependencies but also by statutory provisions and new laws. Still, it is possible to identify factors that characterize an infrastructure as critical. One relevant definition of CIs is given be the German BSI in [100]:

> Critical infrastructures are organisational and physical structures and facilities of such vital importance to a nation's society and economy that their failure or degradation would result in sustained supply shortage, significant disruptions of public safety and security or other dramatic consequences.

Another definition is given by the European Council in its Directive 2008/114/EC [32]:

> [...] 'critical infrastructure' means an asset, system or part thereof located in Member States which is essential for the maintenance of vital societal functions, health, safety, security, economic or social well-being of people, and the disruption or destruction of which would have a significant impact in a Member State as a result of the failure to maintain those functions;

From both definitions we see that a CI is characterized by its high importance for the societal and economic well-being as well as the public health and safety of a nation.

When looking at the various sectors, which are considered to be critical, then we see that the list is different for different countries. For example, the German BSI assigns the following sectors to CIs [100]:

- Transport and traffic (air, maritime, inland waterway, rail, road, logistics)
- Energy (electricity, oil, gas)
- Information technology and telecommunication
- Finance and insurance sector
- Government and public administration
- Food (industry, trade)
- Water (water supply, sewage disposal)
- Health
- Media and culture

The European Union characterizes the following eleven sectors as critical (whereas only "Energy" and "Transport" are mentioned in the directive) [32]:

- Energy
- Nuclear industry
- Information, Communication Technologies (ICT)
- Water
- Food

- Health
- Financial
- Transport
- Chemical Industry
- Space
- Research Facilities

On the other hand, the American government distinguishes 13 types of critical infrastructures as of 2002 [69]:

- Agriculture
- Food
- Water
- Public Health
- Emergency Services
- Government
- Defense Industrial Base
- Information and Telecommunications
- Energy
- Transportation
- Banking and Finance
- Chemical Industry
- Postal and Shipping

From all three lists we can see that the sectors energy, ICT (or communication in general), transportation, finance, water, health, chemical industry can be considered as the most important in terms of their criticality.

2.1.2 Threats

In the last decade, the number of publicly known security incidents in CIs has increased. At the same time, these incident have been triggered by a broader range of events than ever before. Increasing interdependencies among CIs yield to impairment or failure of a CI due to an impairment or failure of other CIs. These cascading effects can be caused, for example, due to natural threats or due to cyber attacks, as the case studies from Sect. 1.4 demonstrate.

These manifold and constantly changing interdependencies challenge CI operators. Limited availability or even failure of one CI affects other CIs depending on it, directly or indirectly. The high complexity and diversity inside the network enables the development of cascading effects. This makes it difficult to assess consequences of an impairment, both for accidental and intentional incidents.

Our goal in this chapter is demonstrating the diversity of the term "security" in the context of CIs, and to see how game theory has been and can further be applied to security problems in this field. To illustrate the various aspects of security of CIs, we here describe some specific risks CI operators face in addition to the general challenges already described in Chap. 1. This list is not intended to be complete but aims at illustrating the diversity of threats and risks faced by CIs.

As briefly mentioned in the previous section, CIs can be found in various sectors. Alas, when hearing the term CI, people may first think about goods and services consumed on a daily basis, such as electricity or water. The awareness is typically lower when it comes to transportation, unless the functionality is limited, e.g., due to bad weather conditions. The importance of food suppliers is even less understood by many people, sometimes including the operators themselves. Further, most operators have a very limited awareness of risks related to IT security. Passwords may be simple to guess, employees may lack training about avoiding malware and/or dangers of social engineering attacks. Between 2015 and 2017, nine case studies on IT security of CIs in Germany have been conducted and analyzed in [60]. Major challenges faced by the participants are the following:

- Remote maintenance of CIs allows a higher level of control and a greater flexibility of working times, but also introduces risks if not implemented correctly and used with care. Threats to a remotely controlled system include cyber attacks that yield disruption of service, unauthorized access to sensitive process data or unauthorized manipulation (e.g., after getting access to credentials).
- Hospitals need to protect personal and health data of patients from both unauthorized access and modification. Prevention focuses especially on employers' awareness since social engineering attacks are still new to this sector.
- Food producers, such as a dairy, need to ensure the quality of their products and prevent attackers from adding chemicals.
- In finance, business processes must be secure. In particular, money transfer should be reliable and protected against unauthorized access or manipulation.
- Enterprises that sell software need to care about cyber attacks first because they need to assure the quality of their products and second due to the customer data they use or process. In case a product is manipulated, it might no longer satisfy the customers wishes and the company suffers from damaged reputation; additionally, there is a risk the product is used for espionage.
- Email communication may be classified depending on the content of the message. If sensitive data is sent without restrictions, consequences range from reputation damage to monetary loss or information for more advanced attacks (such as an APT).
- When the police uses photography for documentation, it is necessary to assure authentication of the images, i.e., no errors or intentional changes should occur at any point. Additionally, secure storage is required and unauthorized access should not be possible.
- Coordination of (disaster) relief forces must still be possible in case IT support fails. In particular, communication should be separated from the classical Information and Communication Technology (ICT) network to ensure emergency calls are possible in any case.

- The German armed forces are not a CI in the classical sense, however, they strongly contribute to the safety of the population and thus need to be protected from targeted (cyber) attacks. Among potential aims of attacks are gain of information, destabilization, air sovereignty or supply with medicine.

The various threats and risks a CI might face can roughly be classified depending on which part of the CI an incident first affects. Thus, we will generally distinguish between the cyber domain in Sect. 2.2, the physical domain in Sect. 2.3, and more advanced attacks that target the cyber-physical system as a compound, in Sect. 2.4 in context with our exposition of existing game theoretic models for security in CIs. However, before investigating security in more detail, we next provide the basic notation as well as a definition of a CI.

Functionality of CIs is of high relevance for a society and thus government, organizations and researchers jointly work on increasing the security of CIs. While there is a decent amount of publications focusing on what has happened and what may happen in the future (e.g., [2] or similar), the literature on analytical models is less rich. Game theory has been frequently applied to investigate the security of critical infrastructures and the corresponding models are often termed *security games* [51]. Such models are quite natural as soon as security is threatened by a strategic attacker, which is the case in many cyber or cyber-physical attacks (at least for the successful attacks). Even for non-intentional threats such as natural disasters, the game theoretic approach may yield some useful insights, if a zero-sum game is set up that assumes a hypothetic (strategic) attacker trying to cause as much damage as possible. While this view is somewhat pessimistic, it yields results that can be interpreted as bounds to the possible payoffs and suggest actions to be on the safe side. Some security games have successfully been put to practice, e.g., by the Los Angeles International Airport [73], by the United States Transportation Security Administration [74] or by the United States Coast Guard [96].

The next three sections review game theoretic approaches to security of CIs, focusing on how to model the most frequent or pressing threats and attacks. The last section of this chapter then gives on overview on existing approaches to simulate consequences of security incidents in CIs.

2.2 Cyber Security

Numerous press reports on recent incidents indicate the vulnerability of infrastructures to cyber attacks. Successful cyber attacks are usually highly strategic and use both profound technical knowledge and information about the targeted CI (e.g., by social engineering). A popular method to learn about potential attacks on a network is the use of *honeypots*. While honeypots look like a ordinary part of the network they do not contain sensitive information but aim at distracting the attacker to learn about his behavior. If the attacker recognizes the target to be a honeypot, he may in turn deceive the defender by choosing different attack strategies than he would use for real network components. This interplay may be modeled as a one-shot Bayesian game of incomplete information between attacker and defender.

In the context of IoT networks, [59] proposed a model where the defender faces an attacker of unknown type, either active or passive. Regardless of the type, the attacker sends a signal "suspicious" or "normal", corresponding to the way his attack looks from the outside. Based on the signal, the defender decides whether to accept the traffic as regular or rerouting it to a honeypot. The existence of a Perfect Bayesian Nash Equilibriums (PBNEs) is shown in the paper and verified using a MATLAB simulation.

Electricity is of crucial importance for operation of any other CI and with the development of smart grids its availability may be reduced by attacks on the communication network. Attackers are typically acting strategically and smart which makes a game theoretic analysis applicable. Assuming the worst case, i.e., that the attacker aims for maximal damage, [18] applies a zero-sum game between attacker and defender to analyze several intelligent attacks, such as vulnerability attack or data injection. The robustness of the network is measured in terms of percolation-based connectivity which allows determination of the defense cost.

2.2.1 Hacking

Incidents such as hacking of the computer controlling containers (enabling drug traffic) in Antwerp [6] or the hacking of the Ukrainian power grid in 2015 and 2016 [20, 48, 106] had significant effects on power grids, hospitals and ports and rose awareness among CI operators. Applying game theoretic models to cyber attacks almost always implicitly assumes an intelligent attacker who acts rationally. This assumption is valid in the case of targeted attacks that involve information gathering as a first step, e.g., through social engineering in combination with publicly available information.

2.2.2 Malware and Ransomware

The Wannacry ransomware [19] and the NotPetya malware [36] (both in 2017) affected many CIs that so far only had a limited awareness of cyber attacks, such as ports [58]. In case of a malware attack, the assumption of a rational attacker may not always be satisfied, e.g., if it is designed in such a way that is reaches as many people as possible. However, the worst case where the attacker aims at causing maximal damage can still be analyzed using zero-sum game models. The effect of malware in a wireless network was studied in [54] through a zero-sum differential game.

Similar to an epidemic model, a network node can be either susceptible to a malware, infected by a malware or already out of use. Differential equations are given that describe the evolution of the number of nodes in each state. Saddle-point strategies are computed. A zero-sum game with stochastic payoffs was applied to the impact of a malware attack on a European Utility network in [57]. In this setting, the payoffs are estimated through a percolation-based model that imitates the spreading of the malware through the utility network. The defender tries to minimize the damage in the sense that he wants to minimize the likelihood for the maximal damage; see also [54]. Strategies to protect a power grid against cyber attacks and ways to mitigate the propagation of malware have in identified in [99] using coloring games.

2.3 Physical Security of Critical Infrastructures

Despite the fact that many recent attacks on CIs start in the cyber domain, it is nonetheless important to keep an eye on the physical security. Besides natural hazards such as flood or earthquake, intentional attacks may focus on the physical domain. Many of these attacks focus on the physical part of communication systems, including ICSs. Protection of a chemical plant is optimized by considering a game between security defender and adversary based on a general intrusion detection model [107]. Relaxing the assumption of simultaneous moves of attacker and defender, a Bayesian Stackelberg game is introduced in [108] and illustrated with a case study.

2.3.1 Eavesdropping

A frequent passive attack is *eavesdropping* where the attacker reads data that has been sent but does not change it. We will skip eavesdropping threats for now as we will revisit this issue in deeper detail in Chap. 11. It is, however, worth mentioning that even sophisticated communication infrastructures like quantum networks may be susceptible to eavesdropping attacks by invalidations of underlying security assumptions such as trusted relay [65, 80], but can be hardened against such threats with help from game theory. We will have to say more about this in Chap. 11.

2.3.2 Jamming

The active counterpart to an eavesdropping attack is *jamming*, where the attacker either blocks data transmission by deleting some data or changes the content and forwards manipulated data. The first type of jamming threatens availability of the systems while the second one challenges integrity. Blocking data exchange between sensors and control units of a Cyber-Physical System (CPS) may result in a DoS attack. In the context of remote state estimation, [63] suggested a zero-sum game

between sensor and attacker to identify optimal times when to send data. The
authors therein prove existence of a Nash equilibrium and propose an algorithm to
compute it. Further game models about jamming are applied in [1, 111].

2.3.3 Terrorist Attacks

Terrorist attacks on oil and gas pipelines are investigated in [82] where the most
attractive targets of an attack as well as optimal defense strategies are determined
through a pipeline security game.

Antagonistic attacks on electric power network have been analyzed in [47] using
a zero-sum game between attacker and defender for the case of a planned attack. For
a random attack, the defender solves an optimization problem trying to minimize
the expected consequences of the attack. In case the attacker tries to maximize
the probability that his payoff is above a predefined level, the two extreme cases
(worst case vs. random case) provide bounds for the expected consequences. The
article illustrates how to evaluate defending strategies using the Swedish national
high voltage system.

2.4 Cyber-Physical Security of Critical Infrastructures

The ongoing digitalization also affects CIs. It enables controlling processes and
smooth operation, e.g., by means of ICSs, but also paves the way for new risks. The
growing interconnectivity of cyber and physical systems increases the vulnerability
of CIs. The interplay between physical and cyber systems typically yields to a
situation where threats are not simply the sum of the threats to the subsystems.
Rather, the two parts may amplify one another during an attack, and this section
focuses on this *hybrid* system. Models of cyber-physical systems used to pay only
limited attention to this interaction in the past.

Interdependencies between two systems affects the effort to protect the two
systems [41] and changes the impact of attacks. Further, it enables new types of
attacks. Sophisticated attackers often exploit combinations of several weaknesses
related to the cyber and the physical domain to reach their goal. This motivated
a joint treatment under the term cyber-physical security, which gained a lot of
attention recently. As the language that describes strategic interaction, game theory
offers a variety of models for such complex attacks. An overview on dynamic game
models for cyber-physical security from the game theoretic perspective is given
in [31], while we here present different models that deal with specific threats for
cyber-physical systems. Coordinated cyber-physical attacks on a power grid have
been modeled through a stochastic game in [104]. As one of the first game theoretic
models that explicitly accounts for interdependencies between cyber and physical
components, the proposed resource allocation stochastic game is used to minimize

the physical impact of the attack. The impact is quantified by an optimal load shedding technique. Existence of equilibria is shown with methods based on the minimax Q-learning algorithm.

Attackers who plan and perform an APTs typically have a lot of information on both the organizational structure as well as technical details of the system they attack [81]. The information ranges from knowledge about vulnerabilities of SCADA server, controllers and network switches to information about employees and their responsibilities. The preparation of an APT is often unnoticed by the targeted organization and this stealthiness makes it very hard to predict the impact to the CI. Game theory models with probability distribution-valued payoffs take into account this unpredictability of the effect of an attack but are still able to find solutions. The price for this higher model accuracy is the task to estimate payoff distributions for each scenario (for each goal, if several are optimized simultaneously). The estimation of such payoffs depends on the considered scenario and on the data available. Payoff estimation may be based on expert assessments and testbed simulations, in case the required resources are available [35]. If such an in-depth analysis is not possible, a model of the CI's dynamics is required. Several classes of models exist that describe the consequences of an incident and take into account cascading effects inside a cyber-physical system, which nowadays CIs are. The currently most frequently used approaches are described in the next section.

2.5   Simulation of Effects of Security Incidents

In the highly interconnected network of CIs, a security incident does not only affect a single CI operator but also others, depending on the affected one (e.g., a shortage in the power supply may affect the functionality of hospitals or water providers). Such cascading effects can be observed in various fields with increasing frequency, including the CI network. While triggers for cascading failures are manifold some characteristics of cascading behavior can be identified, at least in some domains. If available, this information needs to be taken into account in order to model cascading failures adequately.

2.5.1   Network Models

General observations of cascading failures indicate the existence of two phases, namely the slow cascading phase, where things still seem to work properly, and a fast cascading phase, where the system is out of control. Another general feature is the difficulty of measuring the impact of cascading failures that are due to the high complexity and many diverse indirect costs (ranging from monetary cost to working hours of employees or even reputation). Typically, only limited data is available on incidents that caused cascading effects. Whenever possible, historical data is used in combination with expert knowledge to keep the number of assumptions limited. Existing studies as presented in [71] help to identify failure patterns or to identify interdependencies that are not obvious at first glance. Furthermore, uncertainties challenging the interdependencies among infrastructures are discussed in [40].

Despite the strong diversity of models of cascading effects, a rough classification helps to get an overview of the different lines of reasoning. An overview of models and simulation approaches for CI systems is given in [70] and a compact comparison between the different models for cascading failures in power systems is presented in [37]. In the remainder of this section, we group existing approaches into five classes according to their main focus, namely topological models, stochastic models, dynamic simulation models, agent based models and economy based models. This list is not exhaustive since other models exist that do not belong to any of the five groups but the ones discussed here include the most developed approaches.

A formal analysis of cascading effects requires a formal description of the involved systems as a very first step. Typically, this is done through a network model that describes each system through a graph where the components are represented by nodes and connections are represented by edges between them. In the case of a cyber-physical system, a network description of the entire system may be done in two steps. First, both the cyber and the physical system are modeled by a subnetwork (based on the expertise of potentially different experts of the field). Second, interdependencies between the two systems need to be identified and are represented as edges between components of the two subnetworks. While this network representation is quite simple, it helps to get an overview of the situation. Furthermore, properties of the resulting overall graph provide basic information about the system. Additionally, simulations are possible to estimate the impact of failures.

2.5.1.1 Graph-Based Models

Many models of cascading failures are based on physical network topology's properties. For example, *node degrees* are used as weights in [101, 103] and local flow distribution rules in this weighted graph allow analysis of the consequences of an edge failure. Further, *betweenness centrality* has been used to investigate overload breakdowns in scale-free networks [45, 46]. Triggered events and random failure in power grids are compared in [55] by means of graphical topological indices. Additionally, topological properties may be used to define terms such as capacity, e.g. through a *node capacity model* as proposed in [66]. Most of these models assume that a component is either working properly or fails, i.e., its functionality is measured on a binary scale.

Most topological models are very general and work for any complex network, i.e., they are usually not limited to a specific sector. Thus, such models may yield misleading results for specific sectors [44]. Still, they provide a good basis for more advanced models, e.g. to investigate the impact of topology on cascading failures in power grids [22]. Several extensions work with maximum flow theory to model the behavior of power networks, for example, in [34] to identify critical lines and in [33] to model invulnerability of power grids. While these models are generally applicable, they often lack a detailed representation of the actual situation and thus are error-prone when it comes to predictions.

2.5.1.2 Interdependent Network Models

Different networks become more and more interconnected for many reasons (including digitalization) and recent research focuses on such *networks of networks* [94]. This development also affects CIs that depend on one another or are controlled by SCADA networks. These interconnections between formerly independent systems enable advanced attacks [98] in many domains. Recent examples of such incidents include the attack on the Ukrainian power grid in 2015 [61], where the attacker remotely controlled the SCADA system to interrupt operation of the electricity network. While security issues in cyber-physical systems have been recognized [30], research based on interdependent models is still limited. In addition to the cyber-physical dependencies, interdependencies between CIs cause new threats. Such interconnected CIs have been analyzed and classified in [84]. In more detail, consequences of a failure in the electrical grid on a telecommunication network has been investigated in [85] based on an interdependent network model and simulations. Vulnerability and robustness of interconnected networks against cascading failures has been investigated in [10] with the finding that interconnected networks do behave different than single networks. In particular, removal of critical nodes may yield to complete fragmentation of the formerly interdependent networks. Multiple dependencies have further been explored in [93] to analyze the robustness of a power grid. A mitigation strategy against cascading failures in coupled networks has been proposed in [91] and tested with historical data.

2.5.2 Stochastic Models

A core feature of cascading behavior is its unpredictability due to the high complexity and the huge number of factors that may initiate or exacerbate the cascade. Besides the complexity issue, many factors are simply unknown, such as hidden failures in power grids, or cannot be measured at all, e.g., indirect dependencies. In the light of the sheer impossibility of an exact prediction, stochastic models are a natural choice to describe the situation. Additionally, such models allow simulation that may be used to validate the model or make predictions and allow simulation of various possible events. While popular stochastic processes such as Markov chains and branching processes are often used for modeling, other approaches exist as well. For example, the model in [110] describes the failure dynamics of the entire network through a power flow model for the failure propagation combined with a stochastic model for the time between failures. In this way, it provides a simulation of the cascading procedure and investigates the systems robustness.

2.5.2.1 Markov Chain Models

The simplest and best known stochastic model is a *Markov chain*, as it takes into account a certain degree of dependence but is not too complex at the same time as they only have limited memory. Still, the state space may grow exponentially when a more detailed description of the state is needed. On the other hand, there is a rich theory behind Markov chain models, allowing computation of stationary distributions or path lengths. Advanced Markov models with memory exist [105], but these are not (yet) used very frequently due to the more complex mathematics behind the models (in particular, transition matrices need to be replaced by transitions tensors).

A *discrete Markov chain model* that describes the functionality of lines in a power grid on a binary scale has been introduced in [102] (it has to be noted that the time between two state changes is considered as infinitesimal). The main idea there is to deduce the transition probabilities of the Markov chain from a stochastic flow distribution model in order to capture cascading effects. The probabilities of line overloading are computed for a given line flow. A *continuous-time Markov chain model* that describes the system dynamics of power grids is introduced in [79]. The complexity of Markov models is handled by construction of an abstract state space where a state contains all information relevant for the dynamics of the system. This model allows to predict the evolution of the blackout probability as well as an asymptotic analysis on the blackout size. An extension of the model is the *interdependent Markov chain model* to describe cascading failures in interdependent networks [78]. The introduced model showed that interdependencies can highly affect the distributions of failure sizes in the two systems. Simulations demonstrate that systems with exponentially distributed failure sizes seem to be less robust against cascading failures than power-law distributed failure sizes.

2.5.2.2 Branching Process Models

A more complex model of growth and reproduction is a *branching process*. Most branching process models are not very specific and only provide a general estimate of the impact of cascading failure. On the other hand, these models are analytically and computationally tractable and allow direct simulation of the cascading process. Applications to cascading problems have been introduced in [24] where events are grouped into generations and the propagation process is described by a Poisson branching process. Based on this model, the extent of cascading outages in a power grid has been estimated [23] and mitigation strategies have been investigated [28].

A *Galton-Watson branching process* model and the associated estimators have been used to predict the failure distribution in electric power systems [27] and blackout size distributions [75]. Interactions between failures in different components are investigated and simulated in [77]. The influence of interactions on the failure is studied in order to mitigate cascading failures. A multi-type branching process was used to analyze the statistics of cascading outages in [76].

2.5.2.3 High-Level Stochastic Models

A disadvantage of models describing the detailed evolution of cascading failures is the corresponding high running time of simulations based on these models. This is particularly an issue when aiming at real time predictions of the propagation mechanism. In this situation, it is often preferable to use *high-level statistical models* that provide an overview on the cascading process. The CASCADE model [25, 26] initiates a cascading effect by randomly assigning disturbance loads to some components of an electric power transportation system and uses simplified formulas to describe the redistribution of loads, which yields to a cascade. This simplification yields fast estimates of the number of failed components as well as the distribution of blackout size. Based on this model, reliability of power systems has been analyzed in [29]. In some situations, the CASCADE model yields results similar to those resulting from a branching process model [56].

2.5.3 Dynamic Simulation Models

Whenever detailed information about the system at hand is available, more accurate predictions based on more involved simulation models are possible. Since such simulations are more evolved, they are less applicable for real time predictions. Researchers at Oak Ridge National Laboratory (ORNL) proposed a dynamic model describing the cascading process in a power grid (e.g., due to a failing transmission line) based on a linear programming approach in [14]. This *ORNL-PSerc-Alaska* (OPA) model did not provide an accurate simulation of blackouts due to line outage and has thus been extended in [95]. This improved OPA model allows simulation of the patterns of cascading blackouts in power systems as it takes into account the dynamics as well as the potential mitigation actions. A nonlinear dynamic model for cascading failures in power systems is the *Cascading Outage Simulator with Multiprocess Integration Capabilities (COSMIC)* model introduced in [97] that describes many mechanisms by recursive computations of the corresponding differential equations. A dynamic probabilistic risk assessment is used in [42] to describe the coupling between events in cascading failure and the dynamic response of the grid to the perturbation. The model has been extended to capture the fact that cascading outages of power grids often consist of two different phases (a slow and a fast one) [43]. A load flow model that takes into account hidden failures was introduced in [16].

2.5.4 Agent-Based Models

In case the dynamics among the different components of a system of CIs are known, *agent-based models* may provide an adequate description of the behavior of the overall system. Agent-based models can be built from existing system dynamic models or discrete event models [8]. Numerous agent-based models have been developed by Sandia National Laboratories and by Argonne National Laboratory to describe interdependencies between infrastructures. Sandia first developed a model called ASPEN [5] that has been extended and modified several times to model interdependent infrastructures [4, 9].

More recent extensions such as the *NISAC Agent-Based Laboratory for Economics* (N-ABLE) model are used to investigate physical effects of cyber-attacks in infrastructures [53]. In this setting, the agent-based model is applied to simulate interactions between economic agents (e.g., firms) that rely on infrastructures such as electric power or transportation systems. The simulation thus yields an estimate of the consequences of failures in these systems in the real world. Argonne introduced an agent-based model for power systems called *Spot Market Agent Research Tool* (SMART II) in 2000 [67] that takes into account the topology of the network. The model has been extended to investigate natural gas infrastructures and their interdependencies with power infrastructures based on the *Flexible Agent Simulation Toolkit* (FAST) simulation software [68].

Agent-based models are also used to model risk responses in a complex society [12], which in turn provide a basis to simulate responses of society to incidents in critical utilities [11]. Applications include the interaction between communication systems and transportation systems in case of an evacuation [3] or interdependencies between web systems and critical information infrastructures in case of a natural disaster [13]. Further, these models can be used to analyze maintenance strategies in CIs [50]. Impacts of a malicious cyber-attack on a power system have been analyzed in [7], where society is regarded as a group of interacting agents. A simulation model in the context of interdependent CIs was introduced in [15].

Generally, agent-based models provide detailed information about potential consequences of incidents for a given scenario and thus allow for very detailed "What-If" analysis. However, due to the detailed knowledge that is required to build these models, they are typically very specific. If the system changes, agent based models need to be adapted to the new situation. Furthermore, they are built on assumptions about the behavior of the agents. Such assumptions are usually subjective and hard to justify. Especially, if information about the agent's behavior or motivation is missing, agent-based models are not suitable.

2.5.5   *Economy Based Methods*

Since critical infrastructures are crucial for production and economy in general, their interdependencies may as well be investigated with economic models [87]. The two most widely used methods in this context are input-output models and computable general equilibrium models, both described in more detail below.

2.5.5.1 Input-Output Models

Input-output models have been introduced in the early 1950s [62] to describe production and consumption of industry sector. Based on this the *Input-Output Inoperability Model (IIM)* has been proposed in [39] to investigate the risk that a CI is not able to perform its tasks, i.e., the risk of inoperability of an infrastructure. During recent years, the IIM model has been applied to investigate disruptions to economic systems [89], risk analysis in interdependent infrastructures [17, 38], effects on international trade [49] as well as risk propagation in CIs [72].

The extended dynamic IIM [64] describes the development of the inoperability over time including recovery over time. These high-level models are applicable in various situations, ranging from natural disasters [21] to intentional attacks [90] and provide both analytical results and an intuitive understanding of the analyzed dependencies. However, these models are typically not capable of providing detailed information about interdependencies for single components or of taking into account real-time outputs. Even more restrictively, the model assumes linear interdependence and can thus only provide approximate solutions.

2.5.5.2 Computable General Equilibrium Models

Computable general equilibrium models can be seen as extensions of input-output models [86]. Improvements compared to existing input-output models include factors such as non-linearity or consumers responses to prices. Several extensions exist, including a *spatial computable general equilibrium model* as used in [109]. A drawback of these models is that they are trained under normal working conditions (e.g., from historical data) and thus only describe interdependencies in this setting, which might change after extreme events. *Advanced computable general equilibrium models* support investigation of resilience [88]. While computable general equilibrium models capture more details than classical input-output models, they require a lot of data (e.g., to estimate elasticity values for resilience) and on assumptions on the form of production or utility functions. The latter will play a central role in game theoretic models, as goal functions for optimization. For a decision maker, they act as security "scores" to quantify risk and security like a benchmark figure. We will formally introduce these concepts later in Chap. 3.

2.6 Viewing Security as a Control Problem

The complexity of interdependencies between CIs, and the diverse factors of uncertainty induced by external factors beyond our control, as well as the potential of errors from the inside (human failure, software flaws, technical breakdowns, any many others) make the establishment of security a problem of "controlling risk under uncertainty". Figure 2.1 shows an abstraction of the situation, with the CI being the central area of interest (of cloudy shape in the picture), and with external

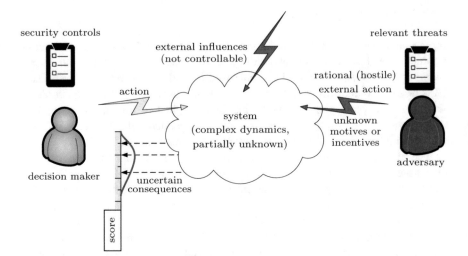

Fig. 2.1 High-level view on security as a control problem

actors taking influence on it. *Decision theory* is herein the search for optimal action
to keep *natural* external factors under control, while *game theory* is about dealing
with rationally acting opponents (shown on the right). This difference is explored
in deeper detail in Chap. 3, with all game models to follow fitting into this bigger
picture in their own particular way. Their use is basically for the "players" outside
the CI, while the inner dynamics of the CI under security control is in many cases
most accurately captured by simulation models. We leave the latter as a separate
challenge outside the scope of the book, since simulation will be of primary use
when it comes to the construction of games, specifically utility functions and loss
distributions. Our main focus in the rest of the book is helping the defending player
in making decisions against both rational and irrational actors as well as taking
influence on a CI from outside or inside.

References

1. Akyol E, Rose K, Basar T (2015) Optimal zero-delay jamming over an additive noise channel.
   IEEE Trans Inf Theory 61(8):4331–4344. https://doi.org/10.1109/tit.2015.2445344
2. BABS (2019) Katalog der Gefährdungen, Katastrophen und Notlagen Schweiz. Technical
   report, Bundesamt für Bevölkerungsschutz (BABS)
3. Barrett C, Beckman R, Channakeshava K, Huang F, Kumar VA, Marathe A, Marathe MV, Pei
   G (2010) Cascading failures in multiple infrastructures: from transportation to communication
   network. In: 2010 5th international conference on critical infrastructure (CRIS). IEEE, pp 1–
   8. https://doi.org/10.1109/CRIS.2010.5617569
4. Barton DC, Eidson ED, Schoenwald DA, Stamber KL, Reinert R (2000) Aspen-EE: an agent-
   based model of infrastructure interdependency. Technical report. SAND2000-2925, 774027,
   Sandia National Labs. https://doi.org/10.2172/774027

5. Basu N, Pryor R, Quint T (1998) ASPEN: a microsimulation model of the economy. Comput Econ 12(3):223–241. https://doi.org/s10.1023/A:1008691115079
6. Bateman T (2013) Police warning after drug traffickers' cyber-attack. www.bbc.com/news/world-europe-24539417. BBC News, retrieved 25 Feb 2020
7. Bompard E, Napoli R, Xue F (2009) Assessment of information impacts in power system security against malicious attacks in a general framework. Reliab Eng Syst Saf 94(6):1087–1094. https://doi.org/10.1016/j.ress.2009.01.002
8. Borshchev A, Filippov A (2004) From system dynamics and discrete event to practical agent based modeling: reasons, techniques, tools. In: The 22nd international conference of the system dynamics society
9. Brown T, Beyeler W, Barton D (2004) Assessing infrastructure interdependencies: the challenge of risk analysis for complex adaptive systems. Int J Crit Infrastruct 1(1):108. https://doi.org/10.1504/IJCIS.2004.003800
10. Buldyrev SV, Parshani R, Paul G, Stanley HE, Havlin S (2010) Catastrophic cascade of failures in interdependent networks. Nature 464:1025. https://doi.org/10.1038/nature08932
11. Busby J, Gouglidis A, Rass S, Konig S (2016) Modelling security risk in critical utilities: the system at risk as a three player game and agent society. In: 2016 IEEE international conference on systems, man, and cybernetics (SMC). IEEE, Budapest, pp 001758–001763. 10.1109/SMC.2016.7844492. http://ieeexplore.ieee.org/document/7844492/
12. Busby JS, Onggo B, Liu Y (2016) Agent-based computational modelling of social risk responses. Eur J Oper Res 251(3):1029–1042. https://doi.org/10.1016/j.ejor.2015.12.034
13. Cardellini V, Casalicchio E, Tucci S (2006) Agent-based modeling of web systems in critical information infrastructures. In: International workshop on complex networks and infrastructure protection (CNIP 2006)
14. Carreras BA, Lynch VE, Dobson I, Newman DE (2002) Critical points and transitions in an electric power transmission model for cascading failure blackouts. Chaos Interdisc J Nonlinear Sci 12(4):985–994. https://doi.org/10.1063/1.1505810
15. Casalicchio E, Galli E, Tucci S (2010) Agent-based modelling of interdependent critical infrastructures. Int J Syst Syst Eng 2(1):60. https://doi.org/10.1504/IJSSE.2010.035381
16. Chen J, Thorp JS, Dobson I (2005) Cascading dynamics and mitigation assessment in power system disturbances via a hidden failure model. Int J Electr Power Energy Syst 27(4):318–326. https://doi.org/10.1016/j.ijepes.2004.12.003
17. Chen P, Scown C, Matthews HS, Garrett JH, Hendrickson C (2009) Managing critical infrastructure interdependence through economic input-output methods. J Infrastruct Syst 15(3):200–210. https://doi.org/10.1061/(ASCE)1076-0342(2009)15:3(200)
18. Chen PY, Cheng SM, Chen KC (2012) Smart attacks in smart grid communication networks. IEEE Commun Mag 50(8):24–29. https://doi.org/10.1109/mcom.2012.6257523
19. Cimpanu C (2017) WannaCry ransomware infects actual medical devices, not just computers. https://www.bleepingcomputer.com/news/security/wannacry-ransomware-infects-actual-medical-devices-not-just-computers/. Bleeping Computer. Retrieved 25 Feb 2020
20. Condliffe J (2016) Ukraine's power grid gets hacked again, a worrying sign for infrastructure attacks, 22 Dec 2016. https://www.technologyreview.com/s/603262/ukraines-power-grid-gets-hacked-again-a-worrying-sign-for-infrastructure-attacks/. Retrieved 26 July 2017
21. Crowther KG, Haimes YY, Taub G (2007) Systemic valuation of strategic preparedness through application of the inoperability input-output model with lessons learned from hurricane katrina. Risk Anal 27(5):1345–1364. https://doi.org/10.1111/j.1539-6924.2007.00965.x
22. Dey P, Mehra R, Kazi F, Wagh S, Singh NM (2016) Impact of topology on the propagation of cascading failure in power grid. IEEE Trans Smart Grid 7(4):1970–1978. https://doi.org/10.1109/TSG.2016.2558465
23. Dobson I (2012) Estimating the propagation and extent of cascading line outages from utility data with a branching process. IEEE Trans Power Syst 27(4):2146–2155. https://doi.org/10.1109/TPWRS.2012.2190112

24. Dobson I, Carreras B, Newman D (2004) A branching process approximation to cascading load-dependent system failure. In: Proceedings of the 37th annual Hawaii international conference on system sciences, 2004. IEEE, 10pp. https://doi.org/10.1109/HICSS.2004. 1265185
25. Dobson I, Carreras BA, Lynch VE, Newman DE (2007) Complex systems analysis of series of blackouts: cascading failure, critical points, and self-organization. Chaos Interdiscip J Nonlinear Sci 17(2):026103. https://doi.org/10.1063/1.2737822
26. Dobson I, Carreras BA, Newman DE (2005) A loading-dependent model of probabilistic cascading failure. Probab Eng Inf Sci 19(1). https://doi.org/10.1017/S0269964805050023
27. Dobson I, Kim J, Wierzbicki KR (2010) Testing branching process estimators of cascading failure with data from a simulation of transmission line outages. Risk Anal 30(4):650–662. https://doi.org/10.1111/j.1539-6924.2010.01369.x
28. Dobson I, Newman DE (2017) Cascading blackout overall structure and some implications for sampling and mitigation. Int J Electr Power Energy Syst 86:29–32. https://doi.org/10. 1016/j.ijepes.2016.09.006
29. Dong H, Cui L (2016) System reliability under cascading failure models. IEEE Trans Reliab 65(2):929–940. https://doi.org/10.1109/TR.2015.2503751
30. Dong P, Han Y, Guo X, Xie F (2015) A systematic review of studies on cyber physical system security. Int J Secur Appl 9(1):155–164. https://doi.org/10.14257/ijsia.2015.9.1.17
31. Etesami SR, Başar T (2019) Dynamic games in cyber-physical security: an overview. Dyn Games Appl. https://doi.org/10.1007/s13235-018-00291-y
32. European Commission (2008) COUNCIL DIRECTIVE 2008/114/EC of 8 December 2008 on the identification and designation of European critical infrastructures and the assessment of the need to improve their protection. Off J Eur Union (L345):75–82. http://eur-lex.europa. eu/legal-content/EN/TXT/PDF/?uri=CELEX:32008L0114&from=EN
33. Fan W, Huang S, Mei S (2016) Invulnerability of power grids based on maximum flow theory. Phys A Stat Mech Appl 462:977–985. https://doi.org/10.1016/j.physa.2016.06.109
34. Fang J, Su C, Chen Z, Sun H, Lund P (2016) Power system structural vulnerability assessment based on an improved maximum flow approach. IEEE Trans Smart Grid 9(2):777–785. https://doi.org/10.1109/TSG.2016.2565619
35. Gouglidis A, König S, Green B, Rossegger K, Hutchison D (2018) Protecting water utility networks from advanced persistent threats: a case study. Springer International Publishing, Cham, pp 313–333. https://doi.org/10.1007/978-3-319-75268-6_13
36. Greenberg A (2018) WIRED: the untold story of NotPetya, the most devastating cyberattck in history. https://www.wired.com/story/notpetya-cyberattack-ukraine-russia-code-crashed- the-world/
37. Guo H, Zheng C, Iu HHC, Fernando T (2017) A critical review of cascading failure analysis and modeling of power system. Renew Sustain Energy Rev 80:9–22. https://doi.org/10.1016/ j.rser.2017.05.206
38. Haimes Y, Santos J, Crowther K, Henry M, Lian C, Yan Z (2007) Risk analysis in interdependent infrastructures. In: Goetz E, Shenoi S (eds) Critical infrastructure protection, vol 253. Springer, pp 297–310. https://doi.org/10.1007/978-0-387-75462-8_21
39. Haimes YY, Pu J (2001) Leontief-based model of risk in complex interconnected infrastruc- tures. J Infrastruct Syst 7(1):1–12. https://doi.org/10.1061/(ASCE)1076-0342(2001)7:1(1)
40. Hasan S, Foliente G (2015) Modeling infrastructure system interdependencies and socioe- conomic impacts of failure in extreme events: emerging R&D challenges. Nat Haz 78(3):2143–2168. https://doi.org/10.1007/s11069-015-1814-7
41. Hausken K (2017) Defense and attack for interdependent systems. Eur J Oper Res 256(2):582–591. https://doi.org/10.1016/j.ejor.2016.06.033

42. Henneaux P, Labeau PE, Maun JC (2012) A level-1 probabilistic risk assessment to blackout hazard in transmission power systems. Reliab Eng Syst Saf 102:41–52. https://doi.org/10.1016/j.ress.2012.02.007

43. Henneaux P, Labeau PE, Maun JC, Haarla L (2016) A two-level probabilistic risk assessment of cascading outages. IEEE Trans Power Syst 31(3):2393–2403. https://doi.org/10.1109/TPWRS.2015.2439214

44. Hines P, Cotilla-Sanchez E, Blumsack S (2010) Do topological models provide good information about electricity infrastructure vulnerability? Chaos Interdisc J Nonlinear Sci 20(3):033122. https://doi.org/10.1063/1.3489887

45. Holme P (2002) Edge overload breakdown in evolving networks. Phys Rev E 66(3). https://doi.org/10.1103/PhysRevE.66.036119

46. Holme P, Kim BJ (2002) Vertex overload breakdown in evolving networks. Phys Rev E 65(6). https://doi.org/10.1103/PhysRevE.65.066109

47. Holmgren A, Jenelius E, Westin J (2007) Evaluating strategies for defending electric power networks against antagonistic attacks. IEEE Trans Power Syst 22(1):76–84. https://doi.org/10.1109/tpwrs.2006.889080

48. ICS-CERT (2016) Cyber-attack against Ukrainian critical infrastructure. https://ics-cert.us-cert.gov/alerts/IR-ALERT-H-16-056-01

49. Jung J, Santos JR, Haimes YY (2009) International trade inoperability input-output model (IT-IIM): theory and application. Risk Anal 29(1):137–154. https://doi.org/10.1111/j.1539-6924.2008.01126.x

50. Kaegi M, Mock R, Kröger W (2009) Analyzing maintenance strategies by agent-based simulations: a feasibility study. Reliab Eng Syst Saf 94(9):1416–1421. https://doi.org/10.1016/j.ress.2009.02.002

51. Kar D, Nguyen TH, Fang F, Brown M, Sinha A, Tambe M, Jiang AX (2016) Trends and applications in Stackelberg security games. In: Handbook of dynamic game theory. Springer International Publishing, pp 1–47. https://doi.org/10.1007/978-3-319-27335-8_27-1

52. Karnouskos S (2011) Stuxnet worm impact on industrial cyber-physical system security. In: IECON 2011 – 37th annual conference of the IEEE industrial electronics society. IEEE. https://doi.org/10.1109/iecon.2011.6120048

53. Kelic A, Warren DE, Phillips LR (2008) Cyber and physical infrastructure interdependencies. Technical report, SAND2008-6192, 945905, Sandia National Laboratories. https://doi.org/10.2172/945905

54. Khouzani M, Sarkar S, Altman E (2012) Saddle-point strategies in malware attack. IEEE J Sel Areas Commun 30(1):31–43. https://doi.org/10.1109/jsac.2012.120104

55. Kim CJ, Obah OB (2007) Vulnerability assessment of power grid using graph topological indices. Int J Emerg Electr Power Syst 8(6). https://doi.org/10.2202/1553-779X.1738

56. Kim J, Dobson I (2010) Approximating a loading-dependent cascading failure model with a branching process. IEEE Trans Reliab 59(4):691–699. https://doi.org/10.1109/TR.2010.2055928

57. König S, Gouglidis A, Green B, Solar A (2018) assessing the impact of malware attacks in utility networks. Springer International Publishing, Cham, pp 335–351. https://doi.org/10.1007/978-3-319-75268-6_14

58. König S, Rass S, Schauer S (2019) Cyber-attack impact estimation for a port. In: Jahn C, Kersten W, Ringle CM (eds) Digital transformation in maritime and city logistics: smart solutions for logistics. In: Proceedings of the hamburg international conference of logistics (HICL), vol 28. epubli GmbH, pp 164–183. https://doi.org/10.15480/882.2496. ISBN 978-3-7502-4949-3

59. La QD, Quek TQS, Lee J (2016) A game theoretic model for enabling honeypots in IoT networks. In: 2016 IEEE international conference on communications (ICC). IEEE. https://doi.org/10.1109/icc.2016.7510833

60. Lechner U, Dännart S, Rieb A, Rudel S (eds) (2018) Case Kritis – Fallstudien zur IT-Sicherheit in Kritischen Infrastrukturen. Logos Verlag, Berlin. https://doi.org/10.30819/4727

61. Lee RM, Assante MJ, Conway T (2016) Analysis of the cyber attack on the Ukrainian power grid. Technical report, E-ISAC, Washington. https://ics.sans.org/media/E-ISAC_SANS_Ukraine_DUC_5.pdf
62. Leontief WW (1951) Input-output economics. Sci Am 185:15–21
63. Li Y, Shi L, Cheng P, Chen J, Quevedo DE (2015) Jamming attacks on remote state estimation in cyber-physical systems: a game-theoretic approach. IEEE Trans Autom Control 60(10):2831–2836. https://doi.org/10.1109/tac.2015.2461851
64. Lian C, Haimes YY (2006) Managing the risk of terrorism to interdependent infrastructure systems through the dynamic inoperability input–output model. Syst Eng 9(3):241–258. https://doi.org/10.1002/sys.20051
65. Mehic M, Fazio P, Rass S, Maurhart O, Peev M, Poppe A, Rozhon J, Niemiec M, Voznak M (2019) A novel approach to quality-of-service provisioning in trusted relay quantum key distribution networks. IEEE/ACM Trans Netw 1–10. https://doi.org/10.1109/TNET.2019.2956079. https://ieeexplore.ieee.org/document/8935373/
66. Motter AE, de Moura APS, Lai YC, Dasgupta P (2002) Topology of the conceptual network of language. Phys Rev E 65(6). https://doi.org/10.1103/PhysRevE.65.065102
67. North MJ (2000) Smart II: the spot market agent research tool version 2.0. Nat Res Environ Issues 8(11)
68. North MJ (2001) Toward strength and stability: agent-based modeling of infrastructure markets. Soc Sci Comput Rev 19(3):307–323. https://doi.org/10.1177/089443930101900306
69. Office of Homeland Security (2002) National strategy for homeland security. Technical report, Department of Homeland Security
70. Oliva G, Panzieri S, Setola R (2012) Modeling and simulation of critical infrastructures. In: Flammini F (ed) WIT transactions on state of the art in science and engineering, vol 1, 1 edn. WIT Press, pp 39–56. https://doi.org/10.2495/978-1-84564-562-5/03
71. Ouyang M (2014) Review on modeling and simulation of interdependent critical infrastructure systems. Reliab Eng Syst Saf 121:43–60. https://doi.org/10.1016/j.ress.2013.06.040. https://linkinghub.elsevier.com/retrieve/pii/S0951832013002056
72. Owusu A, Mohamed S, Anissimov Y (2010) Input-output impact risk propagation in critical infrastructure interdependency. In: International conference on computing in civil and building engineering (icccbe). Nottingham University Press
73. Pita J, Jain M, Ordonez F, Portway C, Tambe M, Western C (2008) ARMOR security for Los Angeles international airport. In: Proceedings of the 23rd AAAI conference on artificial intelligence (2008), pp 1884–1885
74. Pita J, Tambe M, Kiekintveld C, Cullen S, Steigerwald E (2011) GUARDS – innovative application of game theory for national airport security. In: IJCAI 2011, pp 2710–2715. https://doi.org/10.5591/978-1-57735-516-8/IJCAI11-451
75. Qi J, Dobson I, Mei S (2013) Towards estimating the statistics of simulated cascades of outages with branching processes. IEEE Trans Power Syst 28(3):3410–3419. https://doi.org/10.1109/TPWRS.2013.2243479
76. Qi J, Ju W, Sun K (2016) Estimating the propagation of interdependent cascading outages with multi-type branching processes. IEEE Trans Power Syst 1212–1223. https://doi.org/10.1109/TPWRS.2016.2577633
77. Qi J, Sun K, Mei S (2015) An interaction model for simulation and mitigation of cascading failures. IEEE Trans Power Syst 30(2):804–819. https://doi.org/10.1109/TPWRS.2014.2337284
78. Rahnamay-Naeini M, Hayat MM (2016) Cascading failures in interdependent infrastructures: an interdependent Markov-chain approach. IEEE Trans Smart Grid 7(4):1997–2006. https://doi.org/10.1109/TSG.2016.2539823
79. Rahnamay-Naeini M, Wang Z, Ghani N, Mammoli A, Hayat MM (2014) Stochastic analysis of cascading-failure dynamics in power grids. IEEE Trans Power Syst 29(4):1767–1779. https://doi.org/10.1109/TPWRS.2013.2297276

80. Rass S, König S (2012) Turning Quantum Cryptography against itself: how to avoid indirect eavesdropping in quantum networks by passive and active adversaries. Int J Adv Syst Meas 5(1 & 2):22–33

81. Rass S, Konig S, Schauer S (2017) Defending against advanced persistent threats using game-theory. PLoS ONE 12(1):e0168675. https://doi.org/10.1371/journal.pone.0168675

82. Rezazadeh A, Talarico L, Reniers G, Cozzani V, Zhang L (2018) Applying game theory for securing oil and gas pipelines against terrorism. Reliab Eng Syst Saf. https://doi.org/10.1016/j.ress.2018.04.021

83. Rinaldi S (2004) Modeling and simulating critical infrastructures and their interdependencies. In: Proceedings of the 37th annual Hawaii international conference on system sciences, 2004. IEEE. https://doi.org/10.1109/hicss.2004.1265180

84. Rinaldi SM, Peerenboom JP, Kelly TK (2001) Identifying, understanding, and analyzing critical infrastructure interdependencies. IEEE Control Syst 21(6):11–25. https://doi.org/10.1109/37.969131

85. Rosato V, Issacharoff L, Tiriticco F, Meloni S, Porcellinis SD, Setola R (2008) Modelling interdependent infrastructures using interacting dynamical models. Int J Crit Infrastruct 4(1):63. https://doi.org/10.1504/IJCIS.2008.016092

86. Rose A (2004) Economic principles, issues, and research priorities in hazard loss estimation. In: Okuyama Y, Chang SE (eds) Modeling spatial and economic impacts of disasters. Springer, Berlin/Heidelberg, pp 13–36. https://doi.org/10.1007/978-3-540-24787-6_2

87. Rose A (2005) Tracing infrastructure interdependence through economic interdependence. Technical report, CREATE Research Archive. http://research.create.usc.edu/nonpublished_reports/78. Non-published Research Reports, Paper 78

88. Santella N, Steinberg LJ, Parks K (2009) Decision making for extreme events: Modeling critical infrastructure interdependencies to aid mitigation and response planning. Rev Policy Res 26(4):409–422. https://doi.org/10.1111/j.1541-1338.2009.00392.x

89. Santos JR (2006) Inoperability input-output modeling of disruptions to interdependent economic systems. Syst Eng 9(1):20–34. https://doi.org/10.1002/sys.20040

90. Santos JR, Haimes YY, Lian C (2007) A framework for linking cybersecurity metrics to the modeling of macroeconomic interdependencies. Risk Anal 27(5):1283–1297. https://doi.org/10.1111/j.1539-6924.2007.00957.x

91. Schneider CM, Yazdani N, Araújo NAM, Havlin S, Herrmann HJ (2013) Towards designing robust coupled networks. Sci Rep 3(1). https://doi.org/10.1038/srep01969

92. William S (2018) Lessons learned review of the WannaCry Ransomware Cyber Attack. Report NHS, Feb 2018

93. Shao J, Buldyrev SV, Havlin S, Stanley HE (2011) Cascade of failures in coupled network systems with multiple support-dependence relations. Phys Rev E 83(3). https://doi.org/10.1103/PhysRevE.83.036116

94. Shekhtman LM, Danziger MM, Havlin S (2016) Recent advances on failure and recovery in networks of networks. Chaos Solitons Fractals 90:28–36. https://doi.org/10.1016/j.chaos.2016.02.002

95. Mei S, He F, Zhang X, Wu S, Wang G (2009) An improved OPA model and blackout risk assessment. IEEE Trans Power Syst 24(2):814–823. https://doi.org/10.1109/TPWRS.2009.2016521

96. Shieh EA, An B, Yang R, Tambe M, Baldwin C, DiRenzo J, Maule B, Meyer G (2013) PROTECT: an application of computational game theory for the security of the ports of the United States. In: Proceedings of the 26th AAAI conference on artificial intelligence (AAAI'12), pp 2173–2179

97. Song J, Cotilla-Sanchez E, Ghanavati G, Hines PDH (2016) Dynamic modeling of cascading failure in power systems. IEEE Trans Power Syst 31(3):2085–2095. https://doi.org/10.1109/TPWRS.2015.2439237

98. Tazi K, Abdi F, Abbou MF (2015) Review on cyber-physical security of the smart grid: Attacks and defense mechanisms. In: 2015 3rd international renewable and sustainable energy conference (IRSEC). IEEE, pp 1–6. https://doi.org/10.1109/IRSEC.2015.7455127

99. Touhiduzzaman M, Hahn A, Srivastava A (2018) A diversity-based substation cyber defense strategy utilizing coloring games. IEEE Trans Smart Grid 1–1. https://doi.org/10.1109/TSG. 2018.2881672
100. UP KRITIS (2014) Public-private partnership for critical infrastructure protection – basis and goals. Technical report, Bundesamt für Sicherheit in der Informationstechnick (BSI)
101. Wang WX, Chen G (2008) Universal robustness characteristic of weighted networks against cascading failure. Phys Rev E 77(2). https://doi.org/10.1103/PhysRevE.77.026101
102. Wang Z, Scaglione A, Thomas RJ (2012) A Markov-transition model for cascading failures in power grids. In: 2012 45th Hawaii international conference on system sciences. IEEE, pp 2115–2124. https://doi.org/10.1109/HICSS.2012.63
103. Wei DQ, Luo XS, Zhang B (2012) Analysis of cascading failure in complex power networks under the load local preferential redistribution rule. Phys A Stat Mech Appl 391(8):2771–2777. https://doi.org/10.1016/j.physa.2011.12.030
104. Wei L, Sarwat AI, Saad W, Biswas S (2018) Stochastic games for power grid protection against coordinated cyber-physical attacks. IEEE Trans Smart Grid 9(2):684–694. https://doi.org/10.1109/TSG.2016.2561266
105. Wu SJ, Chu MT (2017) Markov chains with memory, tensor formulation, and the dynamics of power iteration. Appl Math Comput 303:226–239. https://doi.org/10.1016/j.amc.2017.01. 030
106. Zetter K (2016) Everything we know about Ukraine's power plant hack | WIRED. https:// www.wired.com/2016/01/everything-we-know-about-ukraines-power-plant-hack/
107. Zhang L, Reniers G (2016) A game-theoretical model to improve process plant protection from terrorist attacks. Risk Anal 36(12):2285–2297. https://doi.org/10.1111/risa.12569
108. Zhang L, Reniers G (2018) Applying a Bayesian Stackelberg game for securing a chemical plant. J Loss Prev Process Ind 51:72–83. https://doi.org/10.1016/j.jlp.2017.11.010
109. Zhang P, Peeta S (2011) A generalized modeling framework to analyze interdependencies among infrastructure systems. Transp Res Part B Methodol 45(3):553–579. https://doi.org/ 10.1016/j.trb.2010.10.001
110. Zhang X, Zhan C, Tse CK (2017) Modeling the dynamics of cascading failures in power systems. IEEE J Emerg Sel Top Circuits Syst 7(2):192–204
111. Zhu Q, Saad W, Han Z, Poor HV, Başalr T (2011) Eavesdropping and jamming in next-generation wireless networks: a game-theoretic approach. In: 2011-MILCOM 2011 military communications conference. IEEE, pp 119–124

Chapter 3
Mathematical Decision Making

I am prepared for the worst, but hope for the best.

B. Disraeli

Abstract Since both, decision- and game theory vitally employ optimization at their core, this chapter will provide the basic ideas, concepts and modeling aspects of optimization. It is intended to provide the mathematical basics for the further chapters. The presentation is to the point of a simple, compact and self-contained description of: (i) what is decision- and game-theory about, (ii) how do the two areas differ, and (iii) how does the practical work with these models look like when we strive for solutions. Specifically, we discuss preference relations, real and stochastic ordering relations and optimization as the most general covering framework, including single- and multi-goal optimization, with applications in being decision theory and game theory. Numeric examples accompany each section and concept. The opening of the chapter will specifically set the notation for all upcoming (mathematical) descriptions, to be consistent throughout the entire presentation (and book).

3.1 Preference and Ordering Relations

Many situations in life, not limited to the context of critical infrastructure protection, call for a choice among a number of options. Depending on the kind of choice, this may be the selection of some item from a finite collection, say, to buy a new security tool, or the determination of the time to implement a certain (fixed) action, say, a security patrol, patch or similar. Decisions about timing are examples of choices made from infinite sets (as the time is a continuous quantity), but both, discrete and continuous choices can be unified under the common denominator of decision theory. So, given two actions a_1, a_2 with a meaning specific to the (here arbitrary) context, how would a decision maker be able to choose between a_1 and a_2, or maybe even decide to refrain from the decision at all? In absence of

S. Rass et al., *Cyber-Security in Critical Infrastructures*, Advanced Sciences and Technologies for Security Applications, https://doi.org/10.1007/978-3-030-46908-5_3

any criterion to judge a_1 relative to a_2, the decision is impossible, unless ad hoc reasons apply. Since relying on coincidental serendipity in all imaginable situations is clearly nonsense, we require a systematic way of inference to guide a decision maker along justified routes to a best decision. The exact meaning of "best" is the key concept to be defined, and decision theory is technically a re-axiomatization of statistical inference towards exactly solving the choice problem in flexible ways.

Ever since humans are involved in this process, decision theory has received a lot of criticism up to rejection by some statisticians, based on the argument that subjective decision making processes are almost always impossible to capture to the accuracy required for mathematics/statistics to work. We will touch some of these issues later in Chap. 6, and postpone a continuation of this discussion until then. For now, we are interested in how actions can be judged against one another, so as to make a choice, even though none of the available options may be "good" in an absolute sense.

Most generally, if a set AS of actions is available to choose from, suppose that the elements of AS admit some relation $\trianglelefteq$, with the understanding that $a_1 \trianglelefteq a_2$ means action a_2 to be *preferable* over action a_1. In mathematical decision theory, it is common practice to quantify actions by assigning a number $g(a)$ to the action a, and to define $a_1 \trianglelefteq a_2$ if and only if $g(a_1) \leq g(a_2)$, provided that such a function g exists. Indeed, the existence of the function g depends on what requirements we impose on the preference relation $\trianglelefteq$. Some properties may be necessary to assure the existence of g, and others may be required so that g is feasible for a mathematical use (e.g., to be linear, continuous, etc.). Note that for the same reason, we clearly cannot call $\trianglelefteq$ an "ordering" relation, since such relations have a precisely defined set of properties, i.e., we call $\leq$ an *ordering relation* , if it is:

- *Reflexive*: every element x satisfies $x \leq x$.
- *Antisymmetric*: if two elements x, y satisfy $x \leq y$ and $y \leq x$ then $x = y$, i.e., they are identical.
- *Transitive*: if $x \leq y$ and $y \leq z$, then also $x \leq z$.

The properties of preference relations are similar but not identical, and we will discuss the complete list in Chap. 5. Generally, neither preference nor ordering relations are special cases of one another. For both kinds of relations, we call them *total* (meaning that any two elements are comparable), if any two elements x, y satisfy the relation as $x \leq y$ and $y \leq x$, or $x \trianglelefteq y$ and $y \trianglelefteq x$, respectively. If there are incomparable elements, we call the respective ordering *partial* . In this case, there exist elements x, y such that neither $x \leq y$ nor $y \leq x$, respectively neither $x \trianglelefteq y$ nor $y \trianglelefteq x$ hold.

Let $(R, \leq)$ be an arbitrary set whose elements are totally ordered (a common choice, though not the only one, is $R = \mathbb{R}$), and let $g : AS \to (R, \leq)$ take values in R. We say that the function g *represents the ordering relation $\leq$ on AS*, if

$$\forall a_1, a_2 : a_1 \trianglelefteq a_2 \iff g(a_1) \leq g(a_2). \tag{3.1}$$

Many practically relevant instances of preference have representations by real-valued, even continuous, functions g, but the representation is nontrivial in general.

The common development of mathematical decision theory starts from a systematic identification of what properties a reasonable preference relation should have, imposes them as axioms, and derives the existence of a continuous and real-valued function g to make it useful with optimization theory. Whether these assumptions accurately reflect decision making processes in reality is a different story. The bottom line of Chap. 5 will be a discussion of empirical findings on when and how the axioms are violated in reality, and how the situation can be fixed again to retain a working decision theory.

For the moment, however, we will continue with the assumption that a function g is available to quantify and thereby rank actions against one another. The value assigned by g be of diverse type, but in almost all cases is either a *score, measure* or *metric*. Even though sometimes understood so, the terms are not synonymous, and it pays to bear in mind the differences:

- A *score* is a number that by itself carries no meaning, and only serves for comparative purposes; say, we may prefer action a_1 over action a_2, if the former scores higher than the latter. The absolute value of the score, however, may be void of semantic. Scores are what risk is typically quantified in, by assigning consecutive loss ranges to consecutive scores. Likewise, security vulnerabilities can be compared in terms of scores, compiled from categorical assessments of a vulnerability's characteristics such as required adversary's skill level, time to exploit, and similar. The Common Vulnerability Scoring System (CVSS) is a well known example and application of this method; with a CVSS number itself having no meaning beyond that of an indicator of severity of some vulnerability.
- A *measure* is similar to a score, but has a meaning on its own, i.e., even in absence of competing action's measures. In the context of risk assessment, two common instances of measures are (among many others):
  - *Probabilities*: each action a may trigger different events to unfold, and the relative frequency of occurrences of such an event (as a consequence of an action a) approximates the *probability* of that event, conditional on the action a. Uncertainty in the consequences of an action then needs reasoning about which effects are probable or unlikely to occur, towards an aid of what action to choose.
  - *Losses, revenues* or *utilities*: A loss, conversely a revenue or utility, generally quantifies (dis)advantage as a consequence of an action a. The terms revenue, utility or payoff are often used synonymously, and a loss is numerically often taken as the same concept only with a negative sign. The quantification of loss in absolute terms can be arbitrary, including monetary units, gain/loss of reputation (say if a company acquires/loses customers or market share), up to saved/lost human lives (with the latter being tied to complex ethical issues which we refrain from discussing any further here).

- A *metric* quantifies the difference between two items. As such, it is similar to a score in the sense of having no semantics (or even being undefined) for a single item, and is only useful to compare objects. Unlike a score, however, a metric usually does not help ranking two objects or actions, but is primarily intended to measure the *similarity* value of a metric with a meaning, as for a measure just explained above.

It should be noted that the terms *measure* and *metric* have rigorous mathematical meanings, and the mathematical objects must not be mixed up with our description above. We will not explore the depths of measure theory or the theoretical applications of metrics in the following, unless this is necessary for our exposition.

3.2   Optimization

The function g now provides us with the means to quantify actions and to mutually rank them, which leads us to the question "Which a^* among all elements $a \in AS$ is the best choice?" and thus to a general optimization problem. Our exposition will evolve around the following generic problem description: let C be a set, and let $g : \Omega \to (R, \leq)$ be a function, whose domain Ω covers C, and whose image space is ordered (to ease matters, we will assume $\leq$ to be a total order on R). In most of what follows, we will take C as some *action space* AS, but for a sound assurance of best decisions, i.e., optima, to exist over the set C, it is often useful to extend C into a larger set $\Omega \supset C$, say, by taking Ω as the convex hull of C, which we formally denote as $\Omega = \Delta(C)$, but other choices are possible too (yet we will not have much use for them). The set $\Delta(AS)$ is defined as $\Delta(AS) = \{(p_1, \ldots, p_n) \in [0, 1]^n : p_1 + p_2 + \ldots + p_n = 1\}$. The simplex extends the set of (deterministic) actions into *randomized* actions, meaning that a decision maker can, whenever the decision comes up, take a different action to behave optimal in a long run. We will later see that the existence of optimal behavior can depend on the freedom to change behavior.

In this notation, we can think of an optimization problem as the task to compute an element x_0 so that

$$[g(x_0) \leq g(x) \quad \forall x \in C] \text{ or } [g(x_0) \geq g(x) \quad \forall x \in C], \tag{3.2}$$

where the left is a *minimization problem* , and the right is a *maximization problem*. We write $\operatorname{argmin}_{x \in C} g(x)$ to reference the *point* x_0 at where g attains a minimum, as opposed to writing $\min_{x \in C} g(x)$ for the minimum *value* at this (herein untold) point x_0 (likewise for maximization problems). Generally, "argmin" will be a set of values in C (the optimum point is generally not unique), while "min" will be a single value in the domain of g. The function g is also called the *goal function*. For mathematical games, the goal will always be some *utility function*, whose particularities and construction are discussed later. The assumption that $\leq$ and $\geq$ are total orders in

(3.2) assures that if an optimal value (not location of) $g(x_0)$ will be unique, if it exists. This uniqueness is lost if the ordering is weakened into a partial one, which unavoidably happens for multicriteria optimization (as discussed in the following Sect. 3.3).

For finite action spaces with n choices, this amounts to finding an optimum of g on the n-dimensional unit cube $[0, 1]^n \subset \mathbb{R}^n$, while for infinite (continuous) action spaces, the simplex Δ becomes the set of all (continuous) distributions defined on AS; turning the sum in the above definition into an integral. Since we will hardly require this more general setting throughout the rest of this book, we leave this remark here for only for the sake of completeness.

The set C indeed can (and will in many cases) contain a multitude up to infinitely many options to choose from (even for a finite set AS, the simplex $\Delta(AS)$ will be a continuum). In these cases, analytic properties of the function g like continuity, differentiability, convexity or others may become important aids to algorithmically perform the optimization. The methods to carry out an optimization in most cases depend on the structure and implied properties of g.

3.3 Multiple Goal Optimization

Many practical situations require a consideration of several scores or measures at the same time. For example, a company may be concerned about financial gains, market share, customer satisfaction, and many other aspects, all of which demand optimized decision making. In this case, we need to consider several goal functions and have to adapt the generic optimization problem defined in (3.2) accordingly.

Let there be $d > 1$ individual goal functions $g_1, \ldots, g_d$ for optimization, then we can put them together as a vector-valued goal function $\mathbf{g} : AS \to \mathbb{R}^d$, written out as $\mathbf{g}(a) = (g_1(a), g_2(a), \ldots, g_n(a))$. Without loss of generality, we may think of the optimization to be all minimizations or all maximizations (by multiplying each goal function by -1, accordingly to change the goal from "min" to "max" or vice versa). Therefore, the formulation of (3.2) needs a redefinition of the $\leq$ and $\geq$ relations between elements of $\mathbb{R}^d$. To this end, we let two vectors $\mathbf{x} = (x_1, \ldots, x_d), \mathbf{y} = (y_1, \ldots, y_d) \in \mathbb{R}^d$ satisfy $\mathbf{x} \leq \mathbf{y}$ if and only if $x_i \leq y_i$ for all coordinates $i = 1, 2, \ldots, d$ (for example, $\mathbf{x} = (1, 2, 3) \leq \mathbf{y} = (1, 2, 4)$). Likewise, $\mathbf{x} \geq \mathbf{y}$ holds if and only if $x_i \geq y_i$ for all coordinates $i = 1, 2, \ldots, d$; the strict versions of both relations is defined in the natural way. The negation of $\mathbf{x} > \mathbf{y}$ is the relation $\mathbf{x} \leq_1 \mathbf{y}$, meaning that there exists some coordinate i_0 for which $x_{i_0} \leq y_{i_0}$, no matter what the other coordinates do. For instance, $\mathbf{x} = (1, 3, 2) \leq_1 \mathbf{y} = (4, 4, 1)$, since the second coordinates in both vectors satisfy the $\leq$ relation (which is enough, although $x_3 \geq y_3$). The relation $\geq_1$ is defined alike, as the complement of the strict $<$ relation between $\mathbf{x}$ and $\mathbf{y}$. The $\leq$ and $\geq$ relations on $\mathbb{R}^d$ are antisymmetric, transitive and reflexive, but unlike $\leq$ on $\mathbb{R}$, the induced order is only partial (for example, neither relation would hold between the two vectors $\mathbf{x} = (1, 3)$ and $\mathbf{y} = (3, 1)$). The strict relations on $\mathbb{R}^d$ behave like their siblings in $\mathbb{R}$, but the $\leq_1$ and $\geq_1$ relations are only reflexive, but neither transitive nor antisymmetric.

Since $\mathbb{R}^d$ is only partially ordered, we cannot hope for a maximum or minimum to be found, but can only find *maximal* and *minimal* values. The difference to a maximum/minimum roots only in the partiality of the order: a minimum is "globally" optimal, meaning that it is $\leq$ to all other elements in the image space of g. A *minimal* element may have incomparable other elements, so that it is only smaller than some, but not smaller than all other elements, namely those to which it has no defined $\leq$-relation (likewise, for a maximal element).

Rewriting (3.2) in terms of $\leq$ thus complicates matters of finding a solution not only because the optimal value $g(x_0)$ is no longer unique, but much less conveniently, we may find different candidate optima $\mathbf{x} = g(x_0)$ and $\mathbf{y} = g(x_1)$ which do not $\leq$- or $\geq$-relate to each other, hence are incomparable.

A choice for a "better" of $\mathbf{x}$ and $\mathbf{y}$ could be made by looking at the coordinates, with different ways of defining a preference: A *lexicographic order* $\mathbf{x} \leq_{\text{lex}} \mathbf{y}$ considers the coordinates from left to right, and puts $\mathbf{x} \leq_{\text{lex}} \mathbf{y}$ if $x_1 < y_1$ and orders them as $\mathbf{x} \geq_{\text{lex}} \mathbf{y}$ if $x_1 > y_1$. Upon a tie $x_1 = y_1$, the process is repeated with the second coordinate and defines the same outcomes, and so on (exactly as one would order words of a language alphabetically, based on the first letter, and upon equality, ordering two words according to the second letter, and so forth). If all coordinates are equal, we have $\mathbf{x} = \mathbf{y}$.

It is worth noting that this ordering can extend *any* ordering on scalars to vectors, with the resulting order being total/partial if the underlying order is total/partial. This is an alternative way of defining even a total order on $\mathbb{R}^d$, but relies on an explicit ranking of coordinates in terms of importance. Looking at our optimization problem, this can be reasonable if the goals themselves can be put in some order of priority, but disregards quantitative trade-offs between goals. For example, goal g_1 may be the most important, but a suboptimal value for it may still be bearable if the second and the third important goal can be made large (or small) to compensate for the suboptimality of goal g_1. This leads to the traditional method of scalarization.

The above problem can be solved using the traditional method of *scalarization*. A scalarization of a vector-valued optimization problem means defining a new goal function by assigning a weight (priorities, importance, etc.) $w_i \in \mathbb{R}$ to the goal function $g_i : \Omega \to \mathbb{R}$ for $i = 1, 2, \ldots, n$, and optimizing the function

$$g^* : \Omega \to \mathbb{R}, \, g^* := \sum_{i=1}^{n} w_i \cdot g_i. \tag{3.3}$$

Without loss of generality, we can additionally assume that all weights are ≥ 0, and that $\sum_i w_i = 1$; for otherwise, if $w_i < 0$, we may replace g_i by $-g_i$, and if $\sum_i \neq 1$, we can just multiply g^* by a some constant to normalize the sum of weights to a unit, which leaves the locations of optima (i.e., the sets "argmin" and "argmax") unchanged, and only amounts to a multiplication of the optimal value by the same constant. Expression (3.3) goes with the untold assumption of the goals to be independent, since the formula does not include any mixed terms. Relaxations of this assumption lead to the *Choquet integral*, whose discussion follows in Sect. 3.6.

In compiling all goals into a single, now scalar-valued, function g^*, we recover the familiar setting of single-criteria optimization. This simplification comes at the price of being necessarily subjective to the extent of how the weights are defined. Additional complications come in if the goals cannot be added meaningfully, say, if g_1 measures monetary gains from installing a new process, while g_2 quantifies efficiency of that process in terms of computational time. Numerically, the two numbers are always addable, but semantically, the sum of the two will have no meaning by itself; in our terminology introduced in Sect. 3.1 above, g^* is thus in any case a *score*.

Nonetheless, the scalarized goal function g^* admits a useful interpretation of optimality, based on the choice of $w_i \geq 0$ and $\sum_{i=1}^{n} w_i = 1$: geometrically, this makes the optimization happen on the *convex hull* of all goals, seeking the optimum at the border of the resulting convex set. This implies that any solution $x_0 = \text{argmax}_{a \in AS}\, g^*(a)$ is *Pareto-optimal*. This means that every alternative point $x_1 \neq x_0$ *might* improve some goal g_i, i.e., $g_i(x_1) \geq g(x_0)$, but necessarily there will be another goal g_j for which $g_j(x_1) < g_j(x_0)$. In our above terminology, a *Pareto-optimum* is thus a point x_0 for which $g(x) \leq_1 g(x_0)$ for all $x \in \Omega$. That is, we cannot simultaneously improve on all goals, since any improvement in one goal requires some additional price to be paid in another goal. Figure 3.1 illustrates this effect with two goals: any movement towards a larger $g(x_1)$ or $g(x_2)$ will necessarily decrease the other goal's value (due to the convexity of the set). The set of all Pareto-optimal points is called the *Pareto front*. Geometrically, it is a part of the boundary of the convex hull induced by the weighted sum (3.3) defining g^*.

Now the question arises, how is the scalarization method better than a lexicographic order? After all, the lexicographic order can easily be made total, has no need for importance *weights* and only requires importance *ranks*. Nevertheless, a flavor of subjectivity in the choices of weights or goal ranks remains in any case. However, if the coordinate functions take only discrete and finitely many values, we can find weights to define g^* such that the numeric order on g^* equals a

Fig. 3.1 2-dimensional example of a Pareto front

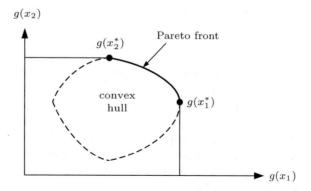

lexicographic order on the underlying vectors. To illustrate this, consider a simple example:

Example 3.1 (Numeric vs. lexicographic order of integers) Let all coordinate functions take values in the range $\{0, 1, 2, \ldots, 9\}$, and define the weights to be $w_i :=$ 10^i. Then, $g^* = \sum_{i=1}^{n} w_i g_i$ will take values being integers in base 10. Attaching leading zeroes to equalize the length of two such integers (within a computer, the bitlengths of any two integers would be equal anyway), it is immediately obvious that the lexicographic order per digit comes up equal to the numeric order of the represented numbers. Obviously, working with more than the 10 numbers above works similarly, only changing the radix of the resulting numeric representative. ◇

3.4   Decision Theory

Decision theory is concerned with choosing optimally from a set of actions, given that the consequences, i.e., *scores*, are either fully determined by the action, or subject to uncertainty induced by nature. This means that whatever uncertainty comes into play has no rational intention to work against the decision maker, though it is the latter's duty to act best in light of this. When looking into the context of security, a decision problem can be formulated on any kind of action set; for example, which anti-malware software to buy given some available offers (this is an optimization over a discrete set), when to conduct the next security training for employees (this is technically an optimization over an infinite, but practically finite, set since it is a choice of a calendar date), or when to expect an attack (this is a choice over continuous time, since the attacker can become active at any point in time, not bound to working hours of its victim).

Some practical situations may allow for the specification of a deterministic function value $g(a)$, ideally computable from a. However, in many practical cases the outcome $g(a)$ from an action a may be better described as a random value, either with a known distribution F, perhaps with a deterministic part in it that we may be able to compute. For example, the consequence of an action a may be likely to be $g(a)$, except for some random variation around it, often expressed by writing

$$X = g(a) + \eta, \tag{3.4}$$

with η having some fixed probability distribution. This may occur if the decision maker *intends* to take action a, but due to circumstances beyond her/his own control, ends up adopting another action a'. If η has zero mean, then $g(a)$ is what the decision maker *expects* to get from action a, while in reality, the result may differ. From a modeling perspective, we may use $g(a)$ as an approximation of the actual consequence, taking uncertainty into account as a random error term $\eta = g(a) - g(a')$, whose distribution may be known. This is one simple way of modeling uncertainty.

A different approach takes the goal function for our optimization as

$$X = g(a + \eta), \tag{3.5}$$

where η may even depend on the action a. We may think of the decision maker acting with a *trembling hand*. This is essentially different to (3.4), and the two coincide only if g is linear in (3.5).

Doing the optimization under such uncertainty is possible in several ways, the simplest of which is just "averaging out" the randomness by replacing the random quantity in (3.4) or (3.5) using the expectation operator. If the error term in (3.4) has zero-mean, taking expectations leaves us with the deterministic expected outcome $g(a)$. Things are more involved in (3.5), where the expectation w.r.t. the distribution of $g(a + \eta)$ amounts to a transformation of the probability densities. The result then depends on the properties of g and the shape of η's distribution. Practically, in absence of either, we may resort to simulations (see Sect. 2.5 for details). If we can acquire an arbitrary lot of stochastically independent (or at least pairwise independent) samples from the random quantity, having the (analytic) assurance that the expectation or at least the variance is finite, the law of large numbers kicks in and for a sequence of samples $X_1, X_2, \ldots$ from (3.4) or (3.5), their long-run average converges to the expectation with probability 1. More precisely, we can compute

$$\overline{X}_n := \frac{1}{n} \sum_{i=1}^{n} X_i, \tag{3.6}$$

over all n sample values that we have, assured that any deviation $\varepsilon > 0$ from the true average $\mathbb{E}(X)$, no matter how small, will happen with vanishing probability in the long run: for all $\varepsilon > 0$, we have $\lim_{n \to \infty} \Pr(|\overline{X}_n - \mathbb{E}(X)| > \varepsilon) = 0$.

A more sophisticated way of handling the uncertainty is improving our information about η by virtue of experience. This is a *Bayesian* approach to decision making, outlined in Sect. 3.4.1. In absence of a hypothesis or the ability to improve it (say, if we cannot observe the adversary because it remains stealthy), a worst-case oriented analysis seeks the best decision under worst possible circumstances. This is the *minimax* type of decision further described in Sect. 3.4.2.

Both methods depend on optimization, which in turn rests on a proper *preference relation* on the action space (as already discussed in previous Sect. 3.2). A decision maker will seek an action with *optimal* consequences, captured by the scores that the utility function assigns, taken as an optimization goal g. The simplest choice is that of a utility function being the expectation of $g(a)$ w.r.t. the uncertainty that goes with the action, either because a is not exactly realized (trembling hand; cf. (3.5)) or because of external random influences (like in (3.4)), or both.

Exploiting the freedom that the goal function g can, but *does not need to*, map into the reals, we gain widely extended possibilities over traditional methods: if we can let an action a's score $g(a)$ be from the set of hyperreals $(*\mathbb{R}, \leq) \supset (\mathbb{R}, \leq)$ [10], we can represent whole probability distributions as a "scalar" hyperreal number. Moreover, and more importantly, the hyperreal space is totally ordered, which lets us do optimization in a restricted yet still useful form. A certain (large) class of probability distributions (specified within Definition 3.1) has a set of unique representatives in the ordered hyperreal space. Returning to our previous intuition, we can then rank two actions based on the probability distribution of their outcomes (a concept that we will later revisit under the name *prospect*). A preference between such distributions can then be derived from the ordering of their representatives in $^*\mathbb{R}$, much similar as to how we would order values from $\mathbb{R}$. This is one example of a so-called *stochastic order*. A simple direct definition of a partial such order calls two random variables X_1, X_2 (say, scores assigned to actions a_1, a_2) as $X_1 \leq_{st} X_2$ if and only if

$$\Pr(X_1 \geq t) \leq \Pr(X_2 \geq t) \quad \text{for all } t \in (-\infty, \infty). \tag{3.7}$$

We call this the *standard stochastic order*. It is easy to construct examples of random variables that do not satisfy this relation in either direction, so $\leq_{st}$ is in any case partial. Upon assuming that the random outcomes are *bounded* and absolutely continuous w.r.t. some measure (Lebesgue or counting), we can fix the partiality to obtain a total ordering as follows:

Definition 3.1 (Stochastic $\preceq$-order) Let two random variables X_1, X_2 with distributions F_1, F_2 be given such that both are ≥ 1, both share the same compact support $\Omega \subset \mathbb{R}$, and both are either categorical or have a continuous density functions f_1, f_2 and satisfying $f_1(a) \neq f_2(a)$ at $a = \max \Omega$ in the continuous case (only). Then, we can put the two into the order $X_1 \preceq X_2$, if and only if there exists some $t_0 \in \Omega^\circ$ (where Ω° is the interior of Ω) such that

$$\Pr(X_1 \geq t) \leq \Pr(X_2 \geq t) \quad \text{for all } t > t_0. \tag{3.8}$$

Definition 3.1 is a special case of the more general definition putting two distributions into $\preceq$-order that equals the hyperreal order of their moment-sequences. It is given in this form here for reasons of similarity to the standard stochastic order (cf. (3.7)), from which we distinguish it by the symbol $\preceq$. Working out the threshold t_0 is rarely needed explicitly, since the ordering is decidable from looking at the tail region (largest risks or losses) from the right end of the support, given that $f_1(a) \neq f_2(a)$.

In many practical cases, the decision about the stochastic $\preceq$-order boils down to deciding a lexicographic order. This holds for categorical as well as for continuous random variables, as long as the support is common between all objects under comparison:

Theorem 3.1 ([10]) *Let* $X \sim F_1, Y \sim F_2$ *be two random variables, with distribution functions F_1, F_2.*

If X, Y are categorical with a common ordered support $\Omega = \{c_1 > c_2 > \ldots > c_n\}$, and with $\mathbf{f}_1, \mathbf{f}_2$ being the respective (empirical) probability mass functions (vectors), then $X \preceq Y \iff \mathbf{f}_1 \leq_{lex} \mathbf{f}_2$, where

$$\mathbf{f}_i = (f_i(c_1), f_i(c_2), \ldots, f_i(c_n)) \in \mathbb{R}^n. \tag{3.9}$$

If F_1, F_2 are continuous with differentiable density functions f_1, $f_2 \in C^\infty(\Omega)$ on a common compact support $\Omega = [a, b] \subset [1, \infty)$, we have $X \preceq Y$ if and only if

$$((-1)^k \cdot f_1^{(k)}(b))_{k \in \mathbb{N}} \leq_{lex} ((-1)^k \cdot f_2^{(k)}(b))_{k \in \mathbb{N}}. \tag{3.10}$$

Remark 3.1 The condition on the smoothness of the density functions in Theorem 3.1 is in fact less restricting than it appears, as we can always find an arbitrarily close approximation from within C^∞ to any given density function, by smoothing with a Gaussian kernel (with a bandwidth made small enough to achieve the desired security by convolution; see Sect. 12.4.3 for more details). Likewise, the condition $f_1(a) \neq f_2(a)$ for continuous distributions is satisfiable by truncating the distributions, while preserving a proximum up to any desired accuracy.

The focus on the tail of a distribution makes the $\preceq$ order interesting for risk management purposes, as it judges on extremal outcomes. In other words, a risk manager would perhaps not bother too much mitigating a risk that is anyway already very low. So, the ranking should be on the high-risk actions, which a tail order like (3.8) is more explicit about; specifically, the threshold t_0 would then draw the line between anticipated disruptions that are acceptable according to business continuity management, and those which need an optimal decision and explicit risk management. Similarly, the upper bound on the losses (maximum value b, largest category c_1 in Theorem 3.1 or compactness of the utility range in Definition 3.1) corresponds to the practical assumption that losses could be infinite, but practically, there will be an upper limit above which larger losses make no difference any more (an infrastructure will at some point be just broken, no matter how much more damage is caused beyond what happened already).

The value b in Definition 3.1 also has a practical meaning as being the anticipated maximum loss that an action may cause, and hence we would not distinguish losses above b or category c_1 any further; any extreme damage of this scale is equally worse than another (for example, if an enterprise runs bankrupt due to a huge financial loss, it does not matter any more if it is because of 100 million or 120 million; the enterprise is in both cases no longer existing).

In the context of a computer system, an excess of this value may be considered as the event of a zero-day exploit, and the probability for losses larger than t_0 is then relevant for risk management. In the terms above, this would be the tail mass of the distribution beyond t_0, which the $\preceq$-ordering minimizes by construction. We will further elaborate on this in Sect. 12.6.2.

For the practical relevance of stochastic orders, suppose that an action a may have consequences on a whole scale of possible outcomes, say $u_1 \geq u_2 \geq u_3 \geq \ldots u_n$, each outcome possibly occurring with some associated (known) probability $p_i = $ Pr(action a has the payoff u_i). The set of pairs $\{(u_i, p_i) : i = 1, 2, \ldots\}$ is nothing else than a discrete probability distribution, elsewhere also called a *prospect*. If the outcome of an action is uncertain in this way, the stochastic $\preceq$-ordering offers a natural and handy tool to rank actions and take optimal decisions.

Indeed, taking the "score" of an action as being just the hyperreal representative of the probability distribution of outcomes, the ordering implied within the space $(^*\mathbb{R}, \leq)$ is exactly the stochastic $\preceq$-order characterized by (3.8).

Using stochastic orders is not generally compatible with Bayesian or minimax decisions, but the specific $\preceq$-ordering *is admissible* for both kinds of decision making.

3.4.1 Bayesian Decisions

The Bayesian method implements the intuition of reducing uncertainty upon data. A bit more formally, if uncertainty comes in via a random error term η (cf. (3.4) or (3.5)), we start from an "initial guess" about how this uncertainty looks like. That is, we define an a priori probability density $f(\eta)$, and start collecting data by observing the utilities after taking actions. This data can take any form (including but not limited to observed utilities), but let us generally assume, without substantial loss of generality, that it is a set of real values $\mathbf{d} \in \mathbb{R}^n$.

Remark 3.2 Practically, most information will not come in a handy numeric form. For example, security incident reports, technical support tickets, etc. are all textual data. For these to fit into the entire numeric world of statistics, decision and game theory, a "conversion" of textual information into representative numerical data is necessary. A first step towards this are document term matrices, basically providing word counts to indicate topics in a text based on frequent occurrence of technical terms. This is already a simple form of representing a text as a (now numeric) matrix, on which subsequent analyses can be conducted. Nevertheless, such simple methods do not say anything about semantics and content of a text – therefore concepts from Natural Language Processing are used. However, we do not go into detail about these data preparation and information retrieval aspects, and rather refer to the vast lot of literature on this topic.

A *Bayesian choice* is an optimal decision in light of data $\mathbf{d} \in \mathbb{R}^n$, and the resulting a posteriori uncertainty about η described by the conditional density $f(\eta|\mathbf{d})$. This latter item is obtained from Bayes' theorem directly via

$$f(\eta|\mathbf{d}) = \frac{f(\mathbf{d}|\eta) \cdot f(\eta)}{f(\mathbf{d})}, \qquad (3.11)$$

with the terms meaning the following:

- $f(\mathbf{d}|\eta)$ is the so-called *likelihood function*. Intuitively, this is the probability of observing exactly the data $\mathbf{d}$, under a given value for η.

    This term is generally obtained from domain knowledge about how the data $\mathbf{d}$ is being generated, given η. If $\mathbf{d}$ is a sample from a random variable with probability density function $f : \mathbb{R}^n \times H \to \mathbb{R}$, $\eta \in H$ is the parameter pinning down the concrete member from the family $f(\mathbf{d}, \cdot)$, the *likelihood function* is exactly $f(\mathbf{d}, \cdot)$, only with "switched roles" of the parameters from $\mathbb{R}^n$ and from H: for a distribution, we usually fix the parameter η (e.g., mean and standard deviation to define a Gaussian distribution) and leave the variable $\mathbf{d}$ to vary. Conversely, the likelihood function *fixes* the data $\mathbf{d}$, and lets the parameter η vary over the set of admissible choices. Each such choice gives a different probability to observe exactly the data $\mathbf{d}$. In the typical application of *maximum likelihood estimation*, we then seek a setting of η that maximizes the chances to observe exactly the given data. Its use in Bayesian decision making is slightly different.

- The term $f(\eta)$ is the (unconditional) distribution of η; this is exactly our a priori distribution, i.e., our so far best hypothesis regarding the uncertainty about η.
- The denominator $f(\mathbf{d})$ is the unconditional distribution of the data, obtained by "integrating out" the parameter η over its whole space of possibilities. Formally, this term comes from the law of total probability and is

$$f(\mathbf{d}) = \int_{\eta \in H} f(\mathbf{d}, \eta) f(\eta) d\eta, \qquad (3.12)$$

where the integral reduces to a sum if H is discrete. Although the denominator is – technically – just there to renormalize (3.11) into a proper probability density, its numerical evaluation imposes the majority of computational overhead tied to Bayesian methods. Designated software aid to help with this task is provided by tools like BUGS ("Bayesian Updating using Gibbs Sampling") [4]. Essentially, these exploit ergodic theorems like the aforementioned law of large numbers, in a highly sophisticated fashion.

In some cases, however, the computational burden and the implied statistical skills necessary to evaluate (3.12) can be avoided by resorting to a proper choice of the prior density $f(\eta)$ relative to the likelihood function: If the two are chosen *conjugate to each other*, then the posterior distribution $f(\eta|\mathbf{d})$ takes the same shape as the prior $f(\eta)$, so that the evaluation of (3.12) can be spared; see Example 3.2. In many other applications, however, the evaluation of (3.12) may reduce to a sum of only a few possible settings for η, which applies to many game theoretic models striving for (perfect) Bayesian equilibria; see Definition 4.3. We will continue this thought later in the context of such equilibria and game theory.

Having computed the posterior distribution $f(\eta|\mathbf{d})$ given our (new) knowledge $\mathbf{d}$, a

$$\text{Bayesian decision} = \underset{a \in AS}{\operatorname{argmax}} \, \mathbb{E}_{f(\eta|\mathbf{d})}[g(a)],$$

$$= \underset{a \in AS}{\operatorname{argmax}} \int_{\eta \in H} g(a) \cdot f(\eta|\mathbf{d}) d\eta$$

that is, we optimize (w.l.o.g. maximize) the average utility quantified by the goal function g, where the average, i.e., expectation, is w.r.t. the posterior distribution. If H is countable or finite, then the integral is replaced by a simple sum over all $\eta \in H$.

Example 3.2 (The β-reputation model) This technique was first described in [2], and later applied for forecasting security incidents in [11]. We illustrate the general technique in the context of intrusion detection.

Consider a security administrator having an Intrusion Detection System (IDS) with an unknown detection rate η. More precisely, we will think of η as the *false-positive alert probability* of the IDS. The choice to be made is whether to become active or not upon an alert by the IDS, since it could be a false alarm, and the security administrator wants to waste as little time and resources as possible (this is the optimization part of the decision making).

The natural behavior of a person would be learning about the false positive rate from experience, and this is exactly what a Bayesian decision is, merely making this intuition rigorous. Hence, in our example the administrator starts from an initial hypothesis on a false positive rate. This is its a priori distribution that we here (for reasons to become clear soon) fix as a β-distribution $\eta \sim \mathcal{B}e(a, b)$ with two parameters $a, b > 0$ to be fixed at the start. The density function of the family $\mathcal{B}e(a, b)$ is $f_\beta(x|a, b) = x^{a-1}(1 - x)^{b-1}/B(a, b)$, where $B(a, b)$ is Euler's beta function. The mean value of this distribution takes a particularly handy form in being

$$\mathbb{E}_{f_\beta(\cdot|a,b)}(\eta) =: \overline{\eta} = \frac{a}{a + b},$$

which will become useful later.

The choice of a, b is arbitrary in first place, but made to reflect the initial belief about the false positive rate. Suppose this to be $\overline{\eta} \approx 0.05$, i.e., 5% false alerts, then we may choose the a priori distribution $\mathcal{B}e(5, 95)$, so that the a priori distribution has the mean $5/(5 + 95) = 0.05$, reflecting our initial belief.

Suppose that incidents come in stochastically independent and at a constant average rate per time unit. Let n be the number of alerts $\mathbf{d} = (d_1, d_2, \ldots, d_n)$ in the last period of data collection, and divide this count into $n = d^+ + d^-$, distinguishing the number of false positives d^- from the number of true positives d^+ (observe that we cannot count the false negatives, as those are naturally not seen).

Now, let us do a Bayes update: for the likelihood function, each false positive comes in with probability η, and the chance to get d^- false positives among a total of n alerts is Binomially distributed with parameters n and η. That is, the likelihood function for n in each time unit is the binomial density

$$f_{bin}(x|n, \eta) = \binom{n}{x}\eta^x(1 - \eta)^{n-x}.$$

This likelihood function is *conjugate* to the β-distribution, which means that the a posteriori density $f(\eta|d)$ will again be a β-distribution and can directly be given (without any need to evaluate (3.12)) as

$$f(\eta|\mathbf{d}) = \mathcal{B}e(a + d^-, b + d^+), \tag{3.13}$$

in a slight abuse of notation, where the right hand side term represents the density function with the given parameters, of course.

Let us consider (3.13) relative to our initial intuition: the Bayes update assigns a simple meaning to the parameters a, b of our chosen prior, namely a counts the negative experience, i.e., false negatives, while b counts the positive experience, i.e., true positives. Consider what happens if we let either experience become overwhelming: if the true positives d^+ tend to infinity (relative to the false alerts), then $E(\eta) \to 0$, reflecting that we gain unconditional trust eventually. Conversely, if the number d^- of false alerts tends to infinity, relative to a (bounded or less largely growing opposite experience count d^+), the distrust converges as $E(\eta) \to 1$, corresponding to the learning effect that the IDS apparently produces almost only false alerts. Both cases are consistent with the intuitive conclusions that we would draw from long run experience, thus subjectively justifying the method in this example context.

Finally, let us return to the Bayesian decision still pending for the security administrator: suppose that going after a correct alert prevents damage of magnitude L_1, as opposed to an unnecessary investment of $-L_2$ being the time and resources wasted on a false alert. A Bayesian decision averages these two outcomes based on the most recent posterior learned so far. Let this be

$$f(\eta|\mathbf{d}) = \mathscr{B}e(a + \sum_i d_i^-, b + \sum_i d_i^+)$$

for the collections $\{d_i^+\}_{i=1}^N$ and $\{d_i^-\}_{i=1}^N$ constituting an entirety of N updates in the past. Then, we find

$$\bar{\eta} = \frac{a + \sum_{i=1}^N d_i^-}{\left(a + \sum_{i=1}^N d_i^-\right) + \left(b + \sum_{i=1}^N d_i^+\right)},$$

and the a posteriori expected utility per action is

$a_1 :=$ "becoming active" has expected utility $= (1 - \bar{\eta}) \cdot L_1 + \bar{\eta} \cdot (-L_2)$
$a_2 :=$ "ignore the alert" has expected utility $= (1 - \bar{\eta}) \cdot L_1$.

$$(3.14)$$

In other words, if the security administrator chooses to do something (become active), then there is a chance of $(1 - \bar{\eta})$ to prevent damage (L_1), but another chance of $\bar{\eta}$ to waste resources ($-L_2$). Conversely, if s/he chooses to ignore the alert, then s/he misses out on a fraction of $(1 - \bar{\eta})$ true positives, causing the proportional lot of damage.

The Bayesian choice minimizes the damage by picking the choice $a \in AS = \{$become active, ignore$\}$ with the smaller expected damage based on (3.14). Since the value of η is based on experience, this opens an attack vector via causing *alert fatigue*. In this case, the adversary may go causing as many false alerts as possible to "train" the Bayesian estimator for η towards complete distrust in the quality of the IDS. Consequently, the security administrator will have stronger and stronger belief (represented by a posterior β-distribution whose one parameter substantially grows over the other), and choosing to ignore more and more alerts. Having "conditioned" the security administrator accordingly, the real attack may start, ironically, exploiting the Bayesian learning mechanism that we just described. A similar concept is known from machine learning, where *adversarial examples* are constructed to trick machine learning models into making false predictions. Minimax decisions, discussed in Sect. 3.4.2 are one way to avoid this possibility, at the price of taking more pessimistic choices, and not necessarily using all available knowledge.   ◇

3.4.2 Minimax-Decisions

Bayesian learning usually hinges on as much data, i.e., incident information, as one can get, but this is in conflict with the usual goal of security being exactly on the *prevention* of incidents. In other words, we should not depend on learning only from our own experience (since we would probably not survive the necessary lot of security incidents to carry the learning to convergence). Threat intelligence and sharing information is thus crucial for learning strategies to be effective, but again, security may incentivize people to keep such information private, as publishing it may damage one's reputation.

Minimax decisions avoid the dependence on continuously incoming new information to learn from, and instead adopts a worst-case hypothesis to make a decision. While a Bayesian decision seeks the best outcome under a hypothesized set of possibilities (the posterior distribution), a minimax decision assumes the worst outcome possible and looks for the best decision then ("hope for the best, but prepare for the worst").

To this end, we formally replace the posterior expectation $\text{argmax}_{a \in AS}$ $\mathbb{E}_{f(\eta|\mathbf{d})}[g(a)]$ in a Bayesian decision by the worst-case setting for η, which governs the (random) consequence $g(a)$. That is,

$$\text{Minimax decision} = \underset{a \in AS}{\text{argmax}}[\min g(a, \cdot)]$$

where the second minimum is over all possible values that g may take upon a given a and some unknown influence here denoted as the second anonymous parameter "·" to g. This means that we consider g in any nontrivial case as a random variable, with a known or unknown distribution. If g is continuous, it may attain its minimum for a least-favourable such distribution f^*, giving $\min g(a) = \mathbb{E}_{f^*}[g(a, \cdot)]$. In this view, a minimax decision appears as a special case of a Bayesian decision.

Remark 3.3 In a more general setting, we would need to replace each min and max by an infimum (i.e., the largest lower bound) and supremum (i.e., the smallest upper bound), which avoids (here omitted) conditions for the optima to be attained. We will have no need to dig into the technicalities to an extent where this becomes relevant, so we will not discuss the general setting any further.

In general, there is no golden rule telling which kind of decision to prefer over the other, and the "meta choice" for the decision making paradigm depends on the availability, reliability, accuracy and usefulness of the data at hand, but also on the possible consequences. Not all kinds of losses are equally comparable on a continuous scale, such as loss of human lives may not be differently bearable whether it is only one or a hundred people dying. In airline security, for example, one would not make much of a difference between terrorist attack letting an airplane crash with two people or two-hundred people in it; the important fact is that either incident shows that the attack was doable and successful. Hence, a minimax decision appears in order. Conversely, when dealing with phishing emails, the minimax paradigm would prescribe to never open an email attachment or hyperlink any more, but for that reason this choice will cause business disruptions in the long run (so, the attack was still sort of successful, only in a different way). Here, a Bayesian decision appears in order, letting the receiver of emails learn to recognize phishing attempts and distinguish them from legitimate traffic by learning from experience.

Example 3.3 (continuation of Example 3.2) If the security administrator cannot update the hypothesis on η, s/he may consider a worst-case scenario instead. With the losses as before, the administrator can choose to ignore the alarm, which gives at most a damage of L_1 if the alarm was actually real. Or s/he can choose to always react on the alarm, which incurs a waste of L_2 resources. Not knowing about what actually is happening, the security officer pessimistically may choose to "always react" if $L_1 > L_2$, since the waste is less than the potential damage. Conversely, if $L_1 < L_2$, then "always ignoring" would be the better choice. If $L_1 = L_2$, then either decision is as good as the other, and all three cases constitute a minimax decision. Systematically, and towards generalizing it to more than two options, we can set up a tableau showing each combination of choice (per row) against the unknown instance of reality (in the column); for the example being, this tableau looks as follows:

$$
\begin{array}{r|c|c|}
 & \text{real} & \text{false} \\
\hline
\text{become active} & 0 & L_2 \\
\hline
\text{ignore} & L_1 & 0 \\
\hline
\end{array}
\qquad (3.15)
$$

Now, a minimax decision should be optimal under the worst case – that worst case being the maximal loss per possibility, which amounts to the per-column maximum. That is $\max\{0, L_1\} = L_1$ and $\max\{L_2, 0\} = L_2$. The minimax choice is for the lesser of the two, hence is for "becoming active" if $L_2 < L_1$, because this action attains the minimum, or for "ignoring" if $L_1 < L_2$, because the minimum of column-maxima is attained for this action under that circumstance.

A Bayesian decision would give the same results, if assign the respective probability 1 to the respective worst case column. That is, a least favourable distribution is a Bernoulli distribution over {real alarm, false alarm} with $\Pr(\text{false alarm}) = \eta$, and given by

$$\eta = \begin{cases} 0, \text{ if } L_2 < L_1 \\ 1, \text{ if } L_1 \le L_2 \end{cases}$$

Imposing the so-defined value into (3.14) reproduces the just described minimax decision. Apparently, an adversary cannot count on alert fatigue, since the security administrator no longer adapts its behavior to incoming information that the adversary could use to influence it. But this advantage is bought at the price of ignoring possible knowledge gained over the lifetime of the system.   ◇

3.5 Game Theory

As mentioned in the previous section, in decision theory the outcome of a decision can depend on uncertainty introduced by nature, which is per definition irrational and has not intention whatsoever to act for or against the decision maker. However, when we now also consider actions of further rational entities, let us call them *players*, which might have their own incentives, we end up with *game theory*.

The general setting of game theory is very similar to that of decision theory, only that we consider several action spaces $AS_1, \ldots, AS_n$, and utility functions $u_1, \ldots, u_n$ hereafter, for a total of n interacting entities (which can, but does not need to include nature). The utility functions are generally mappings $u_i : \prod_{i=1}^{n} AS_i \to (\mathbb{R}, \le)$, where the image set can be replaced by any (totally) ordered set.

The definition of a utility function may, but does not need to, include actions from all other players, up to the possibility of a player's utility to depend not at all on any other's strategy. One example of such a player is nature, which acts irrespectively of whatever the players may do.

3.5.1 Normal Form Games

If all action sets AS_i for $i = 1, 2, \ldots, n$ are finite, then we may directly specify utility values for the i-th player by a tableau, with $|AS_i|$ rows, and $\prod_{j=1, j \ne i}^{n} |AS_j|$ columns, specifying the value of $u_i(a_1, \ldots, a_{i-1}, a_i, a_{i+1}, \ldots, a_n)$ per table cell for the given vector of actions. A game specified like this is said to be in *normal form*. It is common notation to abbreviate this cumbersome specification as $u_i(a_i, \mathbf{a}_{-i})$, in which $\mathbf{a}_{-i} = (a_1, \ldots, a_{i-1}, a_{i+1}, \ldots, a_n) \in AS_{-i}$, where AS_{-i} is the cartesian product of all action spaces, *excluding* AS_i.

Definition 3.2 ((Non-Cooperative) Game) A *game* is a triple $\Gamma = (N, S, H)$, with N being a set of players, S being the family (set) of all action spaces and H the set of all utility functions. To each player $i \in N$, we associate an action space $AS_i \in S$. Further, the set H contains all utility functions $u_i : AS_i \times AS_{-i} \to R$, where R is a totally ordered set of payoffs. If N is finite and all action spaces $AS_i \in S$ are finite, then the game itself is called *finite*.

We will hereafter speak about "games", meaning non-cooperative games without ambiguities. Most of the time, we will put $R := \mathbb{R}$ in Definition 3.2, but other choices are admissible, such as $R :=^* \mathbb{R}$. Observe that the finiteness condition is optional on any of the involved sets, i.e., games can be with infinitely many players and infinite (up to uncountable) action spaces. If the game is finite, we shall hereafter call it a *matrix game*.

Like decision theory, we can then ask for an optimal action of the i-th player, given that the actions of all other players are known. Suppose that all players $\neq i$ have chosen their actions, jointly listed in the vector $\mathbf{a}_{-i}$, then the *best response* BR_i of the i-th player (w.l.o.g.) maximizes the utility:

$$BR_i(\mathbf{a}_{-i}) = \underset{a \in AS_i}{\operatorname{argmax}} \, u_i(a, \mathbf{a}_{-i}). \tag{3.16}$$

This is generally a set-valued mapping, and called the *best response correspondence*. If all players mutually act best against their respective other players, we have an *equilibrium*:

Definition 3.3 (Equilibrium) An *equilibrium* of a game Γ is a vector of strategies $(a_1^*, a_2^*, \ldots) \in \prod_{i \in N} AS_i$ so that

$$\text{for all } i \in N : a_i^* \in BR(\mathbf{a}_{-i}^*). \tag{3.17}$$

Informally, equilibria are behavioral rules in which no player gains any benefit by unilaterally deviating from the equilibrium; in this sense, the equilibrium is an optimal decision for an individual player. However, we stress that it may nonetheless be possible to gain more by teaming up with other players, coordinate actions among the group, and share the joint revenue obtained in this way. This is the route to *cooperative game theory*, which we will not further explore in this book, since our attention will be on non-cooperative games and equilibria (as already mentioned above). The term "non-cooperative" still thus does not imply any conflicts of interests between the players, and exclusively refers to the equilibrium being optimal for a *selfish* player. Games in security naturally have conflicts of interests between the players.

For finite two-player games, the utility functions for each player admit the particular handy specification as a matrix $\mathbf{A}_1$ over the set R, with $|AS_1|$ rows and $|AS_2|$ columns for player 1; respectively transposed as $\mathbf{A}_2 \in R^{|AS_2| \times |AS_1|}$ for player 2. In some cases, a game's description can be simplified by removing actions that cannot be optimal.

Definition 3.4 (Domination) Let a game $\Gamma = (N, S, H)$ be given. A strategy $a \in AS_i$ is said to *weakly dominate* another strategy $b \in AS_i$, if $u_i(a, \mathbf{a}_{-i}) \geq u_i(b, \mathbf{a}_{-i})$ for all $\mathbf{a}_{-i}$ of opponent strategies. If the inequality is strict, we say that the strategy is *strictly dominated*, or just *dominated*.

That is, dominated strategies are those for which there is another strategy giving better utility in all cases. Consequently, we can safely delete such strategies from an action space, since it would in any equilibrium necessarily be played with zero probability.

Example 3.4 (Game with Dominated Strategies)   Consider the following two-person zero-sum matrix game, i.e., a game where the goals of the two players are completely opposite and thus the utilities sum up to zero (see below for a more detailed explanation of zero-sum games). The game has action spaces $AS_1 = \{a_1, \ldots, a_4\}$ and $AS_2 = \{b_1, \ldots, b_4\}$, and payoffs for player 1 defined as

	b_1	b_2	b_3	b_4
a_1	1	3	2	4
a_2	0	3	1	3
a_3	1	4	2	1
a_4	2	1	0	2

If player 1 is maximizing, then strategy a_2 gives the utilities $(0, 3, 1, 3) \leq (1, 3, 2, 4)$ that strategy a_1 would deliver. Thus, strategy a_2 is weakly dominated from player 1's view, and can be deleted. Likewise, for the minimizing player 2, strategy b_3 is better than strategy b_2, since the second and third column satisfy $(3, 3, 4, 1) > (2, 1, 2, 0)$, so the second strategy is strictly dominated for player 2. It follows that both can delete their respective strategies and instead analyze the simpler game now being

	b_1	b_3	b_4
a_1	1	2	4
a_3	1	2	1
a_4	2	0	2

The equilibria, in the sense of actions to be played optimally, are the same, since only actions were excluded that would not have appeared in any equilibrium.   ◇

While the deletion of dominated strategies leaves the set of equilibria unchanged, the existence of equilibria in first place is a different question.

Example 3.5 (APT as "hide-and-seek")   Let us adopt a simplified view on an APT, taking this as an instance of a defender chasing an attacker in some system. Let the defender know that the attacker is present but stealthy, and let the attacker strive for a stealthy takeover. To ease matters further, assume that the system has only two locations 1 and 2 to conquer, with the defender guarding those two. The action spaces for both players are thus $AS_1 = AS_2 = \{1, 2\}$ (the generalization to more than two locations is obvious). Assume that the defender always notices the presence of an attacker (in an intended oversimplification of reality, but sufficient for the illustrative purpose of the example), then we can define the utility functions of the defending player 1 and the attacking player 2 as

$$u_1(a_1, a_2) = \begin{cases} 0, & \text{if } a_1 \neq a_2 \\ 1, & \text{otherwise.} \end{cases} \quad \text{and} \quad u_2(a_1, a_2) = \begin{cases} 1, & \text{if } a_1 \neq a_2 \\ 0, & \text{otherwise.} \end{cases}$$

That is, the defender loses if it checks the wrong place ($a_1 \neq a_2$), in which case the attacker wins, and vice versa.

Is there an equilibrium, in the sense of a best behavior for both actors? The answer is no, since if the defender were to choose $a_1^* = 1$, then the attacker would always win by choosing $a_2^* = 2$. But if $a_2^* = 2$, the defender's best reply is to play $a_1^* = 2$ as well, contrary to our assumption that a_1^* was optimal. The same reasoning holds for the assumption that $a_1^* = 2$, giving $a_2^* = 1$ as optimal for the attacker, and hence rendering $a_1^* = 1$ optimal for the defender. Since the situation is symmetric, the attacker also has no optimal choice within AS_2, so an equilibrium does not exist.

$\diamond$

Example 3.5 gives the crucial hint on that extension, since we just need to allow an optimal action to prescribe "jumping around" between different elements of the action space, if no action is by itself optimal. That is, if choosing one action causes another player to choose a proper action in response, which again forces the first player to revise its behavior, we end up with an infinite sequence of choices, delivering an infinite sequence of utilities per round of the game. Taking the long-run average (fraction) of times when a certain action is chosen defines, if it converges, a probability mass function on the action space of a player. Averaging the utilities in the long run as well, in the same way as for the law of large numbers (3.6), we can ask for optimality of the *long-run average* of utility, induced by a *randomized choice* from the respective action set. This is precisely the way to assure the existence of equilibria in (most) games, including that of Examples 3.4 and 3.5.

Definition 3.5 (Mixed Extension of a Game) For a game $\Gamma = (N, S, H)$, the *mixed extension* is the game (N, S', H'), with the strategy spaces and utility functions being defined as follows:

- $S' = \{\Delta(AS) | AS \in S\}$, where $\Delta(AS)$ is the set of probability distributions defined on the set AS.
- $H' = \{u_i' : \Delta(AS_i) \times \Delta(AS_{-i}) \to R, \text{ with } u_i'(\mathbf{a}_i, \mathbf{a}_{-i}) = \mathbb{E}_{\mathbf{a}_i, \mathbf{a}_{-i}}(u_i)\}$. That is, u_i' gives the *expected utility*, based on the probability distributions of choices of all players.

An equilibrium *in mixed strategies* is a vector of distributions $(\mathbf{a}_i^*)_{i \in N}$ such that for all $i \in N$: $\mathbb{E}_{\mathbf{a}_i^*, \mathbf{a}_{-i}^*}(u_i) \geq \mathbb{E}_{\mathbf{a}, \mathbf{a}_{-i}^*}(u_i)$ for all $\mathbf{a} \in \Delta(AS_i)$.

Example 3.6 (Continuation of Example 3.5) Suppose the defender would choose to check location 1 for 75% of the time, and look after location 2 for only 25% of the time. Then, the following attacker can go for location 1 with probability p, and for location 2 with probability $1 - p$. Writing this down as a tableau, check location

		1	2
		$\frac{1}{4}$	$\frac{3}{4}$
attack location	1 $1-p$	0	1
	2 p	1	0

we find the expected utility for the attacker by applying (3.18) to be

$$u_2((1-p,p),\mathbf{a}_1^* = (1/4,3/4)) = \frac{3}{4} \cdot p \cdot 0 + \frac{3}{4}(1-p) \cdot 1 + \frac{1}{4} \cdot p \cdot 1 + \frac{1}{4} \cdot p \cdot 0$$

$$= \frac{3}{4}(1-p) + \frac{1}{4}p = \frac{3}{4} - \frac{p}{2}$$

The attacker's utility is thus maximal at $p = 0$, giving $\mathbf{a}_2^* = (0, 1)$ and $u_2^* = 3/4$. ◇

Theorem 3.2 (Nash, 1951 [5]) *Every finite game has an equilibrium in mixed strategies.*

For Example 3.5, an equilibrium in mixed strategies would be $\mathbf{a}_1^* = (1/2, 1/2) = \mathbf{a}_2^*$, i.e., both players choose the locations equiprobable.

Theorem 3.2 can be strengthened into assuring equilibria to exist also in infinite games.

Theorem 3.3 (Glicksberg, 1952 [1]) *If for a game in normal form, the strategy spaces are nonempty compact subsets of a metric space, and the utility-functions are continuous w.r.t the metric, then at least one equilibrium in mixed strategies exists.*

The continuity assumption in Glicksberg's theorem is crucial, as for discontinuous utility functions, equilibria may no longer exist (a counterexample is due to [15]). Unfortunately, such cases can become relevant in security, as, for example, disappointment rates in games generally lead to discontinuous utilities (but let us postpone this discussion until Sect. 12.6.1).

Remark 3.4 The literature uses the terms "equilibrium" and "Nash equilibrium" widely synonymous, and so will we do. Historically, the difference originates from the equilibrium concept being introduced by von Neumann and Morgenstern [6] for two person games, and later extended by Nash to n person games allowing also for nonlinear utilities.

Mixed strategies are of particular relevance in security, since it is generally not a good practice to statically implement some security precautions and never update, revise or improve them. In the long run, the adversary will learn about the security mechanism, adapt to it, and ultimately break it. In this context, *mixed strategies* represent nothing else than a repeated change of security precautions to force the attacker to continuously re-adapt to changes of the situation. Or, almost as a proverb: "security is never finished".

If the game has vector-valued payoffs, we can transfer the idea of "loss from unilateral deviation" to the multi-dimensional case by demanding that if a player i unilaterally deviates from the equilibrium will degrade its payoff in at least one coordinate of the goal function $\mathbf{g}_i$, even though a gain in the other coordinates of the player's goal function $\mathbf{g}_i$ may occur. This is the concept of a *Pareto-Nash equilibrium*.

Definition 3.6 (Pareto-Nash Equilibrium) Let $\Gamma = (N = \{1, \dots, n\}, S = \{AS_1, \dots, AS_n\}, H)$ be a (mixed extension) of a game, in which each player has the d-dimensional utility function $\mathbf{u}_i : \prod_{i \in N} \Delta(AS_i) \to (R^d, \leq)$. Let the players be all maximizers (w.l.o.g., since the case of minimizing players is defined analogously). We call a set of elements $(\mathbf{x}_1^*, \dots, \mathbf{x}_n^*)$ a *Pareto-Nash equilibrium* if, for each $i \in N$, we have

$$\mathbf{u}_i(\mathbf{x}, \mathbf{x}_{-i}^*) \leq_1 \mathbf{u}_i(\mathbf{x}_i^*, \mathbf{x}_{-i}^*) \quad \forall \mathbf{x} \in \Delta(AS_i).$$

If AS is finite with, say $k = |AS|$ elements, we call $\Delta(AS)$ the *simplex* constructed over AS, or also the *convex hull*. It is the set of all $(p_1, \dots, p_k) \in [0, 1]$ such that $p_1 + p_2 + \dots + p_k = 1$. Since it creates no ambiguity, we will hereafter by default always consider games to be mixed extensions, and decisions to be randomized in general. This view canonically includes pure strategies by putting all probability mass onto a single element in the action space (thus making the mixed strategies into degenerate distributions). Consequently, we shall not further distinguish utilities defined on action spaces $\prod_{i \in \mathbb{N}} AS_i$ from expected utilities defined over $\prod_{i \in \mathbb{N}} \Delta(AS_i)$ in notation, and write u_i to mean the (expected) utility of the i-th player (in mixed strategies). The two coincide for degenerate distributions.

The expected utility in this setting, with the mixed strategies denoted as $\mathbf{x} \in \Delta(AS_1), \mathbf{y} \in \Delta(AS_2)$ for player 1 and player 2 respectively, is

$$\mathbb{E}_{\mathbf{x},\mathbf{y}}(u_1) = \mathbf{x}^T \cdot \mathbf{A}_1 \cdot \mathbf{y} = \sum_{i=1}^{|AS_1|} \sum_{j=1}^{|AS_2|} x_i \cdot a_{ij} \cdot y_j. \tag{3.18}$$

The formula for player 2 is analogous.

Both, Theorems 3.2 and 3.3 also assure the existence of Pareto-Nash equilibria if the vector-valued utility functions are all scalarized (3.3). The work of [3] then characterizes Pareto-Nash equilibria as (normal) Nash equilibria in certain such scalarized games.

In many cases, it pays to adopt a more accurate model of the conflict than just defining a utility as a consequence of multiple actions chosen at the same time. If the attacker adapts its actions to what the defender does, we can have a leader-follower situation, leading to a Stackelberg equilibrium, whose discussion we postpone until Sect. 4.2.

3.5.2 Zero-Sum Games

The symmetry of utilities, as in Example 3.5 for instance, is a special case, and generally does not need to hold. If a game is such that $\sum_{i \in N} u_i = \text{const}$, then we call it a *constant-sum game*. The utilities in Example 3.5 satisfy this property, since there, $u_1(a_1, a_2) + u_2(a_1, a_2) = 1$, for all a_1, a_2. A game is called *zero-sum*, if the sum of utilities is constant zero. For two players, this corresponds to the setting where the attacker's intentions are exactly opposite to the defender's objectives ("your pain is my gain"). Zero-sum games, however, have two important uses (in theory and practice):

Two games $\Gamma_1 = (N_1, S_1, H_1)$ and $\Gamma_2 = (N_2, S_2, H_2)$ are called *strategically equivalent*, if they have the same set of equilibria. This indeed requires $N_1 \simeq N_2$, and $S_1 \simeq S_2$, where the isomorphism $\simeq$ is interpretable as the sets being identical except for the naming of their elements.

If a game $\Gamma = (N, S, H)$ is constant-sum, say with $\sum_{i \in N} u_i = c \in \mathbb{R}$, then we can find values $\alpha, \beta \in \mathbb{R}$ such that $\alpha \cdot \left(\sum_{i \in N} u_i \right) + \beta = 0$, and define a strategically equivalent game Γ' with N, S and the utilities being $u_i' := \alpha \cdot u_i + \beta$. The affine transformation $x \mapsto \alpha \cdot x + \beta$ apparently leaves all optimum points unchanged, except for the possibility of turning a maximization into a minimization (and vice versa), if $\alpha < 0$. Thus, whenever we have a constant sum game, we may equivalently and without loss of generality, speak about a respective zero-sum game.

Apparently, Examples 3.4 and 3.5 force us to extend Definition 3.2. Let us take a closer look for zero-sum games, as in Example 3.4, but for a general matrix $\mathbf{A} = (a_{ij})$: any row that the maximizing player 1 would choose will cause player 2 to select the column that minimizes the utility for its opponent. Thus, the least that player 1 can achieve is $\underline{v} = \max_i (\min_j (a_{ij}))$. For Example 3.4, this would be $\underline{v} = \max_i (1, 0, 1, 0) = 1$. Likewise, player 1 will choose the row giving the maximum after player 2 has chosen its column; thus, the best possible is then $\overline{v} = \min_j (\max_i (a_{ij}))$. Again, for Example 3.4, this comes to $\overline{v} = \min_j (2, 4, 2, 4) = 2$. The observation that $\underline{v} \leq \overline{v}$ is no coincidence and systematically holds for any function $f : \mathbb{R}^2 \to \mathbb{R}$ (without further hypothesis) whenever a minimum and maximum exists; the *minimax inequality* is

$$\max_x \min_y f(x, y) \leq \min_x \max_y f(x, y) \tag{3.19}$$

The gap that we can compute in this way for a game, which is $\overline{v} - \underline{v} = 1$ for Example 3.4, indicates that the optimum for both players apparently lies somewhere in between the two bounds. The *minimax theorem* (a special case of Theorem 3.2) states that this optimum is achievable, if we extend the game in the proper way.

The practical and important use of zero-sum games, and stems from them being worst-case models: take two games Γ_0, Γ_1, with the same player set N and (mixed) strategy sets $S = \{\Delta(AS_1), \Delta(AS_2)\}$, but with different payoff structures $H_0 = \{u_1, -u_1\}$, and $H_1 = \{u_1, u_2\}$. Writing out the equilibrium Definition 3.3 for the two-player zero-sum setting gives for a utility-maximizing player 1,

$$u_1(\mathbf{x}, \mathbf{y}^*) \leq u_1(\mathbf{x}^*, \mathbf{y}^*) \leq u_1(\mathbf{x}^*, \mathbf{y}), \tag{3.20}$$

where the left inequality follows since player 1 can only lose by unilateral deviation, as can player 2 if it deviates, since this increases the revenue for player 1 (by the zero-sum property of the game). Let us denote the equilibria in Γ_0, Γ_1 by $(\mathbf{x}_0^*, \mathbf{y}_0^*)$ and $(\mathbf{x}_1^*, \mathbf{y}_1^*)$.

Imagine a situation in which player 1 follows an equilibrium strategy $\mathbf{x}_0^*$ in the zero-sum game Γ_0, while player 2 follows an equilibrium strategy $\mathbf{y}_1^*$ in the (nonzero-sum) game Γ_1. Most likely, the equilibria will not be the same, i.e., $\mathbf{x}_0^* \neq \mathbf{x}_1^*$ and $\mathbf{y}_0^* \neq \mathbf{y}_1^*$. Then, player 2 deviates from player 1's perspective, thus increasing player 1's actual payoff over what is expected. Likewise, player 1 deviates from the optimal behavior in Γ_2, thus increasing the payoff for player 2 in this game again by the definition of an equilibrium. It follows that

$$u_1(\mathbf{x}_0^*, \mathbf{y}_0^*) \leq u_1(\mathbf{x}_0^*, \mathbf{y}) \quad \text{for all } \mathbf{y} \in \Delta(AS_2). \tag{3.21}$$

This means that *irrespectively* of how player 2 acts within AS_2, player 1 can at least gain a minimum payoff by (i) presuming a zero-sum game, although this need not be the case in reality, and (ii) acting according to this zero-sum equilibrium (i.e., optimal) strategy. Apparently, the zero-sum assumption may be overly pessimistic, and the attacker's intentions may be far from exactly opposing, but this assumption *spares* the defender from pondering about the actual intentions of its adversary. In this way, the zero-sum assumption simplifies matters in practice, as the defender only needs to be confident about the attacker's *action space*, but does not need to hypothesize about the (hidden) intentions behind an attack.

Although equilibria are generally not unique, the equilibrium *payoff* in a zero-sum game is the same for every equilibrium. This makes the following definition sound:

Definition 3.7 ((Saddle-Point) Value of a Game) Let Γ be a two-player zero-sum game, with any equilibrium $(\mathbf{x}^*, \mathbf{y}^*)$. The *saddle-point value*, or shortly called *value*, of the game is $\mathrm{val}(\Gamma) = u_1(\mathbf{x}^*, \mathbf{y}^*)$.

For a finite game with payoff matrix $\mathbf{A}$ for player 1, the value is given by $(\mathbf{x}^*)^T \cdot \mathbf{A} \cdot \mathbf{y}^*$. The validity of the lower bound (3.21) against *every* behavior of the opponent merits paying special attention to such strategies:

Definition 3.8 (Security Strategy) Let $\Gamma = (N, S, H)$ be a game. A *security strategy* for the i-th player is a (mixed) strategy $\mathbf{a}_i^* \in \Delta(AS_i)$ such that

$$\mathbf{a}_i^* \in \operatorname*{argmax}_{\mathbf{a} \in \Delta(AS_i)} \min_{\mathbf{a}_{-i}} u_i(\mathbf{a}, \mathbf{a}_{-i}), \tag{3.22}$$

That is, a security strategy is a *minimax decision* that bounds the best achievable utility in light of any (mixed) strategy of the opponents of player i. Expression (3.21) asserts a zero-sum equilibrium behavior as an example of a security strategy

in a general (not necessarily zero-sum) game between two players. The existence
of security strategies is a simple consequence of the definition of an equilibrium,
and their computation amounts to merely the respective player taking its opponent
payoff to be precisely opposite to one's own utility, whether or not this assumption
may be accurate in practice. Wrapping up the discussion surrounding expressions
(3.20) and (3.21), we can conclude an existence result:

Theorem 3.4 (Existence of Security Strategies) *Let a two-player game $\Gamma =
(N = \{1, 2\}, S = \{AS_1, AS_2\}, H = \{u_1, u_2\})$ be given. A security strategy
for player i in Γ is given by a Nash-equilibrium strategy in the zero-sum game
$\Gamma_{i,0} = (N, S, H_{i,0} = \{u_i, -u_i\})$.*

Note that the one seeking a security strategy herein intentionally disregards the
payoff structure of the second player (more formally, only u_i but not the opponent's
utility u_{-i} appears in $\Gamma_{i,0}$). From a security practitioner's perspective, this has the
appeal of sparing hypotheses on the opponent's incentives. In other words, security
strategies need *no adversary modeling* in the context of game theory, since all that
is needed is one's own utility or damage assessment.

Example 3.7 ([7, 12]) Consider the two-person non-constant sum game with payoff structure as two matrices, jointly denoted as a tableau of pairs $(u_1(i, j), u_2(i, j))$
for each strategy profile $(i, j) \in AS_1 \times AS_2$ of the two players:

		Player 2			
	(2, 0)	(2, 0)	(1, 4)	(3, 1)	(2, 3)
	(1, 1)	(2, 3)	(2, 1)	(2, 3)	(4, 2)
Player 1	(0, 2)	(3, 2)	(0, 1)	(2, 3)	(2, 1)
	(0, 2)	(4, 2)	(1, 0)	(0, 2)	(1, 2)
	(2, 3)	(2, 1)	(4, 3)	(4, 1)	(3, 0)

This game has multiple equilibria with values $v_1 \in E_1 = \left\{2, 4, \frac{8}{3}, \frac{18}{7}, \frac{9}{4}, \frac{14}{5}\right\}$ for
player 1, and $v_2 \in E_2 = \{2, 3\}$ for player 2.

Now, let us compute a security strategy for the defender, by replacing the utilities
for player 2 by the negative values as for player 1. The resulting game is a zero-sum
matrix game (with the same entries as on the left side for player 1 in the bimatrix
game):

2	2	1	3	2
1	2	2	2	4
0	3	0	2	2
0	4	1	0	1
2	2	4	4	3

The saddle-point value of this game for player 1 is $v_1 = 2 = \min E_1$, so the bound that Definition 3.8 speaks about in (3.22) is even tight (although not generally so).   ◇

If the defender has multiple assets to protect or a player has multiple goals to optimize, the concept of a security strategy is extensible this multi-criteria optimization situation. Essentially, the defender is then playing a *one-against-all game*: it defines a distinct utility function *per goal*, but all utility functions having the same unified action space AS_1 as their domain; only being individually distinct w.r.t. the goal that they quantify. Also, we may allow for different actions for each attack, i.e., not all actions in the common domain AS_1 may be relevant for all goals. For example, an eavesdropping attack on confidentiality may require other actions than a DoS attack on the same system.

Let the utility be defined in terms of loss, so that we have a utility *minimizing* defender, facing a multi goal optimization problem, as in Sect. 3.3. We can model this situation as a multi-objective two-player game: Let the defender have defined functions $u_1^{(1)}, u_1^{(2)}, \ldots, u_1^{(d)}$, corresponding to $d \geq 1$ security goals. To each of these, we can associate an (independent hypothetical) attacker with its own action space $AS_2^{(1)}, \ldots, AS_2^{(d)}$. Without loss of generality, we can unify all these actions in a common action space for all attackers, by taking the union $AS_2 = \bigcup_{i=1}^{d} AS_2^{(d)}$ and letting the adversary's utility be $-\infty$ whenever this value would in connection with a certain action in AS_2. For example, if an action $a \in AS_2$ has some utility for eavesdropping, i.e., $a \in AS_2^{(\text{eavesdrop})}$, but is useless for a DoS (i.e., $u_2^{(\text{DoS})}$ is undefined since $a \notin AS_2^{(\text{DoS})}$), then we just define the utility to be unfavourable against, in fact dominated by, any other action for the DoS attacker, so it would never choose it, just like if it would not be in the action space at all. In this way, we assure that if an action is unavailable for a specific attacker, it will in any case be dominated, and thus will not show up in any equilibrium.

To simultaneously minimize the loss, the defender thus proceeds by defining a weighted sum of utilities for each (hypothetical) adversary and analyzes the two-person game with attacker utility after scalarizing it (cf. (3.3)):

$$u_2 := w_1 \cdot u_2^{(1)} + w_2 \cdot u_2^{(2)} + \ldots + w_d \cdot u_2^{(d)}, \qquad (3.23)$$

for weights $w_1 \cdots w_d$ that the defender needs to choose all ≥ 0 and constrained to satisfy $w_1 + w_2 + \ldots + w_d = 1$. Where do the weights come from? If the defender has information on how likely each type of attacker is, the weights in (3.23) can be set to the corresponding likelihoods. For example, if a DoS attacker is considered twice as likely as an eavesdropper, we have two adversaries, and define the weights $w_{\text{DoS}} = 2/3$ and $w_{\text{eavesdropper}} = 1/3$. If no such hypothesis appears reasonable or is possible, the defender may instead assign goal priorities, and define the weights accordingly to reflect them; cf. Sect. 3.3.

To put this to work, the defender now needs to define the specific utilities per security goal, i.e., for each attacker $j = 1, 2, \ldots, d$. Two situations are herein possible:

1. The defender has an adversary model so as to know the intentions and incentives behind attack type j. In that case, the defender may be able to define an accurate utility function $u_2^{(j)}$ to embody this knowledge.
2. The defender does not know the payoffs for the attacker. Then, the defender may instead model her/his own losses $u_1^{(j)}$ and put $u_2^{(j)} := -u_1^{(j)}$ as a substitute. This is equivalent to going for a minimax decision, i.e., a security strategy against the j-th type of attack(er).

Definition 3.9 (Multi-Goal Security Strategy (MGSS)) Let $\Gamma = (N = \{1, 2\}, S = \{AS_1, AS_2\}, H)$ be a multi-criteria two-player game, with continuous vector-valued payoffs $\mathbf{u}_1 = (u_1^{(i)}, \ldots, u_1^{(d)}) : AS_1 \times AS_2 \to \mathbb{R}^d$ for $d \geq 1$ for the first player. A *multi-goal security strategy (with assurance)* for a maximizing player 1 is a strategy $\mathbf{x}^* \in \Delta(AS_1)$ together with a vector $\mathbf{v} = (v_1, \ldots, v_d)$ such that the following holds:

1. The assurances are the component-wise guaranteed payoffs for player 1, i.e. for all goals $i = 1, 2, \ldots, d$, we have

$$v_i \leq u_1^{(i)}(\mathbf{x}^*, \mathbf{y}) \qquad \forall \mathbf{y} \in \Delta(AS_2), \tag{3.24}$$

with equality being achieved by at least one choice $\mathbf{y}_i^* \in S_2$.
2. At least one assurance becomes void if player 1 deviates from $\mathbf{x}^*$ by playing $\mathbf{x} \neq \mathbf{x}^*$. In that case, some $\mathbf{y}_0 \in \Delta(AS_2)$ exists such that

$$\mathbf{u}_1(\mathbf{x}, \mathbf{y}_0) \leq_1 \mathbf{v}. \tag{3.25}$$

Observe that this definition is agnostic of the second player's utility; in fact, the characteristic of a security strategy is precisely that we do not need to know this information. The existence of MGSS is assured under the same conditions as for Nash equilibria [8]. The existence result is a generalization of Theorem 3.4 that relates to the existence of Pareto-Nash equilibria:

Theorem 3.5 (Existence of MGSS [8]) *Let a security system be modelled as a multi-criteria two-person game Γ. For any such game, there is an auxiliary game Γ', such that $(\mathbf{v}, \mathbf{x}^*)$ is a MGSS, if and only if, it forms a Pareto-Nash equilibrium in the auxiliary game Γ' with the assurances being the saddle-point values of Γ'.*

The computation of an MGSS is possible by Fictitious Play (FP), which we discuss further in Sect. 12.3.4.

The multi goal defense that the defender adopts here can be viewed as a special case of what we call a *Bayesian game*. In such a game, a player is uncertain about the particular *type* of opponent, but knows the set Θ of possible types and individual payoff structures they have. Nature, as an irrational third entity, then decides on the type $\theta \in \Theta$ based on a probability distribution that is known to the defender. This one, in turn, takes an optimal decision based on the knowledge θ, i.e., a Bayesian decision (cf. Sect. 3.4.1). If the attacker is such that it partly exposes its type by sending a signal (even if unintended), the defender may refine its guess about the adversary type into a posterior distribution conditional on the signal, and again takes a Bayesian decision. We will revisit this later in the context of signaling games (see Chap. 4, and Example 3.8 in the next section).

3.5.3 Extensive Form Games

In many situations, it pays to look deeper into the interaction dynamics, since some actions of one player may enable or prevent subsequent actions of other players. Like in many board games, players can take actions in a prescribed order, and may not necessarily be informed about what the other player has done or can do next. The latter are called games with imperfect information (card games are an example, since players usually do not see their opponents deck), and the former are called games without perfect recall (for example, some card games allow exchanging parts of the deck without showing the old and new cards to the others). Security games can exhibit both, for the defender and attacker in similar ways: if the defender does not know where the attacker is located in a system, we have a situation of imperfect information. Likewise, if the defender notices suspicious activity in the system, the attack path over which the adversary came in may be ambiguous, thus the game has no perfect recall.

The examples above induce a couple of apparent ingredients necessary for an accurate description of a game, which we call *extensive form*. Like a normal form game, it has a finite set of players N, each $i \in N$ endowed with an action space AS_i. The game play is in discrete time steps, meaning that the game allows different players or groups of players to move at the same instant of time, and allowing other player (groups) to move afterwards, and so on. A particular instance of the game then gives rise to a *history* $\mathbf{h} = (a_1, a_2, \ldots)$ with elements from $\bigcup_{i \in N} AS_i$ of action, where we assume all action spaces AS_j to be pairwise disjoint, so that the history is unambiguous on which player took what action. The empty sequence of zero length corresponds to no player having moved yet, i.e., the beginning of the game. We denote this special state of the game as $\mathbf{n}$. We say that the move a *continues* a history $\mathbf{h}$, if this move (by whatever player takes it) is allowed by the game's rules to follow at the stage of the game, after the history $\mathbf{h}$, i.e., $\mathbf{h}' = (\mathbf{h}, a)$ is again a valid history for the game.

Given some history $\mathbf{h}$, an extensive form game further defines (i) which player(s) are allowed to take their actions next, and (ii) defines a set of admissible actions at this stage of the game. Both depend on the history $\mathbf{h}$, and for (i), we define a set-valued mapping $p(\mathbf{h}) = \{j \in N : \text{player } j \text{ is to move next}\}$, and for (ii), we define the set $Z(\mathbf{h}) = \{a \in \bigcup_{j \in p(\mathbf{h})} AS_j : \text{player } j \text{ can take action } a \text{ to continue the history } \mathbf{h}\}$. If we restrict our attention only to those actions that the

i-th player can take, we write $Z_i(\mathbf{h})$. The definition of what is possible and who is allowed to move is told by the rules of the game. In cases where multiple players can move simultaneously (not necessarily independent of one another) the value of $p(\mathbf{h})$ is a set with two or more elements, but in many cases will specify exactly one $j \in N$ to move next in the game. When $p(\mathbf{h}) =$, there are no more moves possible, and the game *terminates* by assigning a utility to all players. A history with $p(\mathbf{h}) =$ is called *terminal*. Otherwise, if $p(\mathbf{h}) \neq \emptyset$ (i.e., there is a player that can move), and $Z(\mathbf{h}) \neq \emptyset$, we call the history *finite*. A history that can be continued ad infinitum is called infinite. For example, Chess admits situations with moves to lead into periodicity ("eternal Chess"). Some repeated security games are as well examples of games that never terminate ("security is never done").

The entirety of all histories (including finite, terminal and infinite ones) constitutes the *game tree* B , whose root is $\mathbf{n}$. In this tree, each history $\mathbf{h}$ defines a unique path from the root $\mathbf{n}$ to the last action in the history $\mathbf{h}$, which we will call a *node* v in the tree B. A terminal history is thus a path from the root node $\mathbf{n}$ to a leaf node (final action) v.

The number of v's children in B is the number $|p(\mathbf{h})|$ of players that can move, multiplied by the number of actions $|Z(\mathbf{h})|$ that each player $i \in p(\mathbf{h})$ has in this moment v. If a node is such that no player can move any more, then the game ends, with the *utility function* specifying the payoffs for each player in this terminal state of the game. The game tree B is thus the collection of all possible evolutions of the game. It is generally an infinite graph, since not all instances of the game necessarily need to terminate.

We call a game *finite*, if the graph B is finite, i.e., all sequences of moves terminate sooner or later, in which case a utility is assigned to each player. Normal form games necessarily terminate after one step, since the utility is gained after having taken action. This distinguishes *actions* in the sense that we had so far, from *strategies* in extensive form games: a *strategy* is a prescription of which action to take in any given state of the game. In other words, being at some node v in the game tree B, a strategy tells the moving player which action to take next. Optimizing these strategies is the same as computing equilibria. Indeed, enumerating all terminal paths in the tree B and collecting the utilities gained at their ends, we can convert the game tree into a matrix, i.e., normal form, representation of the game. The action sets that used to label the rows and columns of the matrix (in a two-player setting) then correspond to the action spaces that we had in the chapter up to this section. Now, for extensive form games, we have narrowed the understanding of an "action" into being a possible move at the current state of the game.

Unless the players can observe all other's actions, a player may be unable to determine the exact location of the game in the full tree B of possibilities. That is, from player i's perspective, knowing that player j can move next, player i can partition the set $\{v \in B : j \in p(\mathbf{h})\}$ into sets $I_1, \ldots, I_{j_k}$, with the following property: any two game states v, w in the same set I_ℓ are such that the possible moves of player j are the same, regardless of whether it starts from v or w. We call this an *information set*. If the two possible nodes v, w in B are in different information sets (for player i), then the information gained (hence the name) is the particular set of actions that player j can take from now on. It is often useful to include "nature" as a zeroth player (for example, if a board game prescribes to throw a dice, we may think of this random event being nature choosing a number between 1 and 6).

Example 3.8 (Phishing Game [16]) Consider a simple game of email correspondence, with player 1 being the receiver of emails, and player 2 being any (unknown) sender. The sender of an email can be phishing or honest, and "nature", coming in as a third player 0, decides about the type of sender. The game (tree) thus starts with player 0's move, and then enters an information set for player 1, since this one does not know whether the email was phishing or honest. It thus cannot determine which kind it was, unless there was, say, a digital signature attached to the email. In that case, we know that the email is honest, corresponding to a third child of node $\mathbf{n}$ in the tree, which lies in another, now singleton, information set.

The game specification is directly visible from the tree, and in our previous notation is listed in the following table. Note that the phishing adversary is not even explicitly included:

	$\mathbf{h}$	$\mathbf{n}$	$(\mathbf{n},1)$	$(\mathbf{n},2)$	$(\mathbf{n},3)$
next move function	$p(\mathbf{h})$	$\{0\}$	$\{1\}$	$\{1\}$	$\{1\}$
actions available	$Z(\mathbf{h})$	{phishing, honest+no signature, signature}	{open, delete}	{open, delete}	{open}

When it is player 1's move, this one can decide to open or leave the email unattended if the game is in any state as in information set I_1, or it can only open the email if it is digitally signed (information set I_2). The payoffs could then be assigned as follows: if the receiver opens a phishing mail, then it scores a loss of -1, while the phisher has won (payoff $+1$). If the phishing email gets deleted, the phisher loses as well (payoff -1), and the receiver gains nothing (payoff 0). If an honest email is accidentally considered as phishing and deleted, the receiver suffers a loss -1, while the phisher gains nothing (utility 0, since this email was not sent by her/him). Likewise, the phisher gains nothing in the other case, where the email came from an honest sender, is digitally signed, and gets opened. The respective utilities appear in brackets (receiver utility, phisher's utility) underneath the terminating nodes in the game tree in Fig. 3.2. ◇

"Nature" is generally useful to include as a distinct "player 0", whenever external forces are relevant that no player can influence, but whose circumstances determine what happens in the game. This virtual zero-th player is not assigned any utility, but has an action space in each state of the game, and all players know the probability of each possible action that nature can take when it is its move (i.e., $p(\mathbf{h}) = \{0\}$). When throwing a dice, this is trivial, as any fair dice will equiprobably come up with numbers 1, ..., 6. For the phishing game of Example 3.8, things are more complicated, since player 1 would need to estimate the likelihood to receive a phishing mail.

This was the last ingredient to an extensive form game, which unifies all the components explained so far:

Definition 3.10 (Extensive Form Game) A game in extensive form is a six-tuple consisting of the following:

1. The players, forming a finite set N
2. The possible game plays, specified as a game tree B with root **n**. Let F be the set of terminal or infinite histories in B.
3. A rule who's move it is: This is a function p telling which player(s) can move in each state of the game, i.e., for each node $v \in B$ reached over a history **h** of actions, $p(\mathbf{h}) \subseteq N$ specifies which player(s) move next.
4. Randomness: If there is an element of randomness (not uncertainty), we include a virtual player 0 "Nature" with action set AS_0 and define a probability distribution over AS_0, conditional on each node $v \in B$.
5. Information sets: For all $i \in N$, this is a partition of the set of all $\{v \in B :$ v wascequence **h** of moves, and $p(\mathbf{h}) = i\}$.
6. Utility functions $u : F \to R^{|N|}$, specifying the utility that each player gets when the game finishes.

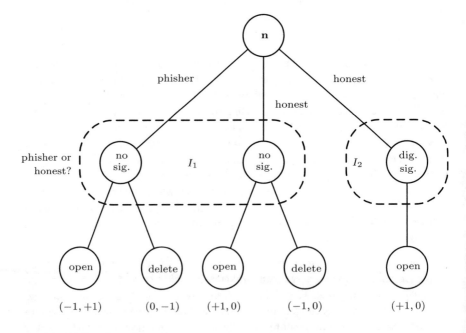

Fig. 3.2 Phishing Game Example

Note that the randomness and information sets are optional in this definition: a "player" 0 is not required, if the game does not involve randomness (for example, Chess). Likewise, the information sets can all be singleton, in case that the current state of the game is always fully known to all players, and we say the game has *perfect information*. This is crucially different from *complete information*: while perfect information means that a player is at every time informed about the current state of the game, complete information means that every player is certain about all details of the game itself, which in particular includes the payoffs that the opponents receive. The two properties are independent, and security has both, games with perfect but incomplete information, and games with imperfect yet complete information (and of course both at the same time):

Example 3.9 (Imperfect but complete information) Consider a network in which Alice wants to send a message to Bob. Let the graph be such that there are $k \geq 2$ node-disjoint paths connecting the two, and Alice can choose any path out of these k at random to transmit her message. The adversary, in turn, can choose any intermediate node in the network to eavesdrop, and hope for the message to pass by. This game has imperfect information, since Alice does not know where the adversary is, but it is a complete information game, since the objectives of all players are clear (as is the rest of the game specification). We will revisit this game for a more detailed analysis in Sect. 11.2.   ◇

Example 3.10 (Perfect but incomplete information) Imagine a quantum network, which is well known to provide perfect point-to-point confidentiality by eaves-dropping detection on the quantum bit exchange. Assume that Alice and Bob are in the quantum network and trying to send messages. They react on each incident of eavesdropping by stopping the transmission and resort to either (i) pausing until the eavesdropper is gone, (ii) switching to a classical encryption to save key material until case (i) occurs, or (iii) switch to an alternative route when they find out that the currently chosen path is insecure. Since the eaves-dropper cannot hide due to the properties of the quantum link, the game has perfect information. The information is nonetheless incomplete, since the actual objective of the attacker remains hidden: it can be a DoS via eavesdropping, as this in any case causes a blockage of the channel or at least a fallback to a weaker form of encryption [14]. It can, however, also be still a sophisticated eavesdropping attempt by enforcing a redirection of traffic through a vulnerable node [9]. We will come back to this application context in Sect. 11.2, where we shall unconditionally quantify the end-to-end security in quantum networks by means of multipath transmission and proving security using game theoretic models.   ◇

Whenever the game has complete information and is finite, we can enumerate all paths from the game tree's root until the leafs (terminal nodes) and record the actions made by all players and assigned utilities (provided that the number of paths is small enough to make a complete list in reasonable computational time). Doing this enumeration for both players naturally creates action sets for them, whose members are then called *strategies*. A strategy (in an extensive form game) is a prescription of what *action* to take every stage (i.e., node) of the game tree.

Strategies are sequences of actions that are determined by the moves of the players as the game proceeds. For a normal form game, a strategy is just a member of the action set: each player picks a member a_i from its action set AS_i and the i-th player $i \in N$ gets the utility as told by $u_i(a_i, \mathbf{a}_{-i})$. The mechanism of the gameplay, however, is not further looked into, since it is all embodied in the utility function. Strategies in extensive form games are exactly different in this aspect, since they describe in detail how the game proceeds from the first moves down until the point when the utilities are being paid to the players.

The conversion from extensive to normal form by enumerating strategies, i.e., paths in the game tree, is generally neither unique nor reversible, since without additional information, the extensive form game cannot be recovered from the normal form. We will come back to specific examples of games in extensive form in Chap. 4, when we describe leader-follower situations in security that naturally lend themselves to extensive form models. Using such a model, Example 4.1 will describe the transition to normal form game using two different extensive form games.

3.6   Extended Concepts: Modeling Goal Interdependence

While scalarizing a multi-objective optimization is the traditional method, more general techniques arise if we explicitly specify how the goals interact, instead of only saying how much weight we put on each goal individually. Formally, let there be n goals, taking specific values g_i. Moreover, let the values be sorted in ascending order $0 \leq g_{(1)} \leq g_{(2)} \leq \ldots \leq g_{(n)}$. The discrete Choquet integral, acting as an aggregation operator in our case, replaces (3.3) by the expression

$$g^* := \sum_{i=1}^{n} w_{(i)} \cdot g_{(i)}, \tag{3.26}$$

in which $w_{(i)} := \mu(A_{(i)}) - \mu(A_{(i+1)})$ with $A_{(i)} := \{i, i + 1, \ldots, n\}$ represents the set of all goals with increasing indices starting at index i. The interesting part is the (so-far) unspecified function μ here, which "valuates" the set of goals in their entirety, and taking their interplay into account. The difference $w_{(i)}$ is thus nothing else than a value (a gain or loss) assigned to the situation when the i-th goal is included (set $A_{(i)}$) or excluded (set $A_{(i+1)}$) from the consideration. Depending on how much the absence of this goal affects the overall aims, the function μ plays the role of the importance weights again. Formally, it is called a *capacity function*, and needs to satisfy $\mu(\emptyset) = 0$ and monotony $[A \leq B] \Rightarrow \mu(A) \leq \mu(B)$. The name Choquet *integral* comes from the similarity to Riemann sums, which the concept can indeed resemble, if we add additional properties to the capacity function to make it a measure.

While appealing, the practical application of this method comes with the price-tag of needing to specify not only the importance *per goal*, but also the importance of all combinations of goals towards a total specification of the capacity μ. That is, (3.26) does only help including the interplay of goals to the extent that we already know and explicitly modeled these beforehand. In the extreme case, we end up specifying μ pointwise on all subsets of the goals $\{1, 2, \ldots, n\}$, which are 2^n many. Thus, practically, we usually confine this to specifying only the importance weights directly (as in (3.3)), and perhaps pairwise or triple-wise interplays (ending up in a number of hand-crafted values for μ that is quadratic or cubic in n).

Fuzzy logic offers yet further possibilities of multi-criteria decision making, all of which boil down to certain (partly heuristic) aggregation functions. While this offers the full flexibility of constructing an aggregation that accounts for the particularities of and mechanisms behind the goal's interdependence, this freedom comes with the price tag of a much more complicated interpretation of the underlying heuristics and mathematical concepts. Although logical connectives like triangular norms and co-norms come out of precise mathematical requirements and enjoy strong connections to statistics (copulae theory), the practical choice from the (continuum) of aggregation operators that fuzzy logic offers, enjoys much less help from theory to make reasonable design choices. Nonetheless, fuzzy theory is remarkably popular and successful in problems of multi-criteria decision making, yet beyond the scope of our account here (see [13] for a comprehensive treatment).

References

1. Glicksberg IL (1952) A further generalization of the kakutani fixed point theorem, with application to Nash equilibrium points. In: Proceedings of the American mathematical society, vol 3, pp 170–174. https://doi.org/10.2307/2032478
2. Jøsang A, Ismail R (2002) The beta reputation system. In: Proceedings of the 15th bled electronic commerce conference
3. Lozovanu D, Solomon D, Zelikovsky A (2005) Multiobjective games and determining Pareto-Nash equilibria. Buletinul Academiei de Stiinte a Republicii Moldova Matematica 3(49):115–122. http://D:\Documents\resources\multiobjective_games_and_optimization\MultiobjectiveGamesandDeterminingPareto-NashEquilibria.pdf
4. MRC Biostatistics Unit, U.o.C. (2008) The BUGS project. https://www.mrc-bsu.cam.ac.uk/software/bugs/
5. Nash JF (1951) Non-cooperative games. Ann Math 54:286–295. http://www.jstor.org/stable/1969529?origin=crossref
6. von Neumann J, Morgenstern O (1944) Theory of games and economic behavior. Princeton University Press, Princeton
7. Rass S (2009) On information-theoretic security: contemporary problems and solutions. Ph.D. thesis, Klagenfurt University, Institute of Applied Informatics
8. Rass S (2013) On game-theoretic network security provisioning. Springer J Netw Syst Manag 21(1):47–64. https://doi.org/10.1007/s10922-012-9229-1

9. Rass S, König S (2012) Turning Quantum Cryptography against itself: how to avoid indirect eavesdropping in quantum networks by passive and active adversaries. Int J Adv Syst Meas 5(1 & 2):22–33
10. Rass S, König S, Schauer S (2016) Decisions with uncertain consequences – a total ordering on loss-distributions. PLoS ONE 11(12):e0168583. https://doi.org/10.1371/journal.pone.0168583
11. Rass S, Kurowski S (2013) On Bayesian trust and risk forecasting for compound systems. In: Proceedings of the 7th international conference on IT security incident management & IT forensics (IMF). IEEE Computer Society, pp 69–82. https://doi.org/10.1109/IMF.2013.13
12. Rass S, Schauer S (eds) (2018) Game theory for security and risk management. Springer, Birkhäuser. http://www.worldcat.org/oclc/1019624428
13. Ross T, Booker JM, Parkinson WJ (2002) Fuzzy logic and probability applications: bridging the gap. ASA SIAM, Philadelphia/U.S.A
14. Schartner P, Rass S (2010) Quantum key distribution and Denial-of-Service: using strengthened classical cryptography as a fallback option. In: Computer symposium (ICS), 2010 international. IEEE, pp 131–136
15. Sion M, Wolfe P (1957) On a game without a value. Princeton University Press, pp 299–306. http://www.jstor.org/stable/j.ctt1b9x26z.20
16. Zhu Q, Rass S (2018) On multi-phase and multi-stage game-theoretic modeling of advanced persistent threats. IEEE Access 6:13958–13971. https://doi.org/10.1109/access.2018.2814481. https://doi.org/10.1109%2Faccess.2018.2814481

Chapter 4
Types of Games

> *Life is not a game. Still, in this life, we choose the games we live to play.*
>
> J.R. Rim

Abstract This chapter introduces the most important classes of games underlying practical security models. These include Stackelberg games, Nash games, signaling games, and games with distribution-valued payoffs. The latter build upon empirical methods and data science to construct games from data, but also reveals theoretic connections to multi-criteria optimization using lexicographic goal priorities (that classical games cannot deal with, but distribution-valued games can handle). Each game description is accompanied by examples from the security domain to motivate and illustrate the use of the individual model. Each class of game is discussed in relation to the other types, highlighting pros and cons, las well as applications, detailed in later chapters.

4.1 Overview

Game theory has been used in a wide variety of security and cyber-security contexts. This lot roots in a few different types of game theoretic models that we discuss in more details in this chapter. Among the application areas are intrusion detection systems [59, 60, 67], adversarial machine learning [30, 51, 53, 54, 56], proactive and adaptive defense [6, 7, 20, 47, 64, 66, 68], cyber deception [18, 28, 34, 36, 58, 61], communications channel jamming [2, 50, 62, 65, 69, 70], Internet of Things [8–11, 24, 28, 32, 33, 35, 72], secure industrial control systems [1, 26, 61, 63] and critical infrastructure security and resilience [3, 3, 16, 17, 21, 22, 41]. In each of these contexts, attackers and defenders attempt to rationally optimize their actions by anticipating the actions of the other players. Hence, game theory captures the fundamental properties of these interactions.

© Springer Nature Switzerland AG 2020
S. Rass et al., *Cyber-Security in Critical Infrastructures*, Advanced Sciences and Technologies for Security Applications,
https://doi.org/10.1007/978-3-030-46908-5_4

Table 4.1 Players, types, actions, and utility functions for three games

Players N	Types Θ	Actions AS	Utility	Duration
Stackelberg game	Typically	$L : a_L \in AS_L$	$L : u_L(a_L, a_F) \in \mathbb{R}$	One-shot
between leader L	uniform	$F : a_F \in AS_F$	$F : u_F(a_L, a_F) \in \mathbb{R}$	leader-follower
and follower F				structure
Nash game	Typically	$V : a_V \in AS_V$	$V : u_V(a_V, a_W) \in \mathbb{R}$	Simultaneous
between semetric	uniform	$W : a_W \in AS_W$	$W : u_W(a_V, a_W) \in \mathbb{R}$	move structure
players V and W				
Signaling game	S has	$S : a_S \in AS_S$	S of each type	One-shot
between sender S	multiple	$R : a_R \in AS_R$	$\theta \in \Theta : u_S^\theta(a_S, a_R) \in \mathbb{R}$	sender-receiver
and receiver R	types $\theta \in \Theta$		$R : u_R(\theta, a_S, a_R) \in \mathbb{R}$	structure
Game with	Multiple types	$S : a_S \in AS_S$	For each player $i \in N$	Various,
distribution valued	possible but	$R : a_R \in AS_R$	$u_i(\mathbf{x}_i, \mathbf{x}_{-i})(t) = \Pr(U \leq t)$ is	including all of
payoffs	not necessary	$u_R(\theta, a_S, a_R)$	a distribution	the above

In Sects. 4.2 and 4.3, we briefly introduce two of the most common game-theoretic models used in defensive deception for cybersecurity and privacy: Stackelberg and Nash games. In Sect. 4.4 we address signaling games, which study scenarios of incomplete information. Section 4.5 describes the more general class of games with uncertainty, where otherwise crisp real-valued payoffs are replaced by random variables that are stochastically ordered. Table 4.1 summarizes the components of the games.

All three games are static games (where we take one-shot interactions to be static although they are not simultaneous). In addition, while Stackelberg and Nash games typically do not include multiple types, Bayesian Stackelberg and Bayesian Nash games do allow multiple types.

4.2   Stackelberg Game

In Stackelberg games [48], the follower moves *after knowing the leader's action*. Often, cybersecurity models take the defender as L and the attacker as F, assuming that the attacker will observe and react to defensive strategies. Their structure is often conveniently described by a game tree, i.a., in extensive form.

Stackelberg games are solved backwards in time (sometimes also called *back-ward induction*). Let $BR_F : AS_L \rightarrow \Delta(AS_F)$ define a *best-response function* of the follower to the leader's action, in mixed strategies to simplify matters here. $BR_F(a_L)$ gives the optimal a_F to respond to a_L. The best-response function could also include a set of equally good actions, i.e., this is why the power set is used. The best response function is defined by

$$BR_F(a_L) = \underset{a_F \in AS_F}{\operatorname{argmax}}\, u_F(a_L, a_F). \tag{4.1}$$

Based on anticipating F's best response, L chooses optimal action a_L^* which satisfies

$$a_L^* \in \underset{a_L \in AS_L}{\operatorname{argmax}}\, u_L(a_L, BR_F(a_L)). \tag{4.2}$$

Then, in equilibrium, the players' actions are $\left(a_L^*, a_F^*\right)$, where $a_F^* \in BR_F\left(a_L^*\right)$.

Definition 4.1 (Stackelberg equilibrium) In a two-player game $\Gamma = (N = \{L, F\},\ S = \{\Delta(AS_L), \Delta(AS_F)\},\ H = \{u_L, u_F\})$, a *Stackelberg equilibrium* is a pair of strategies $(\mathbf{a}_L^*, \mathbf{a}_F^*)$ for the leader and the follower, respectively, if $\mathbf{a}_L^*$ is optimal, given the leader's information, and $\mathbf{a}_F^* \in BR(\mathbf{a}_L^*)$, is the optimal response of the follower (player 2), given that the leader does $\mathbf{a}_L^*$.

The existence of a pure-strategy Stackelberg equilibrium for finite games is guaranteed. The applications of Stackelberg games in security have been pervasive including data privacy [30, 31], infrastructure security [4, 5, 63], virus spreading [14, 19] and cyber insurance [13, 52, 57]. In [5], a three-stage dynamic game is formulated between an attacker and a network designer. The network designer first designs an infrastructure network before the attack and recovers the network from the damages of the attacker. The adversary attacks the infrastructure network once the infrastructure is built with the aim to disconnect the network while the network designer aims to keep the network connected by designing secure and resilient networks before and after the attack, respectively. The two-player game is an extended version of the Stackelberg game where the network designer is the leader in the first stage and the follower in the last stage of the game. The attacker is a follower after the first stage of the game but the leader of the second stage of the game. In [57], the authors have established a bi-level game-theoretic model that boasts the feature of a Stackelberg game where the insurance designer is the leader who announces an insurance policy for the user. The user can be viewed as a follower who plays a zero-sum game with an attacker whose outcome determines the cyber risks for the user. The goal of the insurance is to further mitigate the damage of the attacks on the user. The interactions among the three players can be captured by a two-level game structure in which one game is nested in the other.

Example 4.1 (From extensive to normal form)   Let us rephrase the hide-and-seek game from Example 3.5 in extensive form, letting both players, denoted as "def." and "adv." in Fig. 4.1, take the leader's role. First, let the defender lead, modeling

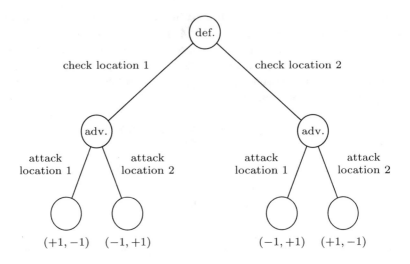

Fig. 4.1 Extensive form game for Example 3.5

the hypothesis that the attacker can observe the defender and react accordingly. Figure 4.1 shows the extensive form game tree, with payoffs denoted as pairs (defender's utility, adversary's utility).

The normal form game is found by systematically enumerating all possible choices of all players, and tabulating them. This leads to a payoff matrix, in which the column player's strategies are written as pairs (X, Y), where X is the action following strategy "check location 1", and Y is strategy following the defender's previous action "check [location] 2". For example, the revenue noted in the cell at location "check [location] 1"/"(attack [location] 1, attack [location] 2)" is obtained as follows: the defender (leader) has chosen to check location 1. The attacker having chosen the first column, will look up the action to follow that particular history, i.e., if the defender's previous action was to check location 1. The prescription of the first column's strategy is to react by attacking location 1, which gives the revenue $+1$ for the defender, and -1 for the attacker. Note that the second prescription in that column would also be to attack location 1, but this action would follow the defender's previous action to check location 2. Thus, this payoff is irrelevant to fill that table cell.

Similarly, if the defender chooses to check location 2 and the attacker picks the first column, that the history to react on would be "check 2", on which the prescription is to attack location 1 (second entry in the column's label, according to our naming convention). In that case, the adversary scores $+1$, and the defender suffers a loss (-1), with the respective cell entry being $(-1, +1)$ accordingly.

Carrying this systematic enumeration of possibilities to the end of considering all possible scenarios in the game tree, we find the normal form representation of the game, as shown in Fig. 4.2. The bold-printed entries therein mark the equilibrium, in fact a Stackelberg equilibrium, for this leader-follower game.

		adversary (follower)			
		attack 1, attack 1	attack 1, attack 2	attack 2, attack 1	attack 2, attack 2
defender	check 1	(1,-1)	(1,-1)	**(-1,1)**	(-1,1)
(leader)	check 2	(-1,1)	(1,-1)	**(-1,1)**	(1,-1)

Fig. 4.2 Normal form of the extensive form game in Fig. 4.1

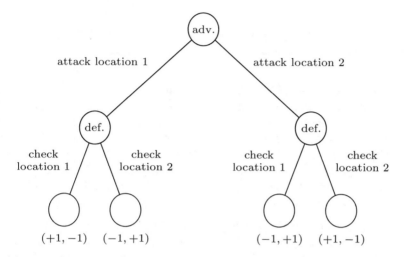

Fig. 4.3 Variant with leading attacker

The expected revenue for the defender in this game is -1, consistently with the fact that the attacker can perfectly adapt to the defender's action, since it can observe the defensive precautions and can act accordingly. This means that the defender's actions should be hidden from the attacker. However, note that this confidentiality does not extend to keeping the action space also confidential. Indeed, the defender *can* let the attacker know what actions are possible in principle, but must keep only the concrete current move as a secret. Risk management, along trust-establishing measures, recommends to publish *what* is done for security, but at the same time prescribes to be confidential about the details. The game theoretic analysis of this particular game teaches the same lesson upon a purely mathematical analysis, which hereby justifies this best practice. We will have to say more about this in Chap. 6.

Let us turn around things to look at the situation where the adversary moves first. Figure 4.3 shows the resulting game tree, with payoff annotation (u_1, u_2) in the leafs referring to the defender's and attacker's payoffs respectively.

The construction of the normal form game works as before, only the shape of the tableau changes into the defender now having follow-up strategies to whatever the adversary did first (note that this version would anyhow be realistic only in a reactive setting, where the defender repairs damage due to adversarial actions, so not all combinations of actions, i.e., strategies, may be practically meaningful).

Fig. 4.4 Normal form of the extensive form game in Fig. 4.3

		adversary (leader)	
		attack 1	attack 2
	check 1, check 1	(+1,-1)	(-1,+1)
defender	check 1, check 2	**(+1,-1)**	**(+1,-1)**
(follower)	check 2, check 1	(-1,+1)	(-1,+1)
	check 2, check 2	(-1,+1)	(+1,-1)

As with the previous game, we highlighted one equilibrium in bold, in Fig. 4.4. It gives the payoff $+1$ to the defender, and -1 to the attacker.

$\diamond$

This conversion assures the existence of equilibria just as for normal form games, although an equilibrium has then a more complex interpretation: it is a *path* in the game tree that reaches the simultaneously optimal utility for all players, or if the equilibrium is in mixed strategies, it is a randomized optimal choice of action for each player in each node of the game tree. Note that this path not necessarily means that each player behaves optimal in each stage of the game. Adding this requirement leads to a refinement of (Nash-)equilibria. We emphasize that this concept applies to general extensive form games, not necessarily only to Stackelberg games.

Definition 4.2 (Subgame Perfect Equilibrium (SPE)) Let a game Γ in extensive form be given, whose game tree is B. A *subgame* in Γ consists of a subtree B' of B, with all other items in Γ applied to the subtree B' only. An equilibrium in Γ is *subgame perfect*, if it induced an equilibrium (of any kind, including Nash, Stackelberg, ...) for every subgame of Γ.

If the game Γ is finite, we call the longest path in B from the root to a leaf the *length* $\ell(\Gamma)$ of the game. An SPE can be found by induction on the games length $\ell(\Gamma)$: if $\ell(\Gamma) = 1$, then starting from the root node **n**, the first move taken by the players $p(\mathbf{n})$ immediately deliver payoffs. This is just as for a finite normal form game, and there is an equilibrium by Nash's Theorem 3.2. Now, suppose that we have found subgame perfect equilibria up to length $\ell(\Gamma) = k$, and we are about to take the $(k + 1)$-st move. Let the history (of length k) up to this point be **h** (generically).

- If nature is to make the next move, then the equilibrium payoffs in the subgame of length $k + 1$ are the weighted average of utilities that all players receive in all actions that nature can play, with weights being the likelihoods for these actions (remember that for player 0, Definition 3.10 requires this information as available).

- If a player $i \neq 0$ is to move, then it selects that action $a \in Z(\mathbf{h}_i, \mathbf{h}_{-i})$ from the available ones, that maximizes this player's outcome among all possibilities, i.e., the subgames until this point in the game, or equivalently, the nodes reachable in k steps from the root. This works only if all information sets are singleton, so it is clear which nodes can be reached, also taking the other player's actions into account.

In each step, we can do this for all (perhaps exponentially many) paths of length k, to find an SPE for a game of length $\ell(\Gamma) = k + 1$, thus proving the following existence result:

Theorem 4.1 ([25]) *Every game in extensive form and with perfect information has a subgame perfect equilibrium.*

The proof of Theorem 4.1 is in fact based on *backward induction* (also used to analyze Stackelberg games): to find a subgame perfect equilibrium, one starts by considering the final mover's action to maximize this player's utility (based on the fact that these utilities are all specified directly) in all possible circumstances, i.e., a minimax decision. Then, one presumes that this action has been taken, and repeats to find the butlast player's action to maximize their utilities, and so on, until the root node is reached. The strategies remaining by that point are the subgame perfect equilibria.

Example 4.2 (Continuation of Example 4.1) Let us apply backward induction on the extensive form game of Fig. 4.1. The last player to move is the attacker, who will choose to attack location 2, whenever the game is in the left branch of the tree, and to attack location 1, when the game is in the right branch of the tree. Assuming those actions to be adopted, the defender has no real choice any more, since it can only choose to lose -1 in the left branch, or lose -1 in the right branch. This delivers the equilibria that we already know, appearing bold-printed in the tableau in Fig. 4.2. Also, it shows that the two equilibria are both subgame perfect.   ◇

Backward induction fails if there are information sets with two or more elements, simply because we would end up in a set whose nodes are indistinguishable. That is, whenever we enter it coming from the root node in B, we have no way of assuring to leave the information set via the intended node.

To fix this, a player needs a *belief* on how likely each element in an information set is. This belief is a probability distribution $\Delta(I)$ supported on the respective information set containing the current stage, and the action is chosen so as to maximize the *expected utility* based on the belief about the current state of the game, conditional on the history **h** of moves up to this point. A Perfect Bayesian Nash Equilibrium (PBNE), which we will discuss in more detail in Sect. 4.4, is therefore based on two hypotheses:

Sequential rationality: Each information set I in the game is associated with an (a priori) belief $\Delta(I)$ assigning likelihoods to all nodes in the information set. Moves are made as Bayesian decisions based on the belief, i.e., maximize the expected utility.

Consistency: Before making the choice for a move out of an information set, the belief $\Delta(I)$ is updated using Bayes' rule, conditional on the history **h** of moves, lasting in I.

The player thus makes a *Bayesian decision* if it currently is inside an information set, and otherwise takes a minimax decision (as in backward induction). Prominent instances of games with PBNE are signaling games (see Sect. 4.4), where the beliefs are imposed on the particular type of adversary that the defender is facing. This is similar to the weights to scalarize a multi goal optimization problem as in Eq. (3.23), where we had multiple virtual adversaries representing different security goals to protect. Signaling games generalize this view.

4.3 Nash Game

While in Stackelberg games players move at different times, in Nash games [27] players move simultaneously. For the comparison, consider a simple Stackelberg game like shown in Fig. 4.5a, relative to the same game formulated as a Nash game in Fig. 4.5b. Nash games are played with *prior commitment*, by which each player commits to his or her strategy before knowing the other player's move. Typically, two-player games of prior commitment are shown in matrix form. Figure 4.5b, however, gives a tree diagram of a two-player game in order to show the difference between this game and a Stackelberg game. Players V and W act simultaneously, or at least without knowing the other player's action. The dashed line connecting the two nodes for W denotes that W does not know which node the game has reached, because she does not know which move V has chosen. This is an instance of an information set (see Definition 3.10).

A Nash equilibrium is a solution concept that describes a point where no player has no incentives to deviate unilaterally. Unlike Expressions (4.1) and (4.2), we now need to satisfy the following conditions simultaneously, and in no preference of one over the other: $\left(a_V^*, a_W^*\right)$ such that

$$a_V^* \in BR_V\left(a_W^*\right), \tag{4.3}$$

$$a_W^* \in BR_W\left(a_V^*\right). \tag{4.4}$$

Equations (4.3) and (4.4) lead to solving for fixed points of the joint map (BR_V, BR_W) and a Nash equilibrium is a point that consistently satisfies the system (4.3) and (4.4). The existence of Nash equilibrium in general requires the mixed extension of a game (Definition 3.5).

In many cybersecurity applications, zero-sum and nonzero-sum games have used to describe security applications such as jamming and eavesdropping in communication networks [50, 69–71], configurations of intrusion detection systems [59, 67], and infrastructure protections [17, 23]. In [59], the interaction between an attacker and an intrusion detector system is captured by a zero-sum game where the attacker chooses a sequence of attacks that can evade the detection while the system administrator aims to configure the intrusion detection system to increase the probability of detection, while reducing the overhead of the detection. In [23], the authors have established a zero-sum game for a large-scale infrastructure network in which each infrastructure node defends itself against an adversary and the risk connected to each node can propagate over the network. The complexity of dealing

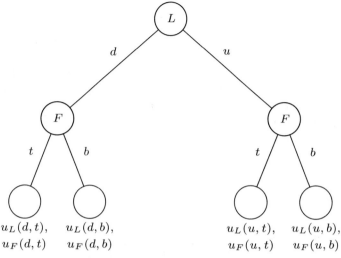

(a) Stackelberg Game (leader-follower)

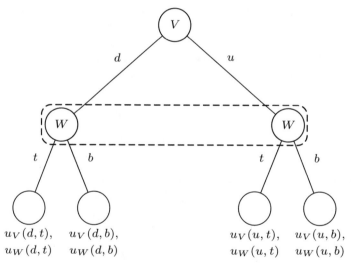

(b) Nash Game (with mutual prior commitment)

Fig. 4.5 Illustration of two game structures: Stackelberg game and Nash game

with the large-scale game can be reduced by developing approximate computational methods and leveraging the sparsity of the network connections; i.e., the nodes are not fully connected. In [69], the authors have formulated a non-cooperative game between the users and the malicious node in the context of relay station-enabled wireless networks. The goal of each wireless user is to choose the serving relay station that can optimize the security and quality-of-service of its transmission, given the presence of interference as well as of a malicious node. Conversely, the malicious node's objective is to decide on whether to eavesdrop, jam, or use a combination of both strategies, in a way to reduce the overall transmission rate of the network. To algorithmically solve this game, FP is applicable, which lets the users and the malicious node iteratively reach a mixed-strategy Nash equilibrium (see Sect. 12.3.4 for details on the general ideas behind FP). The results have also shown that the proposed approach enables the wireless users to significantly improve their average expected utilities which reflect the security and the quality-of-service (in terms of transmission rate) perceived over the chosen relay stations.

4.4   Signaling Game

Continuing the comparison between Stackelberg-, Nash and signaling games, consider Fig. 4.6 as a generalization of the previous two games, where there are now two kinds of player S, and with nature as the zero-th player determining the type at random. Having fixed this choice, each type of sender may act differently and in particular according its own individual incentives. In a general signaling game, the sender S, who has access to private information, transmits a message to a receiver R. The message is not verifiable, so R does not know the underlying information with certainty. Depending on whether the receiver can distinguish different types of senders based on the message (signal) it receives, we distinguish separating from

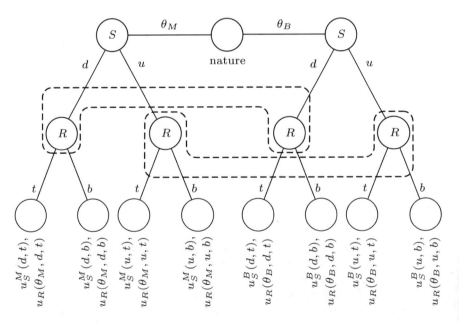

Fig. 4.6 Signaling game example

pooling equilibria (and types in between). Either, the revelation of one's own type or the hiding thereof to the receiver can be in the sender's interest.

Signaling games, like Stackelberg games, are two-player dynamic interactions (Fig. 4.6). Signaling games typically label the players as *sender S* and *receiver R*. The additional component of signaling games is that the sender has access to some information unknown to the receiver. This is called the sender's *type* $\theta \in \Theta$. The receiver only learns about the type based on the sender's action. For this reason, the sender's action (here a_S) is referred to as a *message*. The message need not correspond to the sender's type.

For a generic example, let the set of types of S be $\Theta = \{\theta_B, \theta_M\}$, where θ_B represents a *benign* sender, and θ_M represents a *malicious* sender. Let $p(\theta)$ denote the *prior probability* with which S has each type $\theta \in \Theta$. The utility functions depend on type, i.e., $u_S^M(a_S, a_R)$ and $u_R^B(a_S, a_R)$ give the utility functions for malicious and benign senders, respectively. The function $u_R(\theta, a_S, a_R)$ gives the utility for the receiver when the type of S is θ, the sender message is a_S, and the follower action is a_R. R forms a *belief* $\mu_R(\theta \,|\, a_S)$ that S has type θ given that he sends message $a_S \in AS_S$. To be consistent, this belief should be updated in accord with Bayes' law. A strategy pair in which S and R maximize their own utilities, together with a belief which is consistent, form a PBNE (cf. [12]). In some PBNE, S can cause R to form a specific false belief.

The two-action signal game can be further generalized a signaling game with a finite number of actions. The general setting of the game is presented as follows. First, nature draws a type t_i for the sender a set of feasible types $T = \{t_1, t_2, \cdots, t_I\}$ according to a probability distribution $p(t_i)$, where $p(t_i) > 0$ for every i and $\sum_{i=1}^{I} p(t_i) = 1$. The sender observes t_i, and then chooses a message m_j from a set of feasible messages $M = \{m_1, \cdots, m_J\}$. The receiver observes m_j, and then chooses an action a_k from a set of feasible actions $A = \{a_1, \cdots, a_K\}$. Payoffs are eventually given by the functions $u_S(t_i, m_j, a_k)$ and $u_R(t_i, m_j, a_k)$. The solution concept of signaling games is PBNE. It requires the following conditions to hold consistently:

Condition 1 (Belief Formation) *The receiver forms a belief about the type after observing the message $m_j \in M$ sent by the sender. This belief is denoted by the conditional distribution $\mu_R(t_i | m_j)$, where $\mu_R(t_i | m_j) \geq 0$ for each $t_i \in T$, and*

$$\sum_{t_i \in T} \mu_R(t_i | m_j) = 1, \quad m_j \in M.$$

Note that the sender's belief of the type is obviously represented by the probability measure $p : T \to [0, 1]$.

Condition 2 (Receiver's Problem) *For every message m_j that is received, the receiver maximizes the expected utility, given the belief $\mu_R(t_i | m_j)$, as follows:*

$$\max_{a_k \in A} \bar{u}_R(m_j, a_k) := \sum_{t_i \in T} \mu_R(t_i | m_j) u_R(T_i, m_j, a_k).$$

As a result, the receiver's strategy is given by the mapping $a^* : M \to A$.

Condition 3 (Sender's Problem) *For every type* $t_i \in T$, *the sender's message maximizes the following utility, anticipating receiver's strategy* $a^*(m_j)$:

$$\max_{m_j \in M} u_S(t_i, m_j, a^*(m_j))$$

As a result, the receiver's strategy is given by the mapping $m^* : T \to M$.

This condition is in the same spirit as in the definition of Stackelberg equilibria for leader-follower situations.

Condition 4 (Bayes Rule Consistency) *The strategies* $m_j = m^*(t_i), t_i \in T$, *is used to update the receiver's belief* $\mu_R(t_i | m_j)$:

$$\mu_R(t_i | m_j) = \frac{p(t_i)\sigma(m_j | t_i)}{\sum_{t_i' \in T} p(t_i')\sigma(m_j | t_i')}, \quad if \ \sum_{t_i' \in T} p(t_i')\sigma(m_j | t_i') > 0,$$

$$\mu_R(t_i | m_j) = any \ probability \ distribution, \quad if \ \sum_{t_i' \in T} p(t_i')\sigma(m_j | t_i') = 0,$$

where $\sigma(m_j | t_i)$ *is a distribution induced by the mapping* m^*. *If* m^* *is singleton, then* $\sigma(m_j | t_i) = 1$ *if* $m_j = m^*(t_i)$.

Note that this Bayesian rule depends on the strategy of m^* (and finding m^* depends on a^*). The PBNE requires these conditions to be satisfied at the same time, i.e.,

Definition 4.3 (Perfect Bayesian Nash Equilibrium (PBNE)) A pure-strategy PBNE of the signaling game is a pair of strategies m^*, a^*, μ_R, satisfying the four conditions above.

One way to find a PBNE is to fix μ_R and find a^*, m^* (this process is the same as finding Stackelberg equilibrium), and then find μ_R that will be consistent with a^* and μ_R^* that is obtained. This approach may require some iterations between μ_R and a^*, m^* for complex games.

It is important to stress that not only the receiver can maintain a belief, but generally every player in a signaling game may have some hypothesis on the types of the other players. Our description follows the most common instance of a two-player signaling game, but the generalization to more than two players follows the same ideas.

The application of signaling games in cybersecurity is enormous. One key application domain is cyber deception [29, 35, 36, 49, 55]. The modeling of the asymmetric information between an attacker and a defender captures the incomplete information of the players about the other player's objectives, capabilities, and preferences. In [37], a taxonomy of defensive cyber deception has been proposed and the main literature has recently been summarized. In [36], signaling games are extended by including a detector that gives off probabilistic warnings when the sender acts deceptively. The role of the detector has played an important role in the equilibria of the game. It has been found through equilibrium analysis that the high-quality detectors eliminate some pure-strategy equilibria. Detectors with high true-positive rates encourage more honest signaling than detectors with low false-positive rates, and surprisingly, deceptive senders sometimes benefit from highly accurate deception detectors. In [55], the authors have used a signaling game framework to study GPS signal spoofing where an attacker can deceive a GPS receiver by broadcasting incorrect GPS signals. Defense against such attacks is critical to ensure the reliability and security of many critical infrastructures. The authors have proposed a signal framework in which the GPS receiver can strategically infer the true location when the attacker attempts to mislead it with a fraudulent and purposefully crafted signal. The equilibrium analysis enables the development of a game-theoretic security mechanism to defend against the civil GPS signal spoofing for critical infrastructures. In [16], the signaling games have been extended to a multi-stage framework to address emerging security challenges from APTs which exhibit the stealthy, dynamic and adaptive attack behaviors. The multi-stage game captures the incomplete information of the players' type, and enables an adaptive belief update according to the observable history of the other player's actions. The solution concept of PBNE under the proactive and reactive information structures of the players provides an important analytical tool to predict and design the players' behavior. It has been shown that the benefit of deception of the private attackers' types and motivate defender's use of deception techniques to tilt the information asymmetry.

4.5 Games Over Stochastic Orders

Conventional games require payoffs to come as crisp numbers, and signaling games are one way of bringing in uncertainty by letting there be randomly chosen kinds of opponents (player types). Still, the rewards for each type are quantified in real numbers, and towards a more powerful model of uncertainty, it is natural to ask if the rewards can be something more complex, say whole random variables. This is indeed possible, if we use stochastic orders to replace the natural ordering of the numbers in $\mathbb{R}$.

When people cannot specify precise outcomes like monetary gains or losses, we can often resort to categories describing an uncertain outcome in natural language terms, and then assign the rank of the category as payoff in the game (we will come back to this in Chap. 6 about risk management). For example, if we speak about losses, but cannot be more precise than a subjective estimate of the loss being "negligible", "low", "high" or "very high", then we can assign payoffs 1, ..., 4 to represent the (relative) order position of the categories.

If we can make a reasonable assignment to a fixed category, then we are back at normal games with real-valued (e.g., integer) payoffs. If the uncertainty in this specification, however, extends to ambiguity in what the payoff would be even on a categorical scale, then we need a more powerful object to express what could happen in the current state of the game.

Stochastic orders, which we briefly met in Sect. 3.4, Definition 3.1, allow using whole probability distributions as the payoff. That is, instead of saying that the payoff in any given state of a game (whether it is in normal or extensive form) is "exactly u", we can specify that the payoff is uncertain and a random variable $U \sim F$, where the distribution F can take up all the uncertainty information available. Later, in Chap. 6, we will revisit this concept under the different name of a *prospect*. One way of computing data about possible payoffs as cascading effects may imply, or prospects, is simulation, as we discussed in Sect. 2.5, to understand the direct and indirect effects of security incidents. Empirical game theory can use such simulated or otherwise collected empirical data to define the game models and especially payoff specifications from this information. In doing so, we open the whole spectrum of tools from data science to model uncertain outcomes as probability distributions. In many cases, the models are Gaussian distributions, or more heavy tailed distributions from actuarial science [15], and an easy and quick way back to the familiar real-valued setting of game theory is just taking the average from the data or distributions to work as the payoff value definition. For example, knowing that a payoff is random, say a Gaussian variable $U \sim \mathcal{N}(\mu, \sigma^2)$, the conventional game would simply replace U by its expectation and work with μ as the real-valued revenue to represent U.

Games over stochastic orders, on the contrary, use the full object U and impose a stochastic order to define optimality among random variables. The chosen order should, at least, be total to assure that there is a sound definition of optimality, and the $\preceq$-order given in Definition 3.1 is one possible choice, which is tailored for applications in risk management (as being a stochastic tail order).

The $\preceq$-ordering is defined and constructed as a replacement for $\leq$ on $\mathbb{R}$, so that in particular all so-far described game models transfer almost without change to the setting where their utility functions describe not a numeric payoff but rather a probability distribution over the possible payoffs. We shall therefore refrain from repeating all that has been described so far, since the changes are mostly in the details of the algorithms to solve the game, but not in the underlying concepts [38–40, 42–46].

A point worth mentioning, however, relates to multi-criteria games and the possibility of lexicographically ordering the optimization goals if scalarization is unwanted or not reasonable in practice.

Example 4.3 (Lexicographic Goal Priorities) One example for this is the different importance assigned to the goals of confidentiality, integrity and availability in production-oriented enterprises, vs. companies whose business model roots in information processing: for the latter, the priorities are clearly (C) confidentiality, (I) integrity, and (A) availability. Conversely, a production-oriented company may first strive for availability (A), then for integrity (I) and last (but not least) for confidentiality (C). The lexicographic order in both cases are thus reversed: "C-I-A" for information-processing, vs. "A-I-C" for production. ◇

Somewhat unpleasantly, the $\leq_{lex}$ ordering has no representation by any continuous and real-valued function in the sense of (3.1). It *does have*, however, a representation by functions that take values outside the reals, say $^*\mathbb{R}$. A suitable representation is offered by the stochastic $\preceq$ ordering. Namely, both the ordering of real numbers as well as the lexicographic order on $\mathbb{R}^n$ can be exhibited as special cases of the stochastic $\preceq$-ordering of Definition 3.1 (note that within $\mathbb{R}$, the $\leq_{lex}$ relation equals the canonic ordering of the reals):

Theorem 4.2 *For every two elements* $\mathbf{x}, \mathbf{y} \in \mathbb{R}^n$ *with* $n \geq 1$, *there are categorical distributions* F_X, F_Y *such that* $\mathbf{x} \leq_{lex} \mathbf{y}$ *if and only if* $F_X \preceq F_Y$.

The construction of the representing distributions F_X, F_Y in Theorem 4.2 is indeed simple: let any two vectors $\mathbf{x}, \mathbf{y} \in \mathbb{R}^n$ be given. First, observe that we can add a constant α to both so that $\mathbf{x} + \alpha \cdot (1, 1, \dots, 1) > 0$ where the $>$ holds per component. If we add the same constant α to $\mathbf{x}$ and $\mathbf{y}$, the order between the two remains unchanged. Call the so-shifted vectors $\mathbf{x}', \mathbf{y}'$.

Next, we can pick a value $\lambda > 0$ so that $\lambda \cdot \|\mathbf{x}'\|_1 < 1$ and $\lambda \cdot \|\mathbf{y}'\|_1 < 1$. As before, multiplying by a positive constant leaves the two vectors or reals in the same order.

Since the scaled vectors have a 1-norm < 1, we can add an $(n+1)$-th coordinate to construct the vectors $\mathbf{x}'' := (\mathbf{x}', 1 - \|\mathbf{x}'\|_1)$ and $\mathbf{y}'' := (\mathbf{y}', 1 - \|\mathbf{y}'\|_1)$. By construction, these vectors have (i) all nonnegative entries, and (ii) entries that add up to 1, so that they form probability distributions F_X, F_Y with $\mathbf{x}'', \mathbf{y}''$ as their probability masses. These are exactly the ones that Theorem 4.2 speaks about.

As a special case for $\mathbb{R}^n = \mathbb{R}$, we obtain the ordering of real values preserved if we embed them into Bernoulli distributions: w.l.o.g., take properly scaled and shifted values $0 < a, b < 1$, then $a \leq b$ if and only if the two Bernoulli distributions F_A, F_B with mass vectors $\mathbf{x}'' = (1 - a, a)$ and $\mathbf{y}'' = (1 - b, b)$ satisfy $F_A \preceq F_B$ by Theorem 3.1.

This trick straightforwardly applies to larger sets of real values or vectors, as long as we shift and scale all of them by the *same constants*, which then creates a strategically equivalent game having the same set of equilibria as the original one. As an independent value, Theorem 4.2 enables the use of lexicographic orders for game theory, which is *not* possible in the standard setting of game theory, since the lexicographic order provably has no representation by a continuous and real-valued utility function. The transition to the setting of stochastic orders and hyperreal numbers to do game theory is then even unavoidable. The practical use of this method, especially the application of Theorem 4.2 is made easy by software implementations that we discuss in Sect. 12.4.3 later.

References

1. Alpcan T, Basar T (2004) A game theoretic analysis of intrusion detection in access control systems. In: Decision and control, 2004. CDC. 43rd IEEE conference on, vol 2. IEEE, pp 1568–1573
2. Basar T (1983) The gaussian test channel with an intelligent jammer. IEEE Trans Inf Theory 29(1):152–157
3. Chen J, Zhu Q (2016) Interdependent network formation games with an application to critical infrastructures. In: American control conference (ACC). IEEE, pp 2870–2875. http://ieeexplore.ieee.org/abstract/document/7525354/
4. Chen J, Touati C, Zhu Q (2019) Optimal secure two-layer iot network design. IEEE Trans Control Netw Syst 1–1. https://doi.org/10.1109/TCNS.2019.2906893
5. Chen J, Touati C, Zhu Q (2020) A dynamic game approach to strategic design of secure and resilient infrastructure network. IEEE Trans Inf Forensics Secur 15:462–474. https://doi.org/10.1109/TIFS.2019.2924130
6. Clark A, Zhu Q, Poovendran R, Başar T (2012) Deceptive routing in relay networks. In: Decision and game theory for security. Springer, pp 171–185
7. Farhang S, Manshaei MH, Esfahani MN, Zhu Q (2014) A dynamic bayesian security game framework for strategic defense mechanism design. In: Decision and game theory for security. Springer, pp 319–328
8. Farooq MJ, Zhu Q (2018) A multi-layer feedback system approach to resilient connectivity of remotely deployed mobile Internet of things. IEEE Trans Cogn Commun Netw 4(2):422–432
9. Farooq MJ, Zhu Q (2018) On the secure and reconfigurable multi-layer network design for critical information dissemination in the internet of battlefield things (IoBT). IEEE Trans Wirel Commun 17(4):2618–2632
10. Farooq MJ, Zhu Q (2019) Modeling, analysis, and mitigation of dynamic botnet formation in wireless IoT networks. IEEE Trans Inf Forensics Secur 14:2412–2426
11. Farooq MJ, ElSawy H, Zhu Q, Alouini MS (2017) Optimizing mission critical data dissemination in massive IoT networks. In: Modeling and optimization in mobile, Ad Hoc, and wireless networks (WiOpt), 2017 15th international symposium on. IEEE, pp 1–6. http://ieeexplore.ieee.org/abstract/document/7959930/
12. Fudenberg D, Tirole J (1991) Game theory. MIT Press, Cambridge
13. Hayel Y, Zhu Q (2015) Attack-aware cyber insurance for risk sharing in computer networks. In: Decision and game theory for security. Springer, pp 22–34
14. Hayel Y, Zhu Q (2017) Epidemic protection over heterogeneous networks using evolutionary poisson games. IEEE Trans Inf Forensics Secur 12(8):1786–1800
15. Hogg RV, Klugman SA (1984) Loss distributions. Wiley series in probability and mathematical statistics applied probability and statistics. Wiley, New York. https://doi.org/10.1002/9780470316634. http://gso.gbv.de/DB=2.1/PPNSET?PPN=599519185
16. Huang L, Zhu Q (2018) Analysis and computation of adaptive defense strategies against advanced persistent threats for cyber-physical systems. In: International conference on decision and game theory for security. Springer, pp 205–226
17. Huang L, Zhu Q (2019) Adaptive strategic cyber defense for advanced persistent threats in critical infrastructure networks. ACM SIGMETRICS Perform Eval Rev 46(2):52–56
18. Huang L, Zhu Q (2019) Dynamic Bayesian games for adversarial and defensive cyber deception. In: Al-Shaer E, Wei J, Hamlen KW, Wang C (eds) Autonomous cyber deception: reasoning, adaptive planning, and evaluation of honeyThings. Springer International Publishing, Cham, pp 75–97. https://doi.org/10.1007/978-3-030-02110-8_5
19. Huang Y, Zhu Q (2019) A differential game approach to decentralized virus-resistant weight adaptation policy over complex networks. arXiv preprint arXiv:1905.02237
20. Huang JW, Zhu Q, Krishnamurthy V, Basar T (2010) Distributed correlated q-learning for dynamic transmission control of sensor networks. In: Acoustics speech and signal processing (ICASSP), 2010 IEEE international conference on. IEEE, pp 1982–1985

21. Huang L, Chen J, Zhu Q (2017) A large-scale markov game approach to dynamic protection of interdependent infrastructure networks. In: International conference on decision and game theory for security. Springer, pp 357–376
22. Huang L, Chen J, Zhu Q (2018) Distributed and optimal resilient planning of large-scale interdependent critical infrastructures. In: 2018 winter simulation conference (WSC). IEEE, pp 1096–1107
23. Huang L, Chen J, Zhu Q (2018) Factored Markov game theory for secure interdependent infrastructure networks. In: Game theory for security and risk management. Springer, pp 99–126
24. Kim S (2015) Nested game-based computation offloading scheme for Mobile Cloud IoT systems. EURASIP J Wirel Commun Netw 2015(1):229. 10.1186/s13638-015-0456-5. https://link.springer.com/article/10.1186/s13638-015-0456-5
25. Kuhn H (1953) Extensive games and the problem of information. In: Kuhn H, Tucker A (eds) Contributions to the theory of games II. Annals of mathematics studies 28, vol. II. Princeton University Press, Princeton, pp 193–216
26. Miao F, Zhu Q (2014) A moving-horizon hybrid stochastic game for secure control of cyber-physical systems. In: Decision and control (CDC), 2014 IEEE 53rd annual conference on. IEEE, pp 517–522
27. Nash JF, et al (1950) Equilibrium points in N-person games. Proc Natl Acad Sci 36(1):48–49
28. Pawlick J (2018) A systems science perspective on deception for cybersecurity in the Internet of things. Ph.D. thesis, NY University . http://proxy.library.nyu.edu/login?url=https://search.proquest.com/docview/2051798175?accountid=12768. Copyright – database copyright ProQuest LLC; ProQuest does not claim copyright in the individual underlying works; Last updated – 2018-08-09
29. Pawlick J, Zhu Q (2015) Deception by design: evidence-based signaling games for network defense. arXiv preprint arXiv:1503.05458
30. Pawlick J, Zhu Q (2016) A Stackelberg game perspective on the conflict between machine learning and data obfuscation. In: Information forensics and security (WIFS), 2016 IEEE international workshop on. IEEE, pp 1–6. http://ieeexplore.ieee.org/abstract/document/7823893/
31. Pawlick J, Zhu Q (2017) A mean-field stackelberg game approach for obfuscation adoption in empirical risk minimization. In: 2017 IEEE global conference on signal and information processing (GlobalSIP). IEEE, pp 518–522
32. Pawlick J, Zhu Q (2017) Proactive defense against physical denial of service attacks using poisson signaling games. In: International conference on decision and game theory for security. Springer, pp 336–356
33. Pawlick J, Farhang S, Zhu Q (2015) Flip the cloud: cyber-physical signaling games in the presence of advanced persistent threats. In: Decision and game theory for security. Springer, pp 289–308
34. Pawlick J, Colbert E, Zhu Q (2017) A game-theoretic taxonomy and survey of defensive deception for cybersecurity and privacy. arXiv preprint arXiv:1712.05441
35. Pawlick J, Chen J, Zhu Q (2018) ISTRICT: an interdependent strategic trust mechanism for the cloud-enabled Internet of controlled things. IEEE Trans Inf Forensics Secur 14:1654–1669
36. Pawlick J, Colbert E, Zhu Q (2018) Modeling and analysis of leaky deception using signaling games with evidence. IEEE Trans Inf Forensics Secur 14(7):1871–1886
37. Pawlick J, Colbert E, Zhu Q (2019) A game-theoretic taxonomy and survey of defensive deception for cybersecurity and privacy. ACM Comput Surv (CSUR) 52(4):82
38. Rass S (2015) On game-theoretic risk management (part one)-towards a theory of games with payoffs that are probability-distributions. arXiv:1506.07368v4 [q-fin.EC]
39. Rass S (2015) On game-theoretic risk management (part two) – algorithms to compute Nash-equilibria in games with distributions as payoffs. arXiv:1511.08591v1 [q-fin.EC]
40. Rass S (2017) On game-theoretic risk management (part three) – modeling and applications. arXiv:1711.00708v1 [q-fin.EC]
41. Rass S, Zhu Q (2016) Gadapt: a sequential game-theoretic framework for designing defense-in-depth strategies against advanced persistent threats. In: International conference on decision

and game theory for security. Springer International Publishing, pp 314–326. http://link.
springer.com/chapter/10.1007/978-3-319-47413-7_18
42. Rass S, König S, Schauer S (2015) Uncertainty in games: using probability distributions as
payoffs. In: Khouzani MH, Panaousis E, Theodorakopoulos G (eds) GameSec 2015: decision
and game theory for security, LNCS 9406. Springer, pp 346–357
43. Rass S, König S, Schauer S (2016) Decisions with uncertain consequences – a total ordering on
loss-distributions. PLoS One 11(12):e0168583. https://doi.org/10.1371/journal.pone.0168583
44. Rass S, Konig S, Schauer S (2017) Defending against advanced persistent threats using game-
theory. PLoS One 12(1):e0168675. https://doi.org/10.1371/journal.pone.0168675
45. Rass S, König S, Panaousis E (2019) Cut-the-rope: a game of stealthy intrusion. In: Alpcan
T, Vorobeychik Y, Baras JS, Dán G (eds) Decision and game theory for security. Springer
International Publishing, Cham, pp 404–416
46. Rass S, König S, Alshawish A (2020) HyRiM: multicriteria risk management using zero-sum
games with vector-valued payoffs that are probability distributions. https://cran.r-project.org/
package=HyRiM. Version 1.0.5
47. Van Dijk M, Juels A, Oprea A, Rivest RL (2013) Flipit: the game of "stealthy takeover". J
Cryptol 26(4):655–713
48. Von Stackelberg H (1934) Marktform und gleichgewicht. J. springer, Wien
49. Xu Z, Zhu Q (2015) A cyber-physical game framework for secure and resilient multi-agent
autonomous systems. In: Decision and control (CDC), 2015 IEEE 54th annual conference on.
IEEE, pp 5156–5161
50. Xu Z, Zhu Q (2017) A game-theoretic approach to secure control of communication-based
train control systems under jamming attacks. In: Proceedings of the 1st international workshop
on safe control of connected and autonomous vehicles. ACM, pp 27–34. http://dl.acm.org/
citation.cfm?id=3055381
51. Zhang R, Zhu Q (2015) Secure and resilient distributed machine learning under adversarial
environments. In: 2015 18th international conference on information fusion (fusion). IEEE, pp
644–651
52. Zhang R, Zhu Q (2016) Attack-aware cyber insurance of interdependent computer networks.
SSRN Electron J. https://papers.ssrn.com/sol3/papers.cfm?abstract_id=2848576
53. Zhang R, Zhu Q (2017) A game-theoretic analysis of label flipping attacks on distributed
support vector machines. In: Information sciences and systems (CISS), 2017 51st annual
conference on. IEEE, pp 1–6. http://ieeexplore.ieee.org/abstract/document/7926118/
54. Zhang T, Zhu Q (2017) Dynamic differential privacy for ADMM-based distributed classification learning. IEEE Trans Inf Forensics Secur 12(1):172–187. http://ieeexplore.ieee.org/
abstract/document/7563366/
55. Zhang T, Zhu Q (2017) Strategic defense against deceptive civilian GPS spoofing of unmanned
aerial vehicles. In: International conference on decision and game theory for security. Springer,
pp 213–233
56. Zhang R, Zhu Q (2018) A game-theoretic approach to design secure and resilient distributed
support vector machines. IEEE Trans Neural Netw Learn Syst 29(11):5512–5527
57. Zhang R, Zhu Q, Hayel Y (2017) A bi-level game approach to attack-aware cyber insurance
of computer networks. IEEE J Sel Areas Commun 35(3):779–794. http://ieeexplore.ieee.org/
abstract/document/7859343/
58. Zhang T, Huang L, Pawlick J, Zhu Q (2019) Game-theoretic analysis of cyber deception:
evidence-based strategies and dynamic risk mitigation. arXiv preprint arXiv:1902.03925
59. Zhu Q, Başar T (2009) Dynamic policy-based IDS configuration. In: Decision and control,
2009 held jointly with the 2009 28th Chinese control conference. CDC/CCC 2009. Proceedings
of the 48th IEEE conference on. IEEE, pp 8600–8605
60. Zhu Q, Başar T (2011) Indices of power in optimal IDS default configuration: theory and
examples. In: Decision and game theory for security. Springer, pp 7–21
61. Zhu Q, Başar T (2012) A dynamic game-theoretic approach to resilient control system design
for cascading failures. In: Proceedings of the 1st international conference on high confidence
networked systems. ACM, pp 41–46

62. Zhu Q, Başar T (2013) Game-theoretic approach to feedback-driven multi-stage moving target defense. In: International conference on decision and game theory for security. Springer, pp 246–263
63. Zhu Q, Basar T (2015) Game-theoretic methods for robustness, security, and resilience of cyberphysical control systems: games-in-games principle for optimal cross-layer resilient control systems. Control Syst IEEE 35(1):46–65
64. Zhu Q, Han Z, Başar T (2010) No-regret learning in collaborative spectrum sensing with malicious nodes. In: Communications (ICC), 2010 IEEE international conference on. IEEE, pp 1–6
65. Zhu Q, Li H, Han Z, Basar T (2010) A stochastic game model for jamming in multi-channel cognitive radio systems. In: 2010 IEEE international conference on communications. IEEE, pp 1–6
66. Zhu Q, Tembine H, Başar T (2010) Heterogeneous learning in zero-sum stochastic games with incomplete information. In: Decision and control (CDC), 2010 49th IEEE conference on. IEEE, pp 219–224
67. Zhu Q, Tembine H, Basar T (2010) Network security configurations: a nonzero-sum stochastic game approach. In: American control conference (ACC), 2010. IEEE, pp 1059–1064
68. Zhu Q, Tembine H, Basar T (2011) Distributed strategic learning with application to network security. In: American control conference (ACC), 2011. IEEE, pp 4057–4062
69. Zhu Q, Saad W, Han Z, Poor HV, Başar T (2011) Eavesdropping and jamming in next-generation wireless networks: a game-theoretic approach. In: 2011-MILCOM 2011 military communications conference. IEEE, pp 119–124
70. Zhu Q, Song JB, Başar T (2011) Dynamic secure routing game in distributed cognitive radio networks. In: Global telecommunications conference (GLOBECOM 2011), 2011 IEEE. IEEE, pp 1–6
71. Zhu Q, Yuan Z, Song JB, Han Z, Başar T (2012) Interference aware routing game for cognitive radio multi-hop networks. Sel Areas Commun IEEE J 30(10):2006–2015
72. Zhu Q, Rass S, Schartner P (2018) Community-based security for the Internet of things. arXiv preprint arXiv:1810.00281

Chapter 5
Bounded Rationality

No rational argument will have a rational effect on a man who does not want to adopt a rational attitude.

K. Popper

Abstract This chapter revisits the concept of a utility function, first introduced in Chap. 3, from an axiomatic viewpoint. We review the fundamental principles of decision making as axioms that induce the existence of (continuous) utility functions. Since empirical research of decision situations in real life has shown considerable deviations between mathematical rationality and human behavior, we continue with a series of possible explanations by relaxing or dropping individual axioms from the set of fundamental principles, to explain the difference between human behavior and the utility maximization paradigm. This establishes valuable lessons for the construction of games, say if payoff models are constructed from subjective data (interviews, expert estimates, or similar), but also highlights the need to consider individual risk perception and attitude though the utility function design in a game theoretic model.

5.1 Utility Maximization and Rationality

It is not surprising that game- and decision theory is subject to some criticism since people just not tend to behave like perfectly rational utility maximizers. We shall not speculate about the possible reasons as to why this is the case, but stress that this makes game theory particularly applicable to defenses that a technical system can implement (in a perfectly rational manner), while involving human actors, one needs to account for the possibility that the final decision made by the human is to act arbitrarily. Games that consider uncertainty in the utilities can – to an extent – account for this by stochastically ordering random utilities whose distribution can be modeled more explicitly than just defining an average. However, doing so requires insights and models from psychology. Including aspects of efforts to play

© Springer Nature Switzerland AG 2020
S. Rass et al., *Cyber-Security in Critical Infrastructures*, Advanced Sciences and Technologies for Security Applications,
https://doi.org/10.1007/978-3-030-46908-5_5

equilibria or to avoid disappointments are in any case only auxiliary aspects to be included in the modeling, but eventually, the quality of predicted behavior (of the defender, and more importantly the adversary) will to a significant extent depend on the quality with which the utilities are defined. Security strategies slightly ease life here by sparing models for the attacker's revenues, and letting us rest on a (hopefully accurate) specification of the action spaces. This robustness is bought at a more pessimistic prediction of achievable security (i.e., coarser bounds in (3.22) and (3.24)).

The typical setting of game and decision theory assumes a rational actor seeking to maximize a benefit by choosing an action from a defined set. This presumes unbounded cognitive abilities, since the mathematical optimization implicitly invoked herein may be complex and hard to handle. However, decisions in real life, are additionally constrained by time, available information (reasoning under uncertainty) or not need to be optimal, as long as a desired level of benefit is achieved (satisficing). This may occur when the additional cost towards optimality are perceived as outweighing the additional gains in the optimum. Though such reasoning can in principle – but usually not without severe difficulties – be embodied in a properly designed utility function, reality still shows discrepancies between the hypothesis of a truly rational individual and the actual choices made. Nevertheless, those must not be called irrational or dumb: irrationality is a chosen deviation from reason, which does not preclude hidden different reasons to act like observed. Neither is bounded rationality not equal to dumb, since individuals may also here have hidden, unknown or more complex incentives than what the mathematical optimization (in an oversimplification of matters) may assume. In essence, *bounded rationality* means decision making under limitations of cognitive efforts, available information, or other factors that would determine which option is rationally best.

It will pay to take a systematic look at where the concept of utility maximization (equivalently loss minimization in risk management) originates from. This will help understanding the reasons as to why intelligent and adaptive individuals may nonetheless choose to act seemingly non-utility-maximizing. The first mentioning of utility maximization dates back to Bernoulli in the Saint Petersburg game, and to B. Pascal, with his famous argument why it is rational to believe in god.

Example 5.1 (A Utility-Maximizer's Argument for the Belief in God) Pascal's argument was actually a utility maximizing one: if god does not exist and hence death is the end of everything, then it does not matter whether or not we believe in any religion – there is neither harm nor gain to be expected, we would get zero utility. Otherwise, if god does exist, then not having believed could cause tremendous trouble in the afterlife! If god exists and we believe in religion, then we can expect rewards or punishment depending on how our religious performance has been. In any case, *not* believing in god *always* ends up with an expected negative reward, as opposed to believing in god, which gives us at least the chances for a positive outcome in the afterlife. Hence and logically, it is rational to believe in god (based on a utility maximization argument).   ◇

This appealing connection between science and religion as in Example 5.1 should not be overstressed, since the two satisfy intrinsically different needs. Still, acting according to religious standards may appear as boundedly rational on grounds of this apparently not being utility maximizing (after all, religious ethics should teach not to egoistically pursue ones own welfare on costs borne by others). The point of this is the existence of hidden agendas that may not become visible by the means applied to define utility functionals. Those are, as usual in mathematics, axioms that are regarded as fundamental laws governing the choice behavior. All of these can be put to question, each case directing us to another set of models for bounded rationality.

5.2   The Fundamental Principles of Decision Making

The derivation in the following is a condensed version of the introduction done by C.P. Robert in [17]; we will draw from his account only to the extent necessary for correctness and sufficient to understand why the axiomatic route to model decision making is not necessarily taking us where we want to get.

In Chap. 3, we assumed the existence of some preference relation among the set of choices for a decision problem. This choice set has been an action set AS with elements denoted as $a \in AS$, and a preference relation $a_1 \trianglelefteq a_2$ defined between actions. The more general approach does not consider actions for themselves, but rather assumes decisions to be made towards *rewards* that follow actions as a consequence. Preferences between actions are then determined by the rewards implied by the actions. A reward, in that sense, is thus more or less a quantification of an action, similarly to the utility functions from Chap. 3, but not precisely equivalent to them, since two actions a_1, a_2 were preferred as $a_1 \trianglelefteq a_2$ if and only if some utility function u assigned ordered values $u_1(a_1) \le u_1(a_2)$.

However, the more practical setting is that not the action but rather its consequence would determine the preference. This means that we assign a *reward* $r(a)$ to an action a, and impose the preference relation on the rewards. Conceptually, we may simplify matters by thinking of actions being even *characterized* by their rewards (although there may not be a one-to-one correspondence), but the thinking is that what the decision maker ultimately seeks is a *reward*, and the action itself is only the way of getting it (but as long as the reward is achieved, the decision maker may not care what particular action it may take to get it).

It is thus useful, though not particularly more complicated, to think in terms of rewards, rather than actions, and to let the decision problem be among rewards, with associated actions available to achieve the sought optimal reward. More formally, we will let the ordered set $(R, \trianglelefteq)$ contain all rewards, with some preference (not ordering) relation $\trianglelefteq$ on it. The determination of $\trianglelefteq$ is a matter of a function u that *represents the preference relation* exactly like in (3.1), and our immediate goal for now is finding conditions on the preference relation $\trianglelefteq$ that let us construct a representative function u. Consistent with the literature, and also with Chap. 3, we shall refer to u as a *utility function*, synonymously also a *payoff*.

To start with, a reasonable preference should at least satisfy the following requirements:

Ax_1 (Completeness):   either we have $r_1 \trianglelefteq r_2$ or $r_2 \trianglelefteq r_1$
Ax_2 (Transitivity):   if $r_1 \trianglelefteq r_2$ and $r_2 \trianglelefteq r_3$ then $r_1 \trianglelefteq r_3$.

Both appear the least necessary if we want to speak about a "best" decision. For example, under incompleteness, there would be incomparable rewards and hence actions among which we cannot choose based on the rewards. Likewise, without transitivity, we could have circular preferences like $r_1 \trianglelefteq r_2 \trianglelefteq r_3 \trianglelefteq r_1$, so that every pair of rewards from $\{r_1, r_2, r_3, r_4\}$ is preferable like $r_i \trianglelefteq r_j$, no matter how we compare them. So, no option could ever be "best", and speaking about utility optimization seems senseless. Somewhat surprisingly, transitivity can nonetheless be dropped as a requirement if we move to *regret theory* in Sect. 5.6.

Like for ordering relations, we define *equivalence* between rewards as $r_1 \sim r_2$: $\Longleftrightarrow$ $(r_1 \trianglelefteq r_2) \wedge (r_2 \trianglelefteq r_1)$, i.e., both rewards are $\trianglelefteq$-preferable over another and hence "equally" good.

Actions in real life hardly come with deterministic consequences. Especially in security, investing in additional precautions may or may not work out, or it may be effective only against a specific limited set of known weaknesses. For example, a virus scanner may catch a malware or could miss it, depending on the update status and many other factors. We thus extend the concept of a reward to that of a *prospect*, which is a list of possible rewards associated with individual probabilities. This is similar in spirit to a randomized action, and in fact precisely the continuation of the ideas motivating the use of stochastic orders in Sect. 3.4, and games over such orders (Sect. 4.5).

Definition 5.1 (Prospect)   A *prospect* is a triple $(a, \mathbf{p}, \mathbf{u})$ that associates an action a with a set of possible rewards $\mathbf{u} = (r_1, \ldots, r_k)$ and probabilities $\mathbf{p} = (p_1, \ldots, p_k) = (\Pr(r_1), \ldots, \Pr(r_k))$ to receive them, where $p_1 + p_2 + \ldots + p_k = 1$. Prospects are explicitly denoted as $2 \times n$ matrices

$$\begin{pmatrix} r_1 & r_2 & \ldots & r_n \\ p_1 & p_2 & \ldots & p_n \end{pmatrix},$$

or, whenever the reward set $R = \{r_1, \ldots, r_n\}$ is clear from the context and in a fixed order, we may use the abbreviated notation of a prospect as only the vector of probability masses $\mathbf{p} = (p_1, \ldots, p_n) \in \Delta(R)$.

As an example, think of a patch or system update that *may* but *does not need to* effectively close or remove a vulnerability. It may well happen that, after the patch, certain attacks are no longer possible (this could be a first possible outcome, happening with probability p_1), or remain possible because the patch did not apply for the vulnerability (probability p_2), or the attack remains possible but has only become harder (probability p_3), which is an intermediate reward.

Example 5.2 (Cascading Effects and Prospects from Simulation) More elaborate examples arise when we simulate the impacts of attacks in a complex system. Section 2.5 outlined a variety of models to assess direct and indirect consequences of a security incident, by simulating cascading effects. The bottom line of all these simulation methods is that they come up with a series of scenarios, each of which may correspond to one out of several impact categories (as commonly defined in risk management; see Fig. 6.2 in Chap. 6). Let the impact categories be collected in some (usually ordered) set $R = \{r_1, \ldots, r_n\}$. Upon many repetitions of the simulation, let p_i be the relative frequency of the impact category r_i to come up in the experiments. Then, the pair $((r_1, \ldots, r_n), (p_1, \ldots, p_n))$ is nothing else than an empirical estimate of a prospect, coming as a direct output artefact of a probabilistic simulation. $\diamond$

Prospects generalize the crisp payoffs from game theory to the random ones that we already had in the context of games with probability distributions as payoffs in Sect. 4.5. Under this view, a prospect is nothing else than a payoff that is a distribution and not a number. Ordering prospects then calls for stochastic orders like $\preceq$, like ordering numbers works with the natural numeric ordering $\leq$. The decision challenge is the same in both cases.

Let us assume that all consequences and associated probabilities are known to the decision maker. In choosing among prospects, the decision maker faces a situation of *risk*. This is different from a situation of *uncertainty*, where parts of the prospect (either a consequence or a probability) is unknown to the decision maker.

Formally, let us carry over the $\trianglelefteq$ relation to the so convexified space of prospects, i.e., the set $\Delta(R)$ of probability distributions over the reward space R. In game theory, this would be the action space, but since we are considering a more general view to explain why games may not reflect people's actions in real life, we shall use the more general terminology here. Let us assume that transitivity and completeness carry over to the set $\Delta(R)$, whose elements can be seen as prospects (similar to mixed strategies in the vocabulary of game theory). For example, we may compare two prospects by comparing their mean, but for a general (axiomatic) start, we require completeness and transitivity of $\trianglelefteq$ on $\Delta(R)$ in exactly the form as stated above. For a representation of the preference like in (3.1), we need some more assumptions:

Ax_3 (Conservation of orderings under indifferent alternatives):   for any two prospects $\mathbf{p}, \mathbf{q}$ and any value $0 < \alpha < 1$ call $\alpha \cdot \mathbf{p} + (1 - \alpha) \cdot \mathbf{q}$ the *mixture*. Axiom Ax_3 asks for the condition that given any $\alpha \in (0, 1)$,

$$\text{if } \mathbf{p} \trianglelefteq \mathbf{q} \text{ then } (1 - \alpha) \cdot \mathbf{p} + \alpha \cdot \mathbf{r} \trianglelefteq (1 - \alpha) \cdot \mathbf{q} + \alpha \cdot \mathbf{r}$$

for every $\mathbf{r} \in \Delta(R)$. That is, if two prospects are $\trianglelefteq$-preferable over one another and there is *any* third alternative that may kick in with probability α, then this deviation – since it applies to either option – does not change the overall preference (hence we are indifferent under a change that applies identically to two choices).

Ax_4 (Closedness):   if $\mathbf{p} \trianglelefteq \mathbf{q} \trianglelefteq \mathbf{r}$, then there exist values $\alpha, \beta \in (0, 1)$ such that

$$\alpha \cdot \mathbf{p} + (1 - \alpha) \cdot \mathbf{r} \unlhd \mathbf{q} \unlhd \beta \cdot \mathbf{p} + (1 - \beta) \cdot \mathbf{r}$$

Accepting the axioms $Ax_1, \ldots, Ax_4$ as "fundamental principles of decision making", their consequence (especially using the closedness axiom) is the existence of the sought (so far only presumed to exist) utility function:

Lemma 5.1 ([17, Lem.2.2.2]) *If r_1, r_2 and r are rewards in R with $r_1 \lhd r_2$ and $r_1 \unlhd r \unlhd r_2$ then there is a unique value $0 < v < 1$ such that $r \sim v \cdot r_1 + (1 - v) \cdot r_2$. From this value v, we can define a* utility function *as*

$$u(r) = \begin{cases} v, & \text{if } r_2 \unlhd r \unlhd r_1 \text{ and } r \sim v \cdot r_1 + (1 - v) \cdot r_2; \\ -\frac{v}{1-v}, & \text{if } r_2 \sim v \cdot r_1 + (1 - v) \cdot r; \text{ and} \\ \frac{1}{v}, & \text{if } r_1 \unlhd v \cdot r + (1 - v) \cdot r_2 \end{cases}$$

Lemma 5.1 for the first time semi-explicitly defines a utility function from first principles (being the axioms Ax_1 until Ax_4). Observe that we can allow the rewards to be even infinite. However, infinity is an opaque concept for many individuals, and things can be dramatically simplified if we confine ourselves to *bounded rewards*. That is, there are two values r_1, r_2 such that

$$[r_1, r_2] = \{r : r_1 \unlhd r \unlhd r_2\} \tag{5.1}$$

and $\Pr([r_1, r_2]) = 1$ (i.e., bounded support; an assumption that was also required for the construction of the stochastic $\preceq$-ordering; cf. Definition 3.1). The so-defined utility function preserves the $\unlhd$-ordering on R, and obeys the boundary conditions $u(r_1) = 1$ and $u(r_2) = 0$. Most importantly, by linearization, it extends to finite rewards between the extremes $[r_1, r_2]$ by defining

$$\alpha(r) := \frac{u(r) - u(r_1)}{u(r_2) - u(r_1)}, \text{ and } \beta := \int_{r_1}^{r_2} \alpha(r) d\mathbf{p}(r),$$

where $\mathbf{p}$ is a prospect defined on R (and the integral must hence be understood as of Lebesgue-Stieltjes type). What is interesting is that by virtue of a fifth axiom, the utility function becomes unique up to certain affine transformations:

Ax_5 (Linearization): let δ_{r_1} and δ_{r_2} be the specific prospects corresponding to the extremes of the reward scale $[r_1, r_2]$ (i.e., point masses assigning the full probability mass to the ends of the reward scale).
If a prospect $\mathbf{p}$ is $\sim$-equivalent to $\alpha(r) \cdot \delta_{r_1} + (1 - \alpha(r))\delta_{r_2}$, then this equivalence holds on average.

Under this additional linearization hypothesis, we get the following uniqueness result:

Theorem 5.1 ([17, Thm.2.2.4]) *If $\mathbf{p}, \mathbf{q}$ are prospects from $\Delta(R)$ where R is bounded like in (5.1), then we have:*

Existence: *there is a function u on R so that* $\mathbf{p} \trianglelefteq \mathbf{q}$ *if and only if* $\mathbb{E}_{\mathbf{p}}(u(r)) \leq \mathbb{E}_{\mathbf{q}}(U(r))$.

Uniqueness: *if u' is another utility function satisfying the above equivalence relation, then there are values $a > 0$ and b such that $u'(r) = a \cdot u(r) + b$.*

The uniqueness statement of Theorem 5.1 is easily recognized as the condition of strategic equivalence of strategies in game theory.

So much for the axiomatic groundwork that underlies most of the decision making theory. The whole story so far, implicitly, reveals two other aspects that close the loop back to bounded rationality:

1. The entire development focused on only a *single* measure to be optimized. Namely, we derived the utility function as mapping from the reward space R into the real numbers $\mathbb{R}$. The main reason is that $\mathbb{R}$ is a totally ordered space, but none of the spaces $\mathbb{R}^k$ for $k > 1$ has a natural ordering on it (except for the lexicographic perhaps). Hidden incentives, e.g., avoidance of regret or disappointment counting as further goals besides the explicitly modeled utilities are thus one possible reason of bounded rationality.

2. Each of the axioms (completeness, transitivity, indifference, closedness and linearization) should be consistent with empirical findings for the development to deliver an accurate model of people's decision making. Indeed, there is considerable empirical evidence showing violations of more than one of these assumptions.

Our exposition hereafter partly draws from [21]: bear in mind that we are not looking for occasional random violations here, which we could in any case allow and attribute to imprecise conditions or other errors in the data. Rather, we will look for *systematic* violations of the assumptions underneath the constructions of utility functions. These will not cover all the above axioms, but even one violation would already be sufficient to invalidate the entire construction. On the bright side, violations of these assumptions open directions to new developments and models, some of which we are going to touch briefly hereafter.

5.3 Violations of the Invariance Axiom

The most famous example of a systematic violation of the indifference under equivalent alternatives (axiom $Ax3$) is perhaps *Allais' paradox* or more generally coined the *common consequence effect* [1]. Consider two prospects defined as follows:

$\mathbf{p}_1$: win 1 million with probability 1.

$\mathbf{q}_1$: win 5 million with probability 0.1, win 1 million with probability 0.89, or win nothing with probability 0.01.

Which of the two prospects would you choose? Choosing $\mathbf{p}_1$ is attractive since it makes the person a millionaire instantly and without risk. Choosing $\mathbf{q}_1$ may be even more attractive, since the chances to win the million are only slightly worse, but there is a considerable chance to win even five millions. Based on expected utilities (for $\mathbf{p}$ being 1 million, and for $\mathbf{q_1}$ being $5 \times 0.1 + 1 \times 0.89 = 1.39$ million) we would pick $\mathbf{q_1}$ for the higher expected gains.

Now, consider a slightly modified choice:

$\mathbf{p_2}$:   win 1 million with probability 0.11 or win nothing with probability 0.89.
$\mathbf{q_2}$:   win 5 million with probability 0.1, or win nothing with probability 0.9.

The odds seem quite similar here, but the prizes are considerably different. Again, based on expected utility, for $\mathbf{p_2}$ we have $1 \times 0.11 + 0 \times 0.89 = 0.11$ and $\mathbf{q_2}$ expectedly gives $5 \times 0.1 + 0 \times 0.9 = 0.5$, so a utility-maximizer would go for $\mathbf{q_2}$ here.

Based on a "fully rational" decision making that is consistent with the independence axiom, people would be expected to choose the $\mathbf{q}$-prospectives across this pair of choices. Empirically, however, it turned out that people were picking option $\mathbf{p_1}$ for the certainty of becoming rich, as opposed to picking $\mathbf{q_2}$ for the much higher gains at approximately the same chances for the alternative. This is apparently *not* what utility maximization would predict (but was exactly what M. Allais had predicted). Numerous studies have ever since confirmed his finding, pointing out a systematic violation of axiom Ax_3 apparently.

Where does the violation root? The independence axiom overlooks that people take into account alternatives in their decision making and the "appeal" of an option may strongly depend on what other actions have to offer. The possible disappointment in going for the larger gain may be much more of an incentive to go with the safer option, which is why people would pick the certainty to become rich over the odds to become even richer (or not). On the contrary, if the magnitude of gains is chosen such that the small differences in the probabilities make a significant difference in the utility outcome (and hence the preference), but this difference is not sufficient to convince the decision maker, then the choice is made without regard of the utility function, whose evaluation would require much more cognitive efforts than are required to base the decision just on an approximate comparison of odds.

The overall point of the Allais' paradox is that options *cannot* be valuated (by a utility function) and compared independently from each other – at least this is not what people would do; they prefer a more "holistic" comparison including more alternatives. There is also an empirically verified break-even point in the number of choices: too many choices make the decision over-proportionally hard, so that to reduce cognitive efforts, we may just regard a selection of a few options and abandon perhaps the mathematically optimal ones; this is the well-known *paradox of choice*[12].

A related phenomenon is the *common ratio effect* (see, e.g., [13]): as before, let the choices be among two prospects each, and let us use utility maximization as a predictor, compared to what people were observed to do in reality.

The question of choice is now for the following prospects:

p: either take 10 Dollar for sure or take a gamble to win 16 Dollar with an 80% chance.

q: either win 16 Dollar with chance 20% or win 10 Dollar with a 25% chance.

In **p**, many people would rather take the 10 Dollar than enter a gamble, since the latter has the risk of giving zero revenue. The same people, however, would in **q** choose the gamble for the 16 Dollar, since the odds are approximately similar in the two options. This is indeed incompatible with the independence axiom, since expected utilities point towards a different outcome: let us slightly abuse our notation to write $\mathbb{E}(16)$ and $\mathbb{E}(10)$ to mean the expectations of 16 Dollar or 10 Dollar gains from our choice. Towards a contradiction, let us assume that the individuals from which the finings above were reported mentally were following a utility maximization principle. Then, the preference of the 10 Dollar-option in **q** over the 16 Dollar-option in **q** would mean $0.8 \times \mathbb{E}(16) < 1 \cdot \mathbb{E}(10)$ or equivalently $\mathbb{E}(16) < 1.25 \times \mathbb{E}(10)$. Likewise, the preference of the 16 Dollar-option in **q** over the 10 Dollar-option in **q** would mean $0.2 \times \mathbb{E}(16) > 0.25 \times \mathbb{E}(10)$ or equivalently $\mathbb{E}(16) > \frac{0.25}{0.2} \times \mathbb{E}(10) = 1.25 \times \mathbb{E}(10)$, which is an obvious contradiction [6].

Again, where does the problem arise? The general issue occurs when two revenues $r_1 \trianglelefteq r_2$ are given, whose likelihoods to occur are scaled by the same factor. In the above example, the revenues are $r_1 = 10 \trianglelefteq r_2 = 16$, but the chances to get them have been modified by a factor of 4 in both cases. The empirical evidence shows that people may prefer certainty in one case, but become more willing to take a risk when the chances are approximately similar (as achieved by the common scaling).

5.4 Decision Weights

Recognizing that options are usually ranked relative to the whole set of alternatives, people have modified the expected utility formula for a prospect

$$\mathbf{p} = \begin{pmatrix} r_1 & r_2 & \cdots & r_n \\ p_1 & p_2 & \cdots & p_n \end{pmatrix} = \begin{pmatrix} \mathbf{r} \\ \mathbf{p} \end{pmatrix}$$

from $\mathbb{E}_{\mathbf{p}}(r) = \sum_i p_i \cdot r_i$ into the scalar product $\mathbb{E}_{\mathbf{p}}(r) = w(\mathbf{p})^T \mathbf{r}$ in which the additional *weighting function* $w : [0, 1]^n \to [0, 1]^n$ assigns different importance to the rewards based on what the probabilities are. Depending on how w is chosen, different choice models arise, not all of which may be plausible. An example illustrating this possibility has been given by [21, p. 346f]: let $w : [0, 1] \to [0, 1]$ be a convex function, then over the interior of the probability scale, $w(p) + w(1 - p) < 1$ and there will be an $\varepsilon > 0$ so that a prospect $\mathbf{p} = \begin{pmatrix} x & x + \varepsilon \\ p & 1 - p \end{pmatrix}$ will be rejected in favour of the alternative prospect $\mathbf{q}$ that rewards x with probability 1. Even so, despite the fact that $\mathbf{p}$ clearly dominates $\mathbf{q}$ by our specific choice of ε. The avoidance of this effect can be cast into a requirement on its own: we would certainly like the utility function to be *monotone*, meaning that if a prospect $\mathbf{q}$ is dominated by another prospect $\mathbf{p}$, written as $\mathbf{q} \trianglelefteq \mathbf{p}$, then the utility for the latter should be less than for the dominating prospect; the monotony requirement reflecting dominance in the utility function formally manifests in the familiar form: $\mathbf{q} \trianglelefteq \mathbf{p}$ should imply $u(\mathbf{q}) \leq u(\mathbf{p})$.

First-order dominance among prospects is defined by treating the prospects as distributions and imposing the usual stochastic order (3.7) on them, with the additional condition that the inequality should be strict at least once. Intuitively, a prospect $\mathbf{p}$ dominates the prospect $\mathbf{q}$ if for all possible outcomes, $\mathbf{p}$ rewards at least as much utility than $\mathbf{q}$, and strictly more utility in at least one case of outcome.

Deciding dominance correctly can be an intricate matter for an individual, as Tversky and Kahneman [24] have shown empirically by the following poll: a total of 88 participants were asked to choose between the following two prospects, describing a gamble where a colored ball is drawn from an urn, and gains or losses are tied to the outcome color. The fraction of balls in the urn is different for each colors (and trivially corresponds to the likelihood of drawing the respective ball). The first choice is between prospects

$$
\mathbf{p}_A = \begin{pmatrix} 0\$ & \text{win } 45\$ & \text{win } 30\$ & \text{lose } 15\$ & \text{lose } 15\$ \\ 90\%(\text{white}) & 6\%(\text{red}) & 1\%(\text{green}) & 1\%(\text{blue}) & 2\%(\text{yellow}) \end{pmatrix}
$$

$$
\mathbf{q}_A = \begin{pmatrix} 0\$ & \text{win } 45\$ & \text{win } 45\$ & \text{lose } 10\$ & \text{lose } 15\$ \\ 90\%(\text{white}) & 6\%(\text{red}) & 1\%(\text{green}) & 1\%(\text{blue}) & 2\%(\text{yellow}) \end{pmatrix}
$$

Obviously, $\mathbf{p}_A$ dominates $\mathbf{q}_A$, since the gains under choice $\mathbf{p}_A$ are larger or equal than the rewards under $\mathbf{q}_A$. Quite expectedly, all participants went for $\mathbf{q}_A$. Now, the choice was slightly modified by combining the "loss-cases" (blue and yellow ball) into all yellow balls with a total of 3%, resp., 2% in the urn (and 15 Dollar loss). The gains and losses are left unchanged relative to one another, so that the new choice is between

$$
\mathbf{p}_B = \begin{pmatrix} 0\$ & \text{win } 45\$ & \text{win } 30\$ & \text{lose } 15\$ \\ 90\%(\text{white}) & 6\%(\text{red}) & 1\%(\text{green}) & 3\%(\text{yellow}) \end{pmatrix}
$$

$$
\mathbf{q}_B = \begin{pmatrix} 0\$ & \text{win } 45\$ & \text{win } 45\$ & \text{lose } 15\$ \\ 90\%(\text{white}) & 6\%(\text{red}) & 1\%(\text{green}) & 2\%(\text{yellow}) \end{pmatrix}
$$

Now, conversely to before, $\mathbf{q}_B$ dominates $\mathbf{q}_A$ but this dominance is much harder to notice for a human being. We say that it is *framed* or *intransparent*. In effect, 58 out of the same 88 individuals from before again chose $\mathbf{q}_A$ despite it is arguably the worse option. This outcome is interpreted as people not intentionally acting against dominance, but do so mostly in cases where the dominance relation is not obvious.

A more drastic example is that of the "Asian disease" [23]: people where asked to choose one out of two medications for a (hypothetical) potentially lethal disease. Out of a total of 600 patients, medication A would save 200 lives, and under medication B, 400 patients would die. Obviously, the choices are objectively indifferent, but the way in which they are formulated already creates preferences in the decision maker's minds.

While this framing was done by packing negative attributes into simpler terms (like replacing all blue balls by yellow ones), the converse effect by "unpacking positive attributes" can be used to make an option subjectively more attractive than it actually is. This is the *event-splitting effect*. Empirical studies conducted by [22] and [11] confirm a change of tendency towards choosing options more likely when their positive consequences are more exposed (by splitting them into sub-events).

5.5   Rank-Dependence and Prospect Theory

As we have seen before, approximately equal likelihoods may cause an individual to abandon this information and make the decision depend only on the respective consequence directly (not its expectation). The weighting functions can be used to adjust the probabilities accordingly, say, probabilities that are typically over or underrated can be corrected. Likewise can losses and gains be weighted differently by individuals. This idea is at the core of two more directions to redefine utility functions towards better approximations of real-life decision making, or similarly, explain effects of bounded rationality attributed to violations of the utility maximizing principle.

One type of such modification is *rank-dependent expected utility theory*, originally proposed by [16]: this assumes that the choice does not only depend on the absolute probabilities in the prospect, but also on their relative order (*rank*). Let the prospect be composed from n possible outcomes $r_1 \trianglelefteq r_2 \trianglelefteq \ldots \trianglelefteq r_n$ sorted from worst to best (ascending). In a rank-based approach, we can define decision weights to be $w_i = \pi(p_i + p_{i+1} + \ldots + p_n) - \pi(p_{i+1} + \ldots + p_n)$ for $i = 1, 2, \ldots, n-1$ and $w_n = \pi(p_n)$. The herein new function π is a *probability weighting function*. As suggested by [10], the function π can be chosen to "correct" subjective errors made in the assessment of objective probabilities by individuals. So, an under- or overestimation of probabilities due to optimistic or pessimistic attitudes of an individual could be corrected by a proper choice of π (indeed, connections between the curvature of π and optimism/pessimism have been studied [7, 16, 27]). The choice of w, respectively based on the choice of π, can be made such that the resulting utility function remains monotone (to be consistent with dominated alternatives). See also [26] for a more recent application to scheduling.

Prospect theory [13] extends this idea by embedding the weighting in a two-stage process of decision making: first, prospects are adapted by subjects according to several heuristics, some of which we have already met, i.e., combination of options for simplification, cancellation of approximately equivalent options, etc. In the second phase, the choices among the edited prospects are made by valuating gains and losses relative to a reference point. This amounts to applying again a

Fig. 5.1 Valuation of
outcomes in prospect theory
(example)

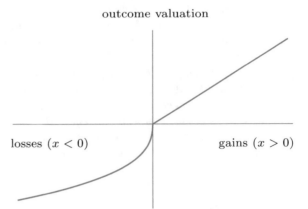

outcome valuation

losses $(x < 0)$ gains $(x > 0)$

Fig. 5.2 Prelec's probability
weighting functions π_α

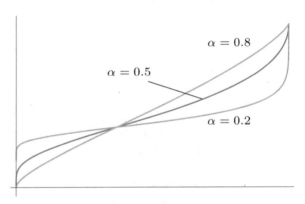

$\alpha = 0.8$

$\alpha = 0.5$

$\alpha = 0.2$

weighting function w and/or a probability weighting function π. This editing is
called *coding*, and the outcome and probability valuations are typically asymmetric
S-shaped functions; examples of which have been given by [15, 25] and are plotted
in Figs. 5.1 and 5.2. Prelec's proposal of the probability weighting function $\pi_\alpha(x) =$
$\exp(-(-\ln x)^\alpha)$ for a parameter $0 < \alpha < 1$ is a popular choice, even in security
applications [19].

5.6 Violations of Transitivity and Regret Theory

Sometimes, choices are not made towards maximal utility, but – subjectively equivalent – towards avoiding disappointment or regret. While we will go into further
detail on disappointment in Sect. 7.4, let us first look at how regret theory [2, 8, 14]
works. Here, the choice is made by comparing consequences of alternatives. Like in
a two-person game, only now played in the mind of the decision maker, the person
first determines the current state s of the world. Then, actions are compared pairwise
depending on the state s; suppose the rewards are $r_{1,s}$ and $r_{2,s}$. The consequences

of the two actions are measured by a function $M(a_s, b_s)$ that $\trianglelefteq$-increases in the first and $\trianglelefteq$-decreases in the second argument. So, if the alternative $r_{2,s}$ is preferable over $r_{1,s}$, then by the downward monotony of M, the alternative can suppress a choice of $r_{1,s}$ upon regret if $r_{2,s}$ is not chosen. Likewise, it can enhance a choice of $r_{1,s}$ as an alternative to $r_{2,s}$ since $r_{1,s}$ is preferable over $r_{2,s}$ and there would be regret again (just being rejoicing in that case). Nonetheless, the individual still maximizes some utility, only that it depends on the state of the world now too. The person would maximize an expectation based on the state of the world given as $\sum_s \Pr(s) \cdot M(r_{i,s}, r_{j,s})$ to compare any two rewards $r_{i,s}$ and $r_{j,s}$. This method of comparison applies to actions only, but can be extended to prospects by using the function M as in a two-person game to define a preference between two prospects $\mathbf{p}, \mathbf{q}$ in the world's state s as

$$\mathbf{p} \trianglelefteq \mathbf{q} \iff \sum_{i,j} p_i \cdot q_i \cdot \psi(i, j) \leq 0, \tag{5.2}$$

with $\psi(i, j) = M(r_{i,s}, r_{j,s}) - M(r_{j,s}, r_{i,s})$. The function ψ herein just tells by its sign which action would be regretted when chosen over the alternative. As usual, we assume stochastic independence of consequences between different actions (otherwise we would make an action's effects depend on an alternative action that has never been taken). The function ψ lets us define a concept of *regret aversion*: let $r \trianglelefteq s \trianglelefteq t$ be three rewards, then regret aversion implies that for these three, we should have $\psi(r, t) \trianglelefteq \psi(r, s) + \psi(s, t)$ (like a triangle inequality). Now, let us suppose regret aversion as the driving force behind decisions, and remember that in regret theory, it matters how consequences are assigned to states. The example of [21] uses a choice between three rewards r_1, r_2 and r_3 and rewards $r_1 \trianglelefteq r_2 \trianglelefteq r_3$. However, the particular assignment of rewards to actions depends on the state of the world. For example, suppose that the choice is in a security game, where the attacker determines the state s (e.g., one out of three possible states s_1, s_2, s_3) of the system and the defender has to make a choice. Let the outcomes under each state and action be given by the matrix (i.e., similar to a matrix game)

	s_1	s_2	s_3
r_1	z	y	x
r_2	x	z	y
r_3	y	x	z

in which we let $\psi(r, s)$ denote the entry in row r and column s. Assuming regret aversion, preference among the actions is determined by the sign of the expression $\psi(r, s) + \psi(s, t) - \psi(r, t)$, and under our assumed preference of rewards, we would get the preference $r_2 \trianglerighteq r_1$ when comparing r_2 to r_1. Comparing the other two pairs in the same way, it is easy to see that we get a cyclic preference structure here being $r_2 \trianglerighteq r_1$, $r_3 \trianglerighteq r_2$ and $r_1 \trianglerighteq r_3$, so transitivity is obviously violated. How does this fit into our imagination of a choice being rational if it maximizes utility? Apparently, the cycles implied by intransitivity preclude a meaningful definition of a "best" action.

In fact, a quick answer is already available to us thanks to game theory: the cyclic structure above is nothing special in games where a Nash equilibrium does not exist in pure strategies, but does exist in mixed strategies. Example 3.6 already illustrated this in a spot checking game. By comparing the payoff structures in Example 3.6, we recognize the cyclic preference structure from the above 3×3-game also in this previous 2×2-payoff matrix: if the defender would check location 1 (all the time), then the attacker would adapt to this by attacking at location 2 (all the time), consequently. Hence, it is better for the defender to check location 2, but if it does so all the time, then the attacker again adapts by attacking at location 1, and so on.

5.7 Border Effects

Similar to the cancellation of approximately equal probabilities (related actions would be considered as nearly identically likely to happen, so the decision would depend only on the quantitative magnitude of the rewards), *border effects* describe a sensitivity of the choice to extreme outcomes. Decision rules derived with this finding in mind can be constructed by introducing proper decision weights, or apply a "correction" of probability values near the two ends of the scale (hence the name "border" effect), similar as done in prospect theory. As a third alternative, the preference $\unlhd$ itself can be designed to account for this sensitivity; a concrete example of which is the stochastic $\preceq$-order introduced in Sect. 3.1.

The $\preceq$-ordering is explicitly sensitive to extreme outcomes, but rather insensitive on the opposite lower end of the loss scale. An additional appeal of stochastic orders in general stems from their natural compatibility to procedural theories of choice, discussed next. In general, an attitude of risk aversion paired with pessimism may over-proportionally weigh possibilities of extreme events, i.e., rational behavior then means just "avoiding the worst", than "seeking the optimal". The former appears equally rational, only under a different definition than a maximization, and we will revisit the idea later in the context of disappointment (Sect. 7.4).

5.8 Procedural Theories

Thinking of your own set of decision heuristics, it may hardly be the case that all decisions are made under strict objectives of maximal utility. Instead, an individual may adapt the decision making depending on the state of the world, current subjective goals, available information and many other factors. Procedural theories embody a potentially large set of rules that may come to application depending on the current situation. Although this is much less mathematically sophisticated than an axiomatic approach starting from first principles (like the axioms above to define a utility function from it), we must bear in mind that we are dealing with *people*, whose behavior is much less exact and deterministic as we may assume. Indeed, though procedural methods may be criticized as cumbersome relative to the more elegant optimization of functions, theories of well-behaved preferences (according to mathematical definitions) do not seem to provide adequate explanations for various effects observed in practice [21, pg.363]. Such deviations *can*, in an abuse of the term, be coined as decisions of bounded rationality, but this also raises the question of whether it might just have been the wrong utility function that was maximized in theory.

The appeal to procedural theories is the freedom to basically encode anything into the decision making mechanism, which opens possibilities to account for all the findings of psychology, sociology and related fields (e.g., [5, 9, 18, 20, 20]) in the design of a decision making framework. Probabilistic elements are easy to include, and simulations of agent based (i.e., procedural) models can be implemented [3, 4]. Indeed, there is no conceptual barrier to use a heterogeneous pool of agents in a prediction of possible choices made by an individual, thus letting the simulation model use different theories of choice that may accurately or incorrectly apply to the given individual at different times or in different situations. In addition, simulation *can* deliver additional information about the "why" of a decision that plain utility maximization cannot. One example is the rate of disappointment that an analysis of pure maximization of utilities cannot deliver directly. Avoidance of disappointment is closely akin to the psychological mechanisms that inspired regret theory, and a general form of a pessimistic choice model.

Wrapping up, bounded rationality may in some cases just be another word for the humble insight that "all models are wrong [. . .]"[1] in explaining rational behavior as "utility maximizing". Rather, there is not necessarily anything wrong with the basic paradigm of defining rationality as guided towards maximal benefit, only this benefit may not be as measurable or crisp as we would like it to be. For example, companies may seek maximal profit, but security generates no *return on investment*. Instead, it is an investment made to protect assets and to enable profits (continuously), and its main purpose is to create trust. This is an intrinsically incommensurable concept, since trust has subjectively distinct meanings.

References

1. Allais M (1953) Le comportement de l'homme rationnel devant le risque: critique des postulats et axiomes de l'ecole americaine. Econometrica 21(4):503. https://doi.org/10.2307/1907921
2. Bell DE (1982) Regret in decision making under uncertainty. Oper Res 30(5):961–981. https://doi.org/10.1287/opre.30.5.961

---

[1] "[. . .] only some models are useful"; a quote that in different formulations is mostly attributed to the statistician George Box.

3. Busby J, Duckett D (2012) Social risk amplification as an attribution: the case of zoonotic disease outbreaks. J Risk Res 15(9):1049–1074. https://doi.org/10.1080/13669877.2012. 670130
4. Busby JS, Onggo B, Liu Y (2016) Agent-based computational modelling of social risk responses. Eur J Oper Res 251(3):1029–1042 https://doi.org/10.1016/j.ejor.2015.12.034
5. Camerer CF (2011) Behavioral game theory: experiments in strategic interaction. The roundtable series in behavioral economics. Princeton University Press, s.l. http://gbv.eblib.com/patron/FullRecord.aspx?p=765287
6. Colman AM (2009) A dictionary of psychology, 3rd edn. Oxford paperback reference. Oxford University Press, Oxford
7. Diecidue E, Wakker PP (2001) On the intuition of rank-dependent utility. J Risk Uncertain 23(3):281–298. https://doi.org/10.1023/A:1011877808366
8. Fishburn PC (1982) Nontransitive measurable utility.s J Math Psychol 26(1):31–67. https://doi.org/10.1016/0022-2496(82)90034-7
9. Gigerenzer G, Selten R (eds) (2002) Bounded rationality: the adaptive toolbox, 1st mit press pbk. edn. MIT Press, Cambridge, MA
10. Gonzalez R, Wu G (1999) On the shape of the probability weighting function. Cogn Psychol 38(1):129–166. https://doi.org/10.1006/cogp.1998.0710
11. Humphrey SJ (1995) Regret aversion or event-splitting effects? more evidence under risk and uncertainty. J Risk Uncertain 11(3):263–274. https://doi.org/10.1007/BF01207789
12. Iyengar SS, Lepper MR (2000) When choice is demotivating: can one desire too much of a good thing? J Pers Soc Psychol 79(6):995–1006. https://doi.org/10.1037//0022-3514.79.6.995
13. Kahneman D, Tversky A (1979) Prospect theory: an analysis of decision under risk. Econometrica 47(2):263. https://doi.org/10.2307/1914185
14. Loomes G, Sugden R (1982) Regret theory: an alternative theory of rational choice under uncertainty. Econ J 92(368):805. https://doi.org/10.2307/2232669
15. Prelec D (1998) The probability weighting function. Econometrica 66(3):497. https://doi.org/10.2307/2998573
16. Quiggin J (1982) A theory of anticipated utility. J Econ Behav Organ 3(4):323–343. https://doi.org/10.1016/0167-2681(82)90008-7
17. Robert CP (2001) The Bayesian choice. Springer, New York
18. Rubinstein A (2002) Modeling bounded rationality, vol 3. Print edn. Zeuthen lecture book series. MIT Press, Cambridge, MA
19. Sanjab A, Saad W, Başar T (2017) Prospect theory for enhanced cyber-physical security of drone delivery systems: a network interdiction game. In: 2017 IEEE international conference on communications (ICC), pp 1–6. https://doi.org/10.1109/ICC.2017.7996862
20. Simon HA (1997) Models of bounded rationality. MIT Press, Cambridge, MA
21. Starmer C (2000) Developments in non-expected utility theory: the hunt for a descriptive theory of choice under risk. J Econ Lit 38(2):332–382. http://www.jstor.org/stable/2565292
22. Starmer C, Sugden R (1993) Testing for juxtaposition and event-splitting effects. J Risk Uncertain 6(3):235–254. https://doi.org/10.1007/BF01072613
23. Tversky A, Kahneman D (1985) The framing of decisions and the psychology of choice. In: Covello VT, Mumpower JL, Stallen PJM, Uppuluri VRR (eds) Environmental impact assessment, technology assessment, and risk analysis, NATO ASI series, series G. Springer, Berlin/Heidelberg, pp 107–129 . https://doi.org/10.1007/978-3-642-70634-9_6
24. Tversky A, Kahneman D (1989) Rational choice and the framing of decisions. In: Karpak B, Zionts S (eds) Multiple criteria decision making and risk analysis using microcomputers, NATO ASI series, series F. Springer, Berlin/Heidelberg, pp 81–126. https://doi.org/10.1007/978-3-642-74919-3_4
25. Tversky A, Kahneman D (1992) Advances in prospect theory: cumulative representation of uncertainty. J Risk Uncertain 5(4):297–323. https://doi.org/10.1007/BF00122574
26. Wang Q, Karlström A, Sundberg M (2014) Scheduling choices under rank dependent utility maximization. Proc Soc Behav Sci 111:301–310. https://doi.org/10.1016/j.sbspro.2014.01.063
27. Yaari ME (1987) The dual theory of choice under risk. Econometrica 55(1):95. https://doi.org/10.2307/1911158

Part II
Security Games

Chapter 6
Risk Management

If you can't measure it, you can't improve it.

P. Drucker

Abstract This chapter embeds game theoretic techniques and models inside the ISO31000 risk management process, as a generic template for the general duty of risk control. We observe similarities between risk management processes and extensive form games, accompanied by the possibility of using game-theoretic algorithms and methods in various steps of a risk management process. Examples include decision making for risk prioritization, choice of best risk mitigation actions or optimal resource allocation for security. To this end, we discuss a variety of systematic methods for adversarial risk analysis (ARA), resilience management (in relation to risk management), level-k thinking, and the assessment of action spaces and utilities for games.

6.1 Steps in a Risk Management Process

Risk management is the duty of a system administration to continuously and permanently maintain awareness, countermeasures and mitigation plans for anything unexpected that could happen in the system's environment and bears potential harm. For enterprise environments, various standards have been compiled from experience and best practices, so as to provide a generic guideline that risk managers can follow. Being widely similar in the overall process phases themselves, and differing mostly in the individual details on what steps to take in each phase, we shall take the ISO31000 as a generic underlying model throughout this chapter. Our exposition draws from [35], which already establishes strong connections between game theory and risk management. This correspondence extends further to other standards [12].

The following phases, taken from the ISO 31000 standard, are essentially present in almost every risk management guideline or standard. The steps build upon one another and are thus practically completed in order of appearance:

© Springer Nature Switzerland AG 2020
S. Rass et al., *Cyber-Security in Critical Infrastructures*, Advanced Sciences
and Technologies for Security Applications,
https://doi.org/10.1007/978-3-030-46908-5_6

1. *Establishing the context*: obtain an understanding of *what* is under risk. This includes gaining a maximum detailed view on the overall system, typically an enterprise, uncovering all its vital components, their interdependency and potential cascading effects upon external influences on parts of the system. The context can be divided into external and internal aspects, which define the scope of the risk management process as such, and the (qualitative or quantitative) management criteria that are relevant to impose.

2. *Risk identification*: the full picture of the system gets further decomposed into assets, threats to these assets and vulnerabilities of these assets; the latter of which are typically properties or characteristics of an asset that allow a threat scenario to occur in reality.

3. *Risk analysis*: for each threat scenario (anticipated in the previous and rigorously defined in this phase), consequences are analyzed and likelihood of occurrence are estimated. In a quantitative risk assessment, the product of the two would give the common risk formula as

$$\text{risk} = \text{impact} \times \text{likelihood}, \tag{6.1}$$

which appears at least at some point in most quantitative and also qualitative risk management frameworks. Observe the similarity of (6.1) to expected utilities: multiplying outcomes of a random variable by the respective likelihoods (and adding them up) means taking *expectations*, so risk as computed via (6.1) can essentially be understood as *expected loss*.

4. *Risk evaluation*: Based on a list of risks with ratings that formulas like (6.1) induce, a selection, ranking and severity level classification is made. This is mainly a matter of prioritizing risks towards subsequent mitigation actions. Understanding risk as expected loss and defining a ranking in the sense of a preference relation $\trianglelefteq$ among risks based on their scoring using formula (6.1), connects this step to Theorem 5.1. Note that a generalized approach replacing (6.1) by a whole distribution object and $\preceq$-stochastically ordering these for a risk evaluation is equally admissible.

5. *Risk treatment*: working through the list of risk in descending order of importance (based on the previous phase), we proceed by identifying threat scenarios, defense strategies (tailored to the threats), mitigation actions (implementing the defenses), and implementation strategies. The latter may fall into one of the following (partly overlapping) classes of action, where any choice herein is always a matter of an a-priori cost-benefit analysis and tradeoffs (e.g., with help from game theory):

   - accepting the risk (if the risk, i.e., the likelihood and impact is below a predefined threshold)
   - preventing the risk (by eliminating vulnerabilities or switching to completely different systems, if possible)
   - relieving risks (by reducing the impact or the likelihood, if the event cannot be prevented)

- transferring the risk (by buying insurance for it. We will have more to say about this in Chap. 7).

The whole process almost naturally allows for a game theoretic perspective on it, such as was done by [26]: the construction of a mathematical two-person game bears striking similarities to the above risk assessment process, since strategies for both players are identified, corresponding to a threat analysis (strategies for one player) and defense actions (strategies for the other player). The definition of payoffs corresponds to the impact analysis, usually done during the identification of risks (step 2) and the analysis of risks (step 3).

The practical process is often part of the business continuity management and disaster recovery, which entail (among others) business impact analyses to first identify critical assets, their degree of potential affection, costs and consequences of outages and recovery.

Finding optimized (equilibrium) strategies for both players manifests itself as the phase of risk treatment. The optimization of opposing interests is necessary at several points in the risk management process, since besides the natural conflict of interest between the defender (strategically the risk manager, operationally its subordinates) and the attacker (being rational as having intentions or even non-rational in being nature itself), we have also the matter of cost-benefit analysis, and – not to forget – matters of conflicting interests in the impact of implementations of mitigation actions. For example, people may accept the need for surveillance since it is necessary for security, but nonetheless may find it unpleasant to be checked themselves (even if it is for the good of the overall community). The accurate quantification of goals is typically necessary for any numeric optimization, but numeric accuracy is precisely what we usually cannot assure in practical applications. Goals like reputation, customer convenience, even monetary impacts are notoriously hard to quantify, and probabilistic models, no matter how sophisticated, at some stage necessarily rely on empirical data. In absence of an empirical data fundament, say, due to the (fortunate) rarity of catastrophic events, getting the numbers into the game theoretic models becomes difficult. This is one reason why game theory has been coined unpopular for risk management matters in the past [2].

In this context, not even the equilibrium concept may obviously be useful in the practical implementation. To illustrate this, we will look at a passenger check at the airport as an example and include "convenience" as a qualitative goal for each passenger (engaging as a player) to apply [13]. Certainly, people find it convenient to know that airport security is there to protect from hazards on the plane, but being under a check oneself is rarely felt comfortable. So, what would be a Nash equilibrium here? If everyone cooperates, then everyone feels uncomfortable by the checks. So, the individually selfish optimum is to refuse the screening and have others contribute. However, if everybody acts selfish in this way, then the threat fully kicks back in since nobody is screened anymore. Mechanisms to implement the optimal balance can be found by game theory, but the typical multitude of equilibria may nonetheless induce implausible or impractical equilibrium strategies [32]. Still, game theory remains a tool to optimize any kind of the balance between conflicting goals as mentioned above.

On the bright side, most of these issues can be addressed by adapting the models accordingly and by changing the perspective hereto. Insofar it concerns qualitative goals, it is a standard procedure to involve experts for their domain knowledge and assessment about qualitative goals. Data collected from such surveys, interviews and other sources (up to social media analysis) can then form a solid data base for building payoff models upon. In turn, it is possible to compile these into values for a game theoretic analysis, or let the games directly work with them (by using stochastic orders as discussed in Sect. 3.4; see also [29]). Other qualitative goals like "convenience" may even admit quantitative measures, e.g., for a passenger screening, we may assume the "degree of annoyance" as proportional to the frequency of a check [31].

6.2 Resilience Analysis

Complementary to risk analysis is *resilience analysis*, with the two being contrastable in various ways. One is viewing risk management as a bottom-up approach that starts with data collection, followed by model building and risk characterization. Resilience management can be viewed as coming from the opposite direction and working top-down: it starts with the goal identification for the management, followed by decision models, specifying which metrics, measures or other ways are useful to quantify values for the decision maker, and culminating in the quantification of actions [19]. Neither can be taken as a cover, substitute or general improvement of the other, since the differences between the two are the angles of view on the same problem, and both may aid one another: resilience management is (among others) concerned with recovery from incidents, potential long-term damages thereof and with rare events. This view can be included in risk management at various stages, say, during a (long-term) impact and likelihood assessment, as well as in the risk treatment stage, when risks are relieved or transferred (thus, enabling a fast recovery by buying insurance and taking precautions beforehand). Likewise, resilience management is not only a matter of bouncing back from an impact, but centers on the integration of risk perception, risk mitigation and risk communication; all three of which are matters that run "laterally" to the generic risk management process (see Fig. 6.1). We leave a deeper discussion to literature related to this book (e.g., [20]), and continue our focus on how game theory fits into both management approaches here. That is, throughout the process, game theory can primarily help with *security resource investment optimization* [10] during risk treatment, but extends even to matters of risk prioritization (e.g., by using a score that is derived from a game model, or at least adopting the concept of orderings on spaces of arbitrary structure; typically the action spaces).

Let us return to formula (6.1) as our starting point to establish of game theory as a tool for risk and resilience management. The product is essentially a numeric

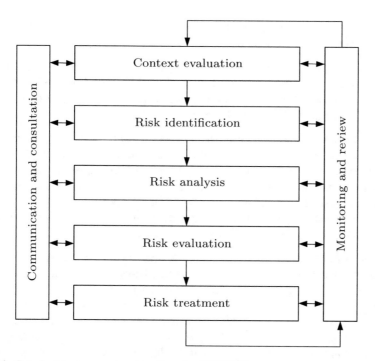

Fig. 6.1 Generic risk assessment process (based on ISO31000)

quantity (even if it is taken over ranks rather than crisp numbers), and as such, game theory can optimize the term as a goal function. This is a challenge on its own [2] and typically calls for sophisticated probabilistic models.

Remark 6.1 Whenever the modeling is based on empirical data, the usual statistical preprocessing is required, including (but not limited to) outlier elimination, normalization of data, debiasing and denoising; basically all that data science has to offer here. If the data is subjective and about risk, then many subjective factors come into play in addition [4, 9, 22, 34, 37, 40, 41], such as personal risk affinity or aversion, background (social, cultural, religious, and others, in addition to expertise). More importantly, risk assessment is nonconstant over time, since the subjective awareness and feeling about risk is highly dependent on recent individual experience and information from personal conversations, social media, news reports, and many other sources.

6.3 Quantifying Security

Security is often understood as a state of a system in which all feasible precautions against all known and relevant threats have been adopted. As such, it is not a quantity arising from some physical process, and cannot be "measured" like other physical values. Nonetheless, reaching a state of security can be thought of as an *event*, to which we can assign probabilities.

Remark 6.2 Conditions under which security is reached may be diverse and depend on the context, application, and range from fully mathematical definitions (examples are given in Chap. 11) up to qualitative indications such as risk standard certification (e.g., ISO, Common Criteria, or similar).

This leads to quantitative *measures* of security, as opposed to security *metrics* or *scores*, as defined in Chap. 3. Scores and metrics are usually for *absolute* assessments, as opposed to metrics, which typically serve a *relative* investigation. Both, scores and measures lend themselves to relative comparisons too, but none of the three necessarily can substitute or even induce the other.

Practical quantifications of impacts and likelihoods are known to be error-prone, as the following Example 6.1 illustrates.

Example 6.1 (risk assessment for lightning strike [21]*)* In the area of Munich (Germany), the chances for a lightning strike are known to be roughly 1.26×10^{-6}, based on meteorological long-term records. Let the impact be estimated to be roughly 10,000 Euros. Then formula (6.1) yields an expected loss of ≈ 0.126 Cent, which does not provide a plausible justification for a countermeasure. Nonetheless, a lightning protection is surely advisable. This is the issue of multiplying large with small numbers, no matter how well-justified those are on statistical grounds. Examples like this lead to the recommendation of using quantitative scales to measure both, impact *and* likelihood, since either may be difficult for an expert to assess.   ◇

To ease the duty for the expert, we can let experts give their answers in terms that are familiar and suitable for the domain, and specify how these map into risk categories for a qualitative assessment (to be subsequently used with optimizations, or simply with formula (6.1)). For an expert assessment, the number of categories on the scales for impact and likelihood should be even to avoid data or answers on the exact middle of the scale, as those are least informative and may be picked by people in lack of any better idea. The beauty of game theory here is its possibility to define payoffs in *any* scale or unit, provided that the resulting quantifications are comparable (totally ordered; see Sect. 5 for a continuation of this point).

One example of such a "catalog based" quantification and definition of security is given in Fig. 6.2a for impacts, and in Fig. 6.2b for the likelihoods. Goals are often *multi-dimensional*, since a decision maker has to care not only for avoidance of monetary losses, but also needs to balance investments against benefits, including effects on the reputation, potential legal consequences, and many more. This points out a general issue with security, since any investment into it bears no direct return. Rather, investments in security are made to *avoid losses*, and acting selfishly towards maximizing the own benefit is not necessarily the same thing as loss minimization.

Categorical quantifications fall in the above category security scores, since the category value is a mere name without a physical meaning behind the number. Security measures usually require more formalism behind them, and in most cases

risk category	loss category			
	loss of intellectual property	damage to reputation	harm to customers	...
negligible (1)	< 500 $	not noticeable	none	...
noticeable (2)	between 500 $ and 10.000 $	noticeable loss of customers (product substituted)	inconvenience experienced but no physical damage caused	...
low (3)	> 10.000 $ and 50.000 $	significant loss of customers	damages to customers' property, but nobody injured	...
medium (4)	> 50.000 $ and 200.000 $	noticeable loss of market share (loss of share value at stock market)	reported and confirmed incidents involving light injuries	...
high (5)	> 200.000 $ and 1 Mio. $	potential loss of marked (lead)	reported incidents involving at least one severe injury but with chances of total recovery	...
critical (6)	> 1 Mio. $	chances of bankrupt	unrecoverable harm caused to at least one customer	...

(a) Example Impact Categories [27]

likelihood category	name	frequency (examples of qualitative descriptions)
1	negligible	occurs roughly once per decade, or every 50 years on average
2	rare	occurs once per year (on average)
3	medium	can be expected once per month
4	likely	may occur several times per month (or even per day)

(b) Example Likelihood Categories

Fig. 6.2   Examples of qualitative security definitions

are defined as likelihoods for the secure state to arise. The exact definition of this event is dictated by the context, and in case of probabilistic security models, is derived from analyses of success chances for the attacker. We will meet examples from cryptography illustrating this in Chap. 11.

At this point, it is important to bear in mind the lessons learned from bounded rationality (Chap. 5): the particular way of how questions are formulated can strongly effect subjective ratings on the given scale. For an example referring to Fig. 6.2a, one could define the categories for the risk dimension "harm to customers" as "x out of y customers affected", or "$p\%$ of customers affected" (with $p = x/y$) or equivalently "every k-th customer on average affected" (with $k \approx y/x$). All three are equivalent statements, but just like as in Sect. 5.4, the way people categorize the risk may be strongly different in the three cases despite the numerical equivalence. Likewise, subjective risk attitudes and individual over- or under-ratings of impacts and likelihoods need to be considered, which is where weighting functions and prospect theory (see Sect. 5.5) provide useful insights and tools.

Concepts "in between" can be operationalized too: suppose that the categories define the units in which the data collection for a risk assessment is carried out. Consequently, each expert in a group of people may utter its subjective rating of impacts and likelihoods. It is straightforward to compile this data into a categorical distribution, supported on the set of risk categories as the definitions (like Fig. 6.2) prescribe. This kind of data processing is particularly flexible, since it provides empirical distributions. These, in turn, can go as priors into a Bayesian analysis, or otherwise be used with game theory directly. The latter method was put to practice in the risk management project "HyRiM" (Hybrid Risk Management for Utility Providers [14]), which specifies a risk management process based on ISO31000, and has each phase therein instantiated by game theory with empirical distributions replacing the usual numerical payoffs (as a practical instance of the games described in Sect. 4.5). The use of systematically, and sometimes partly subjectively, constructed priors receives attention in the next section, where we extend the view on modeling not only the risk as such, but also its origins.

6.4   Adversarial Risk Analysis

Knowing the environment and the enemy is a basic requirement for any effective protection. The famous "Art of War" [38] puts the value of this knowledge in eloquent terms:

> If you know the enemy and know yourself, you need not fear the result of a hundred battles. If you know yourself but not the enemy, for every victory gained you will also suffer a defeat. If you know neither the enemy nor yourself, you will succumb in every battle.

The truth in this statement was known long before game theory was invented to mathematically prove Sun Tzu right in teaching that knowing yourself and the enemy, we do not need to fear the competition. This is what adversarial risk analysis aims at: it adapts the defense to the best hypothesis about upcoming attacks that we can set up based on the information in our hands. If there is no or no reliable information, the next best we can do is substitute the missing data about the enemy by the known information about our own intentions. Zero-sum games are the proper models for this pessimistic attitude. Sun Tzu's insight wisely extends to this situation, since by optimizing worst-case average outcomes, gains and losses will alternatingly occur.

Regardless of whether we model our own or the adversary's intentions, a systematic approach is always advisable, and *influence diagrams* [25] provide one such method. Influence diagrams bear a close similarity to decision trees, Bayesian networks, and also to extensive form representation of games. All three enable a graphical and easy to understand representation of factors relevant for decision making. Let us take a step-by-step approach through the use of influence diagrams towards adversarial risk assessment [32]. An influence diagram comes, in its basic form, as a directed acyclic graph $G = (V, E)$, with every node in V falling into one out of three categories:

Action nodes (rectangles):   these represent a particular action available to the player following up a precursor situation (node) in the graph. These nodes may represent one action or a choice from a set of actions, which may depend on information available a priori. Such information is displayed by incoming edges to the action node (as we will see in Fig. 6.3d).

Uncertainty or *chance* nodes (ovals): The outcome of an action is hardly ever deterministic and in most cases influencing or subject to probabilistic factors, represented by uncertainty nodes. Those are usually tagged with probability distributions conditional on the variables that the parent nodes represent. Formally, if an uncertainty node w has parents $u \rightarrow w$ and $v \rightarrow w$, it would carry a conditional distribution $F_W(w|u, v)$, where all three nodes are considered as random variables. For the example of an action node, say, u, the randomness in u relates to the current choice from the action set. Remember that chance nodes also appear in extensive form games, modeling actions taken by the (irrational) zero-th player "nature".

Utility nodes (hexagons): those model net values for a decision maker, which are functionally or probabilistically dependent on their parents in the graph.

Let us hereafter generally speak about an *item* in an influence diagram mean an object (distribution, function, set) assigned to a node in the graph.

An edge $u \rightarrow v$ between two nodes $u, v \in V$ means that v's state depends on the state of its parent nodes, where the dependency can be probabilistic for chance nodes or functional for utility nodes. Action nodes are where rational choices can be made, such as by utility maximization principles or others.

An influence diagram can thus be used as a graphical representation of relevant factors in a decision and risk management problem. It allows for a systematic decomposition of the situation into factors relevant for the decision making, including their interdependence. The acyclicity of the diagram then enables a systematic approach to identify which data items require specification and data collection. The overall method may (but does not need to) proceed as follows: to any node $v \in V$, we assign a variable with the same name. To ease the notation in the following, let us write v synonymously to mean a node G or the variable relevant in the decision model. For every $v \in V$, call $\mathrm{Pa}(v) := \{u \in V : (u \rightarrow v) \in E\}$ the set of *parents* of the node v in the graph G. Likewise, call $\mathrm{Ch}(v) := \{u \in V : (v \rightarrow u) \in E\}$ the set of v's children in G. Since the diagram is acyclic, there must be some nodes whose parent set is empty, and another set of nodes that have no children. Initialize P as an ordered list of nodes without parents, and follow these steps:

1. For each $v \in P$, specify the following items, depending on v's type:

   - if v is a chance node, construct a distribution for the random variable v, conditional on v's parents. Formally, we need the distribution $F_v(x|\mathrm{Pa}(v)) := \mathrm{Pr}(v \leq x|\mathrm{Pa}(v))$. If $\mathrm{Pa}(v) = \emptyset$, then the distribution is unconditional.
   - if v is a utility node, specify a function f_v that takes all values of v's parents as input to define the value for v (hence, utility nodes would usually not appear without parents, since this would make the utility function constant in this view). In a slight abuse of notation, we may write $f_v : \mathrm{dom}(\mathrm{Pa}(v)) \rightarrow \mathbb{R}$, where dom is the domain in which the parent variables to v take their values.

- if v is an action node, specify a set of values that v may take. These variables are the ones to vary during a subsequent optimization, say game theoretic or other.

2. Since we are now done with v, remove it from the set P and put v's descendant nodes $\mathrm{Ch}(v)$ into P. Then, start over again with Step 1, until the set P has run empty.

We stress that the order in which nodes are processed in the above procedure naturally depends on which items are required for the next node to be properly specified. This lack of specification in the above procedure makes it somewhat incomplete, but as we will see from the examples to follow, it is generally not difficult to iterate through the list of nodes in the proper order, or even resort to a reversed approach, i.e., a recursive bottom-up procedure that starts the above specification from the nodes without children.

We illustrate the above procedure by working through a series of examples of influence diagrams that may generically appear in different situations. The diagrams evolve stepwise, with gray nodes highlighting the difference to the previous picture, respectively. The first diagram in Fig. 6.3a shows a simple situation where an action taken by the decision maker directly leads to a utility. According to the above algorithm, our duty is (iteration 1) to specify the action space, and (iteration 2), to specify the utility function on the action space.

Traditional statistical risk analysis adds an element of randomness to this, which appears as a chance node in Fig. 6.3b. This node, labeled "action" has no parents, and is thus a natural influence on the utility. In addition, the decision maker may choose its actions to influence (say, prevent, mitigate, take general precautions against) a threat. The arrow from the "action" node to the "threat" node expresses this. According to the algorithm, we start with $P = \{\text{action}\}$, and specify an action space (as before), and a distribution for the threat. For natural disasters, such as floods, fires, lightning strikes, or others, we may rely on statistical data for that matter. For example, in domains like insurance distribution models (e.g., extreme value distributions) are known, which may be instantiated and assigned to the "threat"-node here. Armed with both, the procedure prescribes (in its third iteration) the specification of a new utility function that takes the action and the random variable for the threat as input. Its output is thus another random variable, whose distribution is parameterized by the action, and otherwise determined by the transformation of the distribution of the "threat" variable. Since there is no other rational actor in this picture taking actions related to the "threat", we are still in a *non-adversarial* setting here (the only adversaries are nature, resp., chance).

An adversarial risk analysis now adds further action items and players (not necessarily visually distinct) to the picture, shown in Fig. 6.3c. Renaming the natural hazard into a "threat", we may let the threat be under influence of another actor with its own action space.

Remark 6.3 The terms *hazard* and *threats* are often used synonymous; both describing conditions or situations that may cause harm (such as natural disasters,

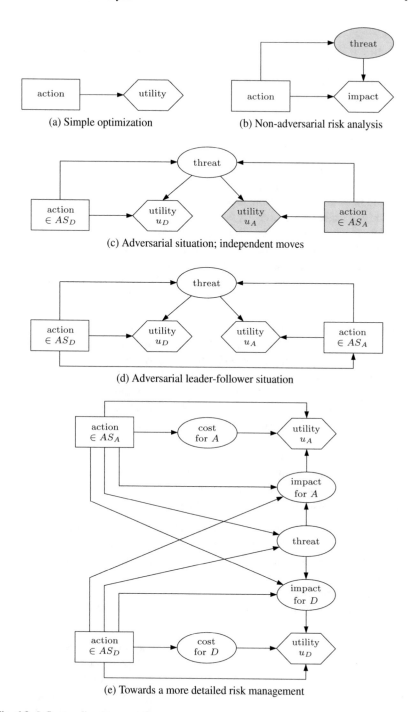

(a) Simple optimization

(b) Non-adversarial risk analysis

(c) Adversarial situation; independent moves

(d) Adversarial leader-follower situation

(e) Towards a more detailed risk management

Fig. 6.3 Influence diagram examples

human errors, or others). Occasionally, a distinction is made between hazard and threat, in the latter being a hazard that is actually upcoming and to be expected. This is already a combination with the likelihood, but not with a vulnerability, and as such not automatically the same as a risk, which is the coincidence of a threat and a condition (the vulnerability) that enables the negative impacts of the threat.

If this actor is rational, it may itself act towards utility maximization. Thus, the influence diagram grows by two new (grayish) items. Note that the labels are now extended by the items to be specified, which can be different (and entirely unrelated) to one another. In this example, let the first player (with white action and utility nodes) be the *defender* opposing an *adversary* who maintains the gray boxes. The items to be specified for both players are subscripted by "D" and "A" respectively, abbreviating the generic names "Daphne" and "Apollo" in the text to follow. The instantiation of the diagram starts with the nodes having no parents, which are now the two action spaces (in any order), and in the next steps would proceed with the distribution of the threat, conditional on both player's actions, and conclude with the specification of utility functions based on the threat and actions (of the respective player).

Figure 6.3c shows no dependency among the two action nodes, which allows the players to take moves simultaneously or asynchronously. In a leader-follower situation, the defender may move first, and the adversary may react upon. This would correspond to an arrow from the white to the gray action node, as displayed in Fig. 6.3d. Note that the sequence of specification following from the above procedure is now different: we start with the defender's action, on which the adversary's action follows (in the second iteration of the procedure). In being a utility maximizer, the adversary will thus observe the current action (from the white action node), and determine its best response to it. For that matter, it requires its own utility function and threat model, so the procedure can for convenience be reversed into a bottom-up specification. Essentially, the influence diagram in Fig. 6.3d requires the following to be specified:

- the action spaces AS_D and AS_A for the defender and the attacker,
- utility functions $u_1 : AS_D \times T$ and $u_2 : AS_A \times T$ for both players, where T is the set of threats that can occur, either naturally or rationally triggered by a rational player,
- a conditional probability distribution $F_T(t|a, d)$, where t is the threat variable (e.g., the impact that it has), and $d \in AS_D, a \in AS_A$ are the actions of both players.

It is possible and often convenient to add further arcs or nodes to the diagram, just as the situation may require. For instance, a two-person game with random payoffs can be modeled with arcs from both action spaces into both utility functions, which effectively models that the utilities of both players depend on their actions (as usual in game theory). Likewise, items appearing in a risk management process like "impacts", "costs", etc. can as well be brought into the diagram, which may then look like as in Fig. 6.3e, or even more detailed or complex. However, the symmetry in these examples is, a mere coincidence, and generally such diagrams need not to be symmetric.

6.4.1 Assessment of Utilities and Chances

Returning to Fig. 6.3d for completing the discussion, how would the decision maker proceed in a leader-follower scenario to get a best decision? In the first step, suppose that all of the items listed above are known to both parties, and that both know that the other party knows (*common knowledge assumption*). As modeled in Fig. 6.3d, Daphne takes a first move d, to which Apollo takes a best response by maximizing the expected utility, i.e., his reply to Daphne's action (indicated by the arc from her action node to Apollos action node) is

$$a^*(d) = \operatorname*{argmax}_a \mathbb{E}_{F_T(t|a,d)}(u_A(a,t)) = \int_\Omega u_A(a,t) f_T(t|a,d) dt, \qquad (6.2)$$

in which f_T denotes the conditional density of the threat variable T on the (fixed) action d of Daphne and (varying) action $a \in AS_A$ that Apollo may take, and the integration range is over the support Ω of f_T. In turn, Daphne will seek her best reply to Apollo's move a^*, which for Daphne means computing

$$d^* = \operatorname*{argmax}_d \mathbb{E}_{F_T(t|a^*,d)}(u_A(a,t)) = \int_\Omega u_D(d,t) f_T(t|a^*,d) dt.$$

The pair $(d^*, a^*(d^*))$ is a (Stackelberg) equilibrium in this sequential game, provided that all items in the influence diagram are common knowledge to both parties. What if this is not the case? The lack of knowledge can be addressed in various ways, so let us assume that Daphne knows only AS_D, u_D and the conditional threat distribution F_T, but lacks information on u_A and AS_A.

In a Bayesian approach, Daphne would make up a hypothesis about the unknown items, essentially in form of an a priori distribution F_A over the pair (u_A, AS_A). The precise choice of this prior is the typically hardest part in a Bayesian analysis, and some (suggested) choices include uniform distributions, non-informative priors, or level-k thinking (discussed later). In any case, Daphne needs to empathize with Apollo, and, in putting herself in Apollo's shoes, hypothesize about what his options and utilities could be. Given ideas about these, she can compile the prior distribution F_A. Then, she fixes a decision $d = d_0$ and samples from F_A (using Monte-Carlo simulation) to get set of n random instances of u_A and AS_A. Each of these can then be used with (6.2) to get a sequence of best replies $a_1^*, a_2^*, \ldots, a_n^*$, to Daphne's action d. Using this simulated data, Daphne can then come up with another empirical probability density $\hat{p}_D(a|d_0)$ reflecting how she thinks that Apollo will reply to her action d_0. The whole procedure is of course repeated for all her actions accordingly, to define a probability density $\hat{p}_D(a|d)$ to model Apollo's random replies a to Daphne's random actions d. She then goes to compute her best reply under what she expects from Apollo, i.e., she computes

$$d^* = \operatorname*{argmax}_{d \in AS_D} \sum_{a \in AS_A} \hat{p}_D(a|d) \mathbb{E}_{F_T(t|a,d)}(u_A(a,t)). \qquad (6.3)$$

So far, Daphne's agenda is straightforward to carry out, but the specification of the prior distribution F_A remains unclear yet. In many situations, Daphne can reasonably ad hoc presume something about Apollo's skill level, technical or monetary resources, inner and extrinsic motivations, and other factors [5, 18]. Assumptions on bounded resources are common in cryptography (see Chap. 11), but limited budget or knowledge may be reasonable to assume elsewhere too. Standards like CVSS explicitly account for required time to exploit, necessary background skills (up to insider knowledge) and many other aspects that may be relevant and useful to model the attacker's possibilities (i.e., the set AS_A) or capabilities.

However, the issue becomes much more involved if the players take actions independently, such as in Fig. 6.3c: now, for Daphne to reason about Apollo's behavior, she must consider that Apollo, by symmetry, will do the same to assess her behavior. If we adopt a common knowledge assumption again, then Daphne and Apollo may both have a probabilistic idea about the other's behavior (a *common prior* from which both, Daphne's and Apollo's items in the influence diagram are sampled). Under this common prior assumption, Harsanyi [11] proposed the use of PBNEs. Essentially, Daphne (likewise Apollo) assumes that Apollo (resp. Daphne) will behave like in one of n known ways or *types*, each of which has its individual utility function u_A. Each type occurs with likelihood p_i for $i = 1, 2, \ldots, n$, and this distribution is known to Daphne and Apollo, so that she can analogously rewrite (6.3) into (a minimax decision problem)

$$d^* = \operatorname*{argmax}_{d \in AS_D} \sum_{i=1}^{n} \hat{p}_i \cdot \mathbb{E}_{F_T(t|a^*(i),d)}(u_A(a^*(i), t)), \qquad (6.4)$$

where $a^*(i)$ is the best reply under each of the potential types of Apollo (which is computable, since we know the individual payoffs and other parameters). In carrying the Bayesian approach further, both could then update their beliefs based on what they observe from the other player. A *robust* alternative hereto [1] is Daphne assuming a worst-case type (disregarding any likelihood herein) and change (6.4) into

$$d^* = \operatorname*{argmax}_{d \in AS_D} \left[\min_i \mathbb{E}_{F_T(t|a^*(i),d)}(u_A(a^*(i), t)) \right],$$

assuming that she will always face the worst possible Apollo (among all known kinds $i = 1, 2, \ldots$).

Alas, this does not ultimately settle the issue of requiring a prior distribution, either on the influence diagram items, or on player types together with their respective descriptions (corresponding to what we have in the influence diagram). Without a common knowledge assumption, Daphne could start pondering about how Apollo would behave, and by symmetry, Apollo would do the same reasoning about Daphne. From Daphne's view, Apollo is rational and hence will adapt his behavior to what he thinks Daphne does. So, in addition to her own (known) behavior, she would need a hypothesis about what Apollo thinks she does. Likewise, Apollo would try to anticipate Daphne's behavior based on what he thinks she thinks that he does. This thinking-about-what-the-other-is-thinking-about-me reasoning can be continued ad infinitum, and would impute an unrealistic level of strategic thinking on both sides.

Level-k thinking systematizes the idea [33] of strategic sophistication in a clever way to offer a reasonable stopping point in the above regress: it puts players into classes of ascending strategic sophistication, starting with the simplest players being at level 0. Those act entirely random and without any strategic preference among their actions. Formally, they randomize their choices uniformly over their action spaces. A level 1 player assumes his opponent to be a level 0 player, thus acts best against an opponent who takes action uniformly at random. A level 2 player assumes a level 1 opponent, and so on. For Daphne and Apollo, they can easily utilize this idea to construct priors iteratively: Daphne starts with presuming Apollo to play at level 0. Hence, she uses a uniform prior distribution $p_D^{(0)}(a|d)$ over all parameters that are unknown to her. Based on this hypothesis, she uses Monte-Carlo sampling from the prior, empirical distributions therefrom, and (6.3) to compute the best replies to a level 0 Apollo, which puts herself into level 1. Then, turning into Apollo's role, Daphne computes Apollo's best response against a level 0 version of herself (Daphne), using a uniform prior for level 0, sampling, and optimization as in (6.3) (only from Apollo's perspective). The optimal solutions for both, Daphne and Apollo, then relate to level 1 players. In turn, Daphne can use Apollo's level 1 response as a prior, do the sampling, and compute her best reply to the level 1 Apollo. This puts her on level 2. Likewise, Apollo can use Daphne's level 1 behavior as a prior distribution and compute his best reply to a level 1 Daphne. Now, we have level 2 priors for both, Daphne and Apollo, and both may restart until the desired level of strategic sophistication is reached.

The benefit of this method lies in the fact that experimental evidence underpins level-k thinking as explanation for human behavior, so choosing the priors in this way enjoys empirical support [6–8, 17]. Moreover, a strategic sophistication above level $k = 3$ is unusual in practice, which is already a practical a suggestion on where to stop the recursion.

Example 6.2 (Level-k Defender in Cyber Deception) Imagine a cyber deception defense based on honeypots. A level-1 defender would assume a level-0 adversary, which comes as the assumption of the attacker *not* recognizing whether or not it has hit the honeypot or the real system. In that sense, the attacker is "unaware" of the defender's strategy, or equivalently, a level below the defender.

A level-2 defender, in turn, assumes a level-1 attacker. This adversary thus has the same knowledge as a level-1 defender from before, so it knows about the honeypot and will recognize it. As a strategy, the attacker may thus deliberately attack the honeypot as a decoy action, so as to trick the defender into thinking that it hit the honeypot only. In a sense, the attacker thus plays its own deception game with the defender. The defender thus knows that the attacker knows about the honeypots being in place. It can thus decide on its own whether or not to "believe" that the activity on the honeypot was the only sign of an attack, or that it possibly was a decoy action to hide the real adversarial activity.

A level-3 defender would assume a level-2 adversary, which in turn means that the attacker knows that the defender will be skeptic even if there is suspicious activity on the honeypot. The defender thus thinks about what the attacker thinks about what the defender may do, and so on.   ◇

Bayesian reasoning about the attacker is not the only possibility, yet from a stochastic viewpoint, it appears as the most natural approach to consider the maximal lot of information available, when modeling the adversary. An alternative that leads Daphne to a more minimax-like reasoning is her assuming Apollo's intentions to be exactly opposite to hers, in absence of better knowledge. In the influence diagram, this would correspond to substituting the unknown utility function u_A by the known $-u_D$, effectively creating a something like a zero-sum game model. However, we cannot substitute action spaces accordingly, and zero-day exploits are on example of unexpected attacks that have obviously been contained in Apollo's action space, but were unknown to Daphne. So, she could not prepare for it (despite any strategic sophistication).

6.4.2   Assessment of Action Spaces

Systematic approaches to the assessment of action spaces differ in the target entity: for the defender (Daphne), large catalogues are available in various standards on security risk management (e.g., [3, 15, 24] to name a few). Those are also available to Apollo, helping him to assess Daphne's action space. For Daphne, to reason about Apollo's action spaces, she can resort to known catalogues of vulnerabilities and ethical hacking resources [23, 36] and Topological Vulnerability Analysis (TVA) [16]. A popular tool here are *attack graphs*, in which nodes correspond to logical points inside a system (machines, places, or similar), which are related by arcs that show how a child node can be reached from a parent node during an attack. Such trees are typically annotated by pre- and post-conditions, telling which exploits, vulnerabilities and services are available to traverse from one node to the next. Enumerating all paths from initial entry nodes of the attacker down to the target asset gives the action space model for the attacker. An example instance of this process is given in Sect. 9.4 in the context the Cut-The-Rope game.

Zero-day exploits can be included in these attack graphs by resorting to their general characteristics, in lack of any possible estimate on impact or likelihood. We will look into heuristic risk assessments for zero-day attacks in Sect. 12.6.2.2. Specifically, the only safe assumption on a zero-day exploits regard network connections from the source to the victim, services offered by the victim and existing privileges on the source to exploit the service. These correspond to worst-case pre- and post-conditions (say, if the zero-day exploit is assumed to create alleviated privileges on the target machine), from which we can derive attack paths (scenarios) as elements of the action space estimates (say, by path enumeration, Bayesian network analysis [39], subjective expert surveys [30] or other means).

Viewed from a different angle, zero-day exploits can be thought of attacks yet outside the known action spaces. The risk of these is thus a matter of how likely the action space is *incomplete*. In some specialized situations [28], this incompleteness can, upon discovery, be easily fixed, since for some communication games (see Sect. 11.2), it suffices to counteract the zero-day exploit only in a *single* out of the many possibilities in AS_A, to completely restore security. Attack trees offer similar possibilities, since if the connectivity of the attack graph is low, it suffices to add protection to only a few nodes in the network, perhaps. The precise condition uses domination between strategies, and follows later as Theorem 12.1 in Sect. 12.6.2.1, where Example 12.6 is given to illustrate the effect of extending the action space.

References

1. Aghassi M, Bertsimas D (2006) Robust game theory. Math Program 107(1–2):231–273. https://doi.org/10.1007/s10107-005-0686-0
2. Bier VM, Cox LA (2007) Probabilistic risk analysis for engineered systems. In: Edwards W (ed) Advances in decision analysis. Cambridge University Press, Cambridge, pp 279–301
3. CC Consortium (2018) Common criteria for information technology. https://www.commoncriteriaportal.org
4. Chauvin B, Hermand D, Mullet E (2007) Risk perception and personality facets. Risk Anal 27(1):171–185. https://doi.org/10.1111/j.1539-6924.2006.00867.x
5. Clemen RT, Reilly T (2014) Making hard decisions with decision tools, 3rd rev. edn. South-Western/Cengage learning, Mason. Reilly, Terence (VerfasserIn)
6. Costa-Gomes MA, Crawford VP (2006) Cognition and behavior in two-person guessing games: an experimental study. Am Econ Rev 96(5):1737–1768. https://doi.org/10.1257/aer.96.5.1737
7. Crawford VP, Iriberri N (2007) Level-k auctions: can a nonequilibrium model of strategic thinking explain the winner's curse and overbidding in private-value auctions? Econometrica 75(6):1721–1770. https://doi.org/10.1111/j.1468-0262.2007.00810.x
8. Crawford V, Gneezy U, Rottenstreich Y (2008) The power of focal points is limited: even minute payoff asymmetry may yield large coordination failures. Am Econ Rev 98(4):1443–1458
9. Dohmen T, Falk A, Huffman D, Sunde U, Schupp J, Wagner GG (2011) Individual risk attitudes: measurement, determinants, and behavioral consequences. J Eur Econ Assoc 9(3):522–550. https://doi.org/10.1111/j.1542-4774.2011.01015.x
10. Fielder A, König S, Panaousis E, Schauer S, Rass S (2018) Risk assessment uncertainties in cybersecurity investments. Games 9(2):34. https://doi.org/10.3390/g9020034. http://www.mdpi.com/2073-4336/9/2/34/pdf

11. Harsanyi JC (1973) Games with randomly disturbed payoffs: a new rationale for mixed-strategy equilibrium points. Int J Game Theory 2(1):1–23
12. He W, Xia C, Zhang C, Ji Y, Ma X (2008) A network security risk assessment framework based on game theory. Futur Gener Commun Netw 2:249–253. https://doi.org/10.1109/FGCN.2008.166.
13. Heal G, Kunreuther H (2005) You can only die once: interdependent security in an uncertain world. In: Richardson HW, Moore JE, Gordon P (eds) The economic impacts of terrorist attacks. Edward Elgar, Cheltenham/Northampton. https://doi.org/10.4337/9781845428150.00008
14. HyRiM Consortium (2015) Hybrid risk management for utility providers. https://hyrim.net/. EUAFP7 Project No. 608090, project from 2015–2017
15. Informationstechnik, B.f.S.i.d. (2008) BSI-Standard 100–2: IT-Grundschutz methodology. https://www.bsi.bund.de/SharedDocs/Downloads/EN/BSI/Publications/BSIStandards/standard_100-2_e_pdf.pdf?__blob=publicationFile&v=1
16. Jajodia S, Noel S, O'Berry B (2005) Massive computing: topological analysis of network attack vulnerability. Springer, Berlin/New York
17. Kawagoe T, Takizawa H (2009) Equilibrium refinement vs. level-k analysis: an experimental study of cheap-talk games with private information. Games Econ Behav 66(1):238–255. https://doi.org/10.1016/j.geb.2008.04.008
18. Keeney RL, Raiffa H (1976) Decisions with multiple objectives: preferences and value tradeoffs. Wiley series in probability and mathematical statistics. Wiley, New York. Raiffa, Howard (VerfasserIn)
19. Linkov I, Palma-Oliveira JM (2017) An introduction to resilience for critical infrastructures. In: Linkov I, Palma-Oliveira JM (eds) Resilience and risk. Springer Netherlands, Dordrecht, pp 3–17. https://doi.org/10.1007/978-94-024-1123-2_1. http://link.springer.com/10.1007/978-94-024-1123-2_1
20. Linkov I, Palma-Oliveira JM (eds) (2017) Resilience and risk: methods and application in environment, cyber and social domains. NATO science for peace and security series. Series C, environmental security. Springer, Dordrecht
21. Münch I (2012) Wege zur Risikobewertung. In: Schartner P, Taeger J (eds) DACH security 2012. SysSec, pp 326–337
22. Nicholson N, Soane E, Fenton-O'Creevy M, Willman P (2006) Personality and domain–specific risk taking. J Risk Res 8(2):157–176. https://doi.org/10.1080/13669870320001238556
23. NIST (2018) National vulnerability database. https://nvd.nist.gov/
24. Organisation IS (2009) ISO/IEC 31000 – risk management – principles and guidelines. http://www.iso.org/iso/home/store/catalogue_tc/catalogue_detail.htm?csnumber=43170. Accessed 11 Apr 2016
25. Pearl J (2005) Influence diagrams—historical and personal perspectives. Decis Anal 2(4):232–234. https://doi.org/10.1287/deca.1050.0055
26. Rajbhandari L, Snekkenes EA (2011) Mapping between classical risk management and game theoretical approaches. In: Decker BD, Lapon J, Naessens V, Uhl A (eds) Communications and multimedia security: 12th IFIP TC 6/TC 11 international conference, CMS 2011, Ghent, 19–21 Oct 2011. Proceedings. Springer, Berlin/Heidelberg, pp 147–154
27. Rass S (2017) On game-theoretic risk management (part three) – modeling and applications. arXiv:1711.00708v1 [q-fin.EC]
28. Rass S, Schartner P (2011) Information-leakage in hybrid randomized protocols. In: Lopez J, Samarati P (eds) Proceedings of the international conference on security and cryptography (SECRYPT). SciTePress – Science and Technology Publications, pp 134–143
29. Rass S, König S, Schauer S (2016) Decisions with uncertain consequences – a total ordering on loss-distributions. PLoS One 11(12):e0168583. https://doi.org/10.1371/journal.pone.0168583
30. Rass S, Konig S, Schauer S (2017) Defending against advanced persistent threats using game-theory. PLoS One 12(1):e0168675. https://doi.org/10.1371/journal.pone.0168675
31. Rass S, Alshawish A, Abid MA, Schauer S, Zhu Q, de Meer H (2017) Physical intrusion games – optimizing surveillance by simulation and game theory. IEEE Access 5:8394–8407. https://doi.org/10.1109/ACCESS.2017.2693425

32. Rios Insua D, Rios J, Banks D (2009) Adversarial risk analysis. Risk Anal 104(486):841–854
33. Rothschild C, McLay L, Guikema S (2012) Adversarial risk analysis with incomplete information: a level-K approach. Risk Anal 32(7):1219–1231. http://doi.wiley.com/10.1111/j.1539-6924.2011.01701.x
34. Rubio VJ, Hernández JM, Márquez MO (2012) The assessment of risk preferences as an estimation of risk propensity. In: Assailly JP (ed) Psychology of risk, psychology research progress. Nova Science Publishers, Inc, New York, pp 53–81
35. Schauer S (2018) A risk management approach for highly interconnected networks. In: Rass S, Schauer S (eds) Game theory for security and risk management. Springer, Birkhäuser, pp 285–311
36. Shema M (2014) Anti-hacker tool kit, 4th edn. McGraw-Hill/Osborne, New York
37. Skotnes R (2015) Risk perception regarding the safety and security of ICT systems in electric power supply network companies. Saf Sci Monit 19(1):1–15
38. Sun Tzu, Giles L (2015) The art of war. OCLC: 1076737045
39. Sun X, Dai J, Liu P, Singhal A, Yen J (2016) Towards probabilistic identification of zero-day attack paths. In: 2016 IEEE conference on communications and network security (CNS). IEEE, Piscataway, pp 64–72. https://doi.org/10.1109/CNS.2016.7860471
40. Weber CS (2014) Determinants of risk tolerance. Int J Econ Financ Manag Sci 2(2):143. https://doi.org/10.11648/j.ijefm.20140202.15
41. Weber EU, Blais AR, Betz NE (2002) A domain-specific risk-attitude scale: measuring risk perceptions and risk behaviors. J Behav Decis Mak 15(4):263–290. https://doi.org/10.1002/bdm.414

Chapter 7
Insurance

> *Insurance: An ingenious modern game of chance in which the*
> *player is permitted to enjoy the comfortable conviction that he is*
> *beating the man who keeps the table.*
>
> A. Bierce

Abstract Cyber insurance provides users a valuable additional layer of protection to transfer cyber data risks to third-parties. An incentive-compatible cyber insurance policy can reduce the number of successful cyber-attacks by incentivizing the adoption of preventative measures in return for more coverage and the implementation of best practices by pricing premiums based on an insured level of self-protection. This chapter introduces a bi-level game-theoretic model that nests a zero-sum game in a moral-hazard type of principal-agent game to capture complex interactions between a user, an attacker, and the insurer. The game framework provides an integrative view of cyber insurance and enables a systematic design of incentive-compatible and attack-aware insurance policy. The chapter also introduces a new metric of disappointment rate that measures the difference between the actual damage and the expected damage.

7.1 Why Cyber-Insurance?

Cyber risks for infrastructures are of growing concerns, not only due to the advent of APTs (see Chap. 1). The cyber risk will not only create cyber incidents including identity theft, cyber extortion, and network disruption, but also can lead to the malfunction of the entire infrastructure and its key services to users and customers. It becomes a critical issue for operators to safeguard infrastructures from the intentional and unintentional actions that would inflict damage on the system. Conventional countermeasures include installing IDS, blacklisting malicious hosts, filtering/blocking traffic into the network. However, these methods cannot guarantee perfect security and can be evaded by sophisticated adversaries despite the advances

© Springer Nature Switzerland AG 2020
S. Rass et al., *Cyber-Security in Critical Infrastructures*, Advanced Sciences and Technologies for Security Applications,
https://doi.org/10.1007/978-3-030-46908-5_7

in technologies. Therefore, cyber risks are inevitable and it is essential to find other means to mitigate the risks and impact.

Cyber insurance is an important tool in risk management to transfer risks [52, 53]; cf. Chap. 6. Complement to the technological solutions to cybersecurity, cyber insurance can mitigate the loss of the targeted system and increase the resiliency of the victim by enabling quick financial and system recovery from cyber incidents. Such scheme is particularly helpful to small and medium size infrastructure systems that cannot afford a significant investment in cyber protection. At the time of writing this book, the market of cyber insurance is still in its infancy. U.S. penetration level of the insured is less than 15%. Promisingly, the market is growing fast at a 30% annual growth rate since 2011. The key challenge with cyber insurance lies in the difficulty to assess different types of cyber risks and impact of the cyber incidents that are launched by resourceful adversaries who are stealthy and purposeful. The design of cyber insurance also needs to take into account moral hazards and adverse selection problems. The insured tend to lack in incentives to improve their security measures to safeguard against attacks. As the nodes in the cyber space are increasingly connected, the unprotected cyber risks can propagate to other uninsured nodes. With asymmetric information of the insured, the insurer also has tendency to increase the premium rates for higher risks, making the cyber insurance less affordable to end users.

In this chapter, we aim to provide a baseline framework to understand the interactions among the players in the cyber insurance market and leverage it to design optimal cyber insurance for the infrastructure services. One key application of the framework is to provide assurance to the infrastructure managers and users and transfer their risks when the attack on power grid fails to provide electric power to a food processing plant, when cloud servers break down and fail to provide airline customer check-in information, and when the trains collide due to the communication systems fail. It is clear that cyber insurance plays a key role in mitigating the cyber risks that interconnect the communications and information systems of an infrastructure with their physical impact on the infrastructure or linked infrastructures. The interdependencies among (critical) infrastructures, ICT, and their operators and users can propagate the cyber risks and exacerbate the damages on the critical infrastructures. To this end, the understanding of the cyber insurance of interconnected players is the key to the holistic understanding of the risk management of interdependent infrastructures.

This chapter will first present a principal-agent game-theoretic model to capture the interactions between one insurer and one customer. The insurer is deemed as the principal who does not have incomplete information about customer's security policies. The customer, which refers to the infrastructure operator or the customer, implements his local protection and pays a premium to the insurer. The insurer designs an incentive compatible insurance mechanism that includes the premium and the coverage policy, while the customer determines whether to participate in the insurance and his effort to defend against attacks. The chapter will also focus on an attack-aware cyber insurance model by introducing the adversarial behaviors into the framework. The behavior of an attacker determines the type of cyber threats, e.g. denial of service (DoS) attacks, data breaches, phishing and spoofing. The distinction of threat types plays a role in determining the type of losses and the coverage policies. The data breaches can lead to not only financial losses but also damage of the reputations. The coverage may only cover certain agreed percentage of the financial losses.

7.2 Background

The challenges of cyber security are not only technical issues but also economic and policy issues [3]. Recently, the use of cyber insurance to enhance the level of security in cyber-physical systems has been studied [22, 23]. While these works deal with externality effects of cyber security in networks, few of them take into account in the model the cyber attack from a malicious adversary to distinguish from classical insurance models. In [28], the authors have considered direct and indirect losses, respectively due to cyber attacks and indirect infections from other nodes in the network. However, the cyber attacks are taken as random inputs rather than a strategic adversary. The moral hazard model in economics literature [15, 16] deal with hidden actions from an agent, and aims to address the question: How does a principal design the agent's wage contract to maximize his effort? This framework is related to insurance markets and has been used to model cyber insurance [6] as a solution for mitigating losses from cyber attacks. In addition, in [1], the authors have studied a security investment problem in a network with externality effect. Each node determines his security investment level and competes with a strategic attacker. Their model does not focus on the insurance policies and hidden-action framework. In this work, we enrich the moral-hazard type of economic frameworks by incorporating attack models, and provide a holistic viewpoint towards cyber insurance and a systematic approach to design insurance policies. The network effect on security decision process has been studied in [27]. The authors have considered a variation of the linear influence networks model in which each node represents a network company and directed links model the positive or negative influence between neighbor nodes.

Game-theoretic models are natural frameworks to capture the adversarial and defensive interactions between players [11, 17, 18, 24, 25, 37, 50, 54, 56, 57]. Game theory can provide a quantitative measure of the quality of protection with the concept of Nash equilibrium where both defender and an attacker seek optimal strategies, and no one has an incentive to deviate unilaterally from their equilibrium strategies despite their conflict for security objectives. The equilibrium concept also provides a quantitative prediction of the security outcomes of the scenario the game model captures. In most cases, this will be an expectation of loss, while in other (generalized) cases, it can also be the likelihood for extreme events (such as for zero-day exploits, and games over stochastic tail orders). Later, in Sect. 7.4, we will also discuss how to handle excesses over what is expected.

There are various types of game models that can capture different class of adversaries. For example, games of incomplete information are suitable for understanding cyber deception [17, 29, 32, 33, 50, 57]; dynamic games are useful for modeling cyber-physical system security [8, 18, 25, 43–47]; and zero-sum and Stackelberg games for security risk management (see Chap. 6, [30, 31, 41, 48, 49, 51]. In this work, we build a zero-sum game between an attacker and a defender to quantify the cybersecurity risks associated with adversarial behaviors. This game model is nested in the principle-agent game models to establish an integrated framework that captures the defender, the attacker, and the insurer.

7.3   Three-Person Game Framework for Cyber Insurance

In this section, we introduce a principal-agent model for cyber insurance that involve users and insurers. Users here can refer to an infrastructure operator that manages cyber networks that face threats from an attacker, making users vulnerable to data breaches, task failures, and severe financial losses. The objective of the users is to find an efficient way to mitigate the loss due to the cyber attacks. To this end, there are two main approaches. One is to deploy local protections, such as firewalls and IDSs [4, 35], frequent change of passwords [36], timely software patching and proactive moving target defenses [19]; see also Sect. 12.5. These defense mechanisms can reduce the success rate of the attacks, but cannot guarantee perfect network security for users. There are still chances for the users to be hacked by the attackers. The other approach is to adopt cyber-insurance. The users pay a premium fee so that the loss due to cyber attacks can be compensated by the insurer. This mechanism provides an additional layer of mitigation to reduce the loss further that the technical solutions of the first approach cannot prevent. To capture the two options in our framework, we allow users to decide their protection levels as well as their rational choice of participation in the insurance program.

Attackers are the adversaries who launch cyber-attacks, such as node capture attacks [39] and denial of services (DoS) attacks [20], to acquire private data from users or cause disruptions of the network services. Hence, the objective of the attacker is to find an efficient attack strategy to inflict as much damage to the users as possible. We use attack levels to represent different attack strategies to capture various types of attacks of different levels of severity. A higher attack level is more costly to launch, but it will create more severe damage. Since the loss of the users not only depends on the attack strategies but also insurance policies. The optimal strategy of the attacker will also be influenced by the coverage levels of an insurance policy.

An insurer is a person or company that underwrites an insurance risk by providing users an incentive compatible cyber-insurance policy that includes a premium and the level of coverage. The premium is a subscription fee that is paid by the users to participate in the insurance program while the coverage level is the proportion of loss that will be compensated by the insurer as a consequence of successful cyber attacks. The insurers have two objectives. One is to make a profit from providing the insurance, and the other one is to reduce the average losses of the users, which is also directly related to the cost of the insurer. An insurer's problem is to determine the subscription fee and the coverage levels of the insurance. Note that the average losses depend on both users' local protection levels and attackers' attack levels. Moreover, the rational users will only enroll in the insurance when the average reduction in the cost is higher than or equal to the premium he paid to the insurer.

The objectives of users, attackers, and insurers, and the effects of their actions are all intertwined. We use a three-player game to capture the complex interactions among the three parties. The conflicting objectives of a customer and an attacker can be captured by a local game at each node in which the customer determines a defense strategy while the adversary chooses an attack strategy. The outcome of the local interactions at each node determines its cyber risk and the cyber insurance is used as an additional method to further reduce the loss due to the cyber risk. The insurers are the leaders or principals in the framework who design insurance policies for the users while the users can be viewed as followers or agents who determine their defense strategies under a given insurance policy.

7.3.1 Attack-Aware Cyber Insurance

We first formulate the game between the customer and the attacker, then we describe the insurer's problem under the equilibrium of the customer and the attacker's game. An illustration of the cyber-insurance model is shown in Fig. 7.1: in this game, the action pair (p_c, p_a) chosen by the customer and the attacker results in a risk level not directly observable by the insurer. The insurer designs an insurance policy that includes a premium subscription fee and the coverage level to cover part of the loss due to the cyber attack.

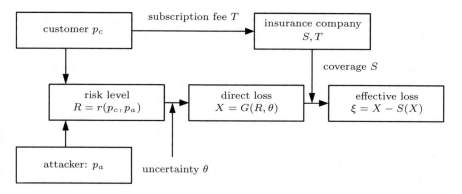

Fig. 7.1 Illustration of three-player cyber insurance game

Let $p_c \in [0, 1]$ and $p_a \in [0, 1]$ denote the local protection level of the customer and the attack level of the attacker. On one hand, a large p_c indicates a cautious customer while a small p_c indicates that the customer is reckless. A reckless customer may click on suspicious links of received spam emails, fail to patch the computer system frequently, and leave cyber footprints for an adversary to acquire system information. On the other hand, a large p_a indicates a powerful attacker, and a small p_a indicates a powerless attacker. The abstraction of using p_c and p_a captures the effectiveness of a wide range of heterogeneous defense and attack strategies without a fine-grained modeling of individual mechanisms. This will allow us to focus on the consequence of security issues and the choice of a mechanism that induces the result.

The action pair of the customer and the attacker (p_c, p_a) determines the risk level of the customer $R \in \mathbb{R}_{\geq 0}$. A larger p_c and a smaller p_a indicate a higher risk level of the customer. We use the following risk function r to denote the connections between the customer's and the attacker's actions and the risk level of the customer.

Definition 7.1 Function $r(p_c, p_a) : [0, 1]^2 \to \mathbb{R}_{\geq 0}$ gives the risk level R of the customer with respect to the customer's local protection level p_c and the attack's attack level p_a. Moreover, it is assumed to be continuous on $(0, 1]^2$, convex and monotonically decreasing on $p_c \in [0, 1]$, and concave and monotonically increasing in $p_a \in [0, 1]$.

Note that the monotonicity in $p_c \in [0, 1]$ indicates that a larger local protection level of customer leads to a smaller risk level of customer while the monotonicity in $p_a \in [0, 1]$ indicates that a larger attack level of attacker leads to a larger risk level of customer. Since r is convex on p_c, the risk decreases smaller when the customer adopts larger local protection level. Since r is concave on p_a, the risk increases faster when the attacker conducts a higher attack level. Without loss of generality, we use the following risk function,

$$r(p_c, p_a) = \log \left(\frac{p_a}{p_c} + 1 \right). \tag{7.1}$$

Similar types of functions have also been widely used in jamming attacks in wireless networks [2, 24] and rate control problems [21, 55]. Under the risk level of R, the economic loss of the customer can be represented as a random variable X measured in monetary units (e.g., US Dollar, Euro, etc.), which can be expressed as $X = G(R, \theta)$, where θ is a random variable with probability density function g that captures the uncertainties in the measurement or system parameters. For example, a data breach due to the compromise of a server can be a consequence of low security level at the customer end. The magnitude of the loss depends on the content and the significance of the data, and the extent of the breach. The variations in these parameters are captured by the random variable θ. Since the risks of being attacked cannot be perfectly eliminated, the customer can transfer the remaining risks to the third party, the insurer, by paying a premium or subscription fee T for a coverage of $S(X)$ when he faces a loss of X, where $S : \mathbb{R}_{\geq 0} \to \mathbb{R}_{\geq 0}$ is the payment function that reduces the loss of the customer if he is insured. Thus, the effective loss ξ to the customer becomes $\xi = X - S(X)$.

Given the attacker's action p_a and the insurer's coverage function S, the customer aims to minimize the average effective loss by finding the optimal local protection level p_c^*. Such objective can be captured by the following optimization problem

$$\min_{p_c \in [0,1]} \mathbb{E}_\theta[H(\xi)] = \mathbb{E}_\theta[H(X - S(X))], \tag{7.2}$$

where $H : \mathbb{R}_{\geq 0} \to \mathbb{R}_{\geq 0}$ is the loss function of the customer, which is increasing on ξ. The subscription fee T is not included in this optimization problem, as the fee is a constant decided by the insurer.

The loss function $H(\xi)$ indicates the customer's risk attitude or propensity. A convex $H(\xi)$ indicates that the customer is risk-averse, i.e., the customer cares more about the risk, while a concave $H(\xi)$ indicates that the customer is risk-taking, i.e., he cares more about the cost, rather than the risk. A linear $H(\xi)$ in ξ indicates that the customer is risk-neutral. In this paper, we consider a risk-averse customer, and use a typical risk-averse loss function that $H(\xi) = e^{\gamma \xi}$ with $\gamma > 0$, where γ indicates how much the customer cares about the loss.

Note that the cost function in (7.2) can be expressed explicitly as a function of X. Thus, problem (7.2) can be rewritten by taking expectations with respect to the sufficient statistics of X. Let f be the probability density function of X. Clearly, f is a transformation from the density function g (associated with the random variable θ) under the mapping G. In addition, g also depends on the action pair (p_c, p_a) through the risk variable R. Therefore, we can write $f(x|p_c, p_a)$ to capture the parameterization of the density function. Without loss of generality, we assume that X follows an exponential distribution, i.e., $X \sim \mathscr{E}x(\frac{1}{R})$, where $R := r(p_c, p_a)$ is the risk level of the customer. The exponential distribution has been widely used in risk and reliability analysis[5, 9, 12, 26]. Thus the density function can be written as

$$f(x|p_c, p_a) = \frac{1}{R} e^{-\frac{1}{R}x} = \frac{1}{r(p_c, p_a)} e^{-\frac{1}{r(p_c, p_a)}x}$$

$$= \frac{1}{\log(\frac{p_a}{p_c} + 1)} e^{-\frac{1}{\log(\frac{p_a}{p_c} + 1)}x}, \forall x \in \mathbb{R}_{\geq 0}.$$

The average amount of loss given actions p_c and p_a is $\mathbb{E}(X) = R = r(p_c, p_a) = \log(\frac{p_a}{p_c} + 1)$; cf. (7.1). For small p_c and large p_a, the risk level of the customer R tends to be large, which leads to a large average loss of the customer. We further assume that the insurance policy $S(X)$ is linear in X, i.e., $S(X) = sX$, where $s \in [0, 1]$ indicates the coverage level of the insurance. Hence, the effective loss is given by $\xi = (1-s)X$. The average effective loss given the insurance coverage level s and the action pair (p_c, p_a) is $\mathbb{E}(\xi) = \mathbb{E}((1-s)X) = (1-s)\mathbb{E}(X) = (1-s)\log(\frac{p_a}{p_c} + 1)$. When s is large, the effective loss is small. As a result, we arrive at

$$
\begin{aligned}
\mathbb{E}[H(\xi)] &:= \int_{x_i \in \mathbb{R}_{>0}} H(x - S(x)) f(x|p_c, p_a) dx \\
&= \frac{1}{R} \int_0^\infty e^{[\gamma(1-s) - \frac{1}{R}]x} dx \\
&= \frac{1}{1 - \gamma(1-s)R} = \frac{1}{1 - \gamma(1-s)\log(\frac{p_a}{p_c}+1)}.
\end{aligned}
\qquad (7.3)
$$

The loss is finite if $\gamma(1 - s) - \frac{1}{R} < 0$, i.e., when

$$
1 - \gamma(1 - s) \log\left(\frac{p_a}{p_c} + 1\right) > 0. \qquad (7.4)
$$

Otherwise, the loss expectation will eventually become infinite, i.e., $\mathbb{E}[H(\xi)] \to \infty$. In this regime, no insurance scheme can be found to mitigate the loss. Condition (7.4) gives a feasible set of parameters under which cyber insurance is effective and provides a fundamental limit on the level of mitigation. Note that minimizing (7.3) is equivalent as minimizing $\gamma(1 - s) \log(\frac{p_a}{p_c} + 1)$ under the feasibility condition (7.4). The customer's problem can be rewritten as follows:

$$
\begin{aligned}
\min_{p_c \in [0,1]} \quad & J_u(p_c, p_a, s) := \gamma(1-s)R = \gamma(1-s)\log(\tfrac{p_a}{p_c}+1) \\
\text{s.t.} \quad & 1 - \gamma(1-s)\log(\tfrac{p_a}{p_c}+1) > 0.
\end{aligned}
\qquad (7.5)
$$

Problem (7.5) captures the customer's objective to minimize average effective loss given the attack level p_a and the insurance coverage level s. On the other hand, the attacker aims to find the optimal attack level p_a^* that maximizes the average loss of the customer given its local protection level and insurer's coverage level s. Such conflicting interests of the customer and the attacker constitutes a zero-sum game, which takes the following minimax or max-min form,

$$
\begin{array}{ccc}
\min_{p_c \in [0,1]} \max_{p_a \in [0,1]} K(p_c, p_a, s) & \text{or} & \max_{p_a \in [0,1]} \min_{p_c \in [0,1]} K(p_c, p_a, s) \\
\text{s.t. } p_c, p_a, \text{ and } \gamma \text{ satisfy (7.4)} & & \text{s.t. } p_c, p_a, \text{ and } \gamma \text{ satisfy (7.4)}
\end{array}
\qquad (7.6)
$$

where

$$
\begin{aligned}
K(p_c, p_a, s) &:= \gamma(1-s)R + c_u p_c - c_a p_a \\
&= \gamma(1-s)\log(\frac{p_a}{p_c} + 1) + c_u p_c - c_a p_a,
\end{aligned}
\qquad (7.7)
$$

The first term of the objective function K captures the average effective loss given insurance coverage level s, the local protection level p_c and the attack level p_a. The second and third terms indicate the cost of the customer and the attacker, respectively. $c_u \in \mathbb{R}_{>0}$ is the cost parameter of the customer. A larger c_u indicates that local protection is costly. $c_a \in \mathbb{R}_{>0}$ denotes the cost parameter of the attacker to conduct an attack level of p_a. A larger c_u indicates that a cyber-attack is costly. Note that c_u and c_a can be interpreted as the market price of local protections and cyber-attacks, and they are known by the insurer. The constraint indicates the feasible set of the customer. Note that if s, p_c, and p_a are not feasible, K is taken to be an infinite cost. Minimizing $K(p_c, p_a, s)$ captures the customer's objective to minimize the average effective loss with the most cost-effective local protection level. Maximizing $K(p_c, p_a, s)$ captures the attacker's objective to maximize the average effective loss of the customer with least attack level. Note that the minimax form of (7.6) can be interpreted as a worst-case solution for a customer who uses a security strategy by anticipating the worst-case attack scenarios.

Furthermore, problem (7.6) yields an equilibrium with saddle-point to the insurance coverage level s comes as an instance of Definition 3.3 for the two-player special case:

Definition 7.2 (Insurance Game) Let $AS_u(s)$, $AS_a(s)$ and $AS_{u,a}(s)$ be the action sets for the customer and the attacker given an insurance coverage level s. Then the strategy pair (p_c^*, p_a^*) is an equilibrium of the zero-sum game defined by the triple $\Gamma_z := (\{User, Attacker\}, \{AS_u(s), AS_a(s), AS_{u,a}(s)\}, K)$, if for all $p_c \in AS_u(s)$, $p_a \in AS_a(s)$, $(p_c, p_a) \in AS_{u,a}(s)$,

$$K(p_c^*, p_a, s) \leq K(p_c^*, p_a^*, s) \leq K(p_c, p_a^*, s), \tag{7.8}$$

where K and $AS_{u,a}(s)$ is the objective function and feasible set defined in (7.7) and by (7.4).

Note that under a given insurance coverage level s, (p_c^*, p_a^*) must satisfy the feasible constraint (7.4). Thus, we aim to look for a constrained saddle-point of the zero-sum game with coupled constraints on the strategies of the players.

Proposition 7.1 *Given an insurance coverage level s that satisfies*

$$1 - \gamma(1 - s)\log\left(\frac{c_u}{c_a} + 1\right) > 0, \tag{7.9}$$

there exists a unique equilibrium of the zero-sum game Γ_z from Definition 7.2, given by

$$p_c^* = \frac{\gamma(1-s)}{c_u + c_a}, \quad p_a^* = \frac{c_u \gamma(1-s)}{c_a(c_u + c_a)}. \tag{7.10}$$

Proposition 7.1 shows that the saddle-point of the zero-sum game between the customer and the attacker is related to the insurer's policy s. Note that when s is large, both the p_c^* and p_a^* is small, indicating that both the customer and the attacker will take weak actions. Moreover, we have the following observations regarding the saddle-point, i.e., security strategy.

Remark 7.1 (Peltzman Effect) When the insure provides higher coverage level s, the saddle-point risk of the customer p_c^* tend to be smaller, i.e., the customer takes a weaker local protection. Such risky behavior of the customer in response to insurance is usually referred as Peltzman effect [34].

Corollary 7.1 (Invariability of the saddle-point ratio) *The saddle-point satisfies* $\frac{p_a^*}{p_c^*} = \frac{c_u}{c_a}$, *i.e., the ratio of the actions of the customer and the attacker is only related to* c_u *and* c_a, *and it is independent of the insurer's policy s. In particular, when* $c_u = c_a$, $\frac{p_a^*}{p_c^*} = 1$, *i.e., the saddle-point strategy becomes symmetric, as* $p_c^* = p_a^* = \frac{\gamma(1-s)}{c_u+c_a} = \frac{\gamma(1-s)}{2c_u} = \frac{\gamma(1-s)}{2c_a}$.

Remark 7.2 (Constant Cost Determined saddle-point Risk) The customer has a constant saddle-point risk level $R^* = r(p_c^*, p_a^*) = \log\left(\frac{p_a^*}{p_c^*} + 1\right) = \log\left(\frac{c_u}{c_a} + 1\right)$ at the equilibrium, which is determined by the costs of adopting protections and launching attacks. The ratio is independent of coverage level s.

Corollary 7.2 *At the saddle-point, the average direct loss of the customer is* $\mathbb{E}(X) = R^* = \log\left(\frac{c_u}{c_a} + 1\right)$, *the average effective loss of the customer is* $\mathbb{E}(\xi) = \mathbb{E}((1-s)X) = (1-s)\mathbb{E}(X) = (1-s)R^* = (1-s)\log\left(\frac{c_u}{c_a} + 1\right)$, *the average payment of the insurer to the customer is* $\mathbb{E}(sX) = s\mathbb{E}(X) = sR^* = s\log\left(\frac{c_u}{c_a} + 1\right)$.

Corollary 7.1 indicates the constant saddle-point strategy ratio of the customer and the attacker, which is determined only by the cost parameters c_u and c_a, i.e., the market prices or costs for applying certain levels of protections and attacks, respectively. As a result, the saddle-point risk level of the customer is constant, and only determined by the market as shown in Remark 7.2. Thus, the average direct loss is constant as shown in Corollary 7.2. However, when the insurance coverage level s does not satisfy (7.9), the insurability of the customer is not guaranteed, which is shown in the following proposition.

Proposition 7.2 (Fundamental Limits on Insurability) *Given an insurance coverage level s that* $1 - \gamma(1-s)\log\left(\frac{c_u}{c_a} + 1\right) \leq 0$, (p_c^*, p_a^*) *does not satisfy the feasible inequality (7.4), thus, the average direct of the customer is* $\mathbb{E}(X) \to \infty$, *and the zero-sum game* Γ_z *from Definition 7.2 does not admit a finite equilibrium loss. Thus, the customer is not insurable, as the insurance policy cannot mitigate his loss. The insurer will not also provide insurance to the customer who is not insurable.*

From (7.4) and (7.9), we also get:

Proposition 7.3 *Under an insurable scenario, the cost parameter of the customer must satisfy* $c_u < c_a(1 - e^{-\gamma(1-s)})$, *and the local protection level of the customer must satisfy* $p_c > \frac{\gamma(1-s)}{c_a}e^{\gamma(1-s)}$.

It is important to note that the customer must pay a subscription fee $T \in \mathbb{R}_{\geq 0}$ to be insurer. The incentive for the customer to buy insurance exists when the average loss at equilibrium under the insurance is lower than the loss incurred without insurance.

If the amount of the payment from the insurer is low, then the customer tends not to be insured. In addition, if the payment is low, then the risk for the insurer will be high and the customer may behave recklessly in the cyber-space.

7.3.2   Insurer's Problem

The insurer announces the insurance policy $\{s, T\}$, where s indicates the coverage level, T indicates the subscription, and then the customer's and the attacker's conflicting interests formulates a zero-sum game, which yields a unique solution as shown in Proposition 7.1, with the corresponding equilibrium loss as shown in Corollary 7.2. Note that T is the gross profit of the insurer as he charges it from the customer first, but when the customer faces a loss X, the insurer must pay sX to the customer. The operating profit of the insurer can be captured as $T - sX$. The insurer cannot directly observe the actions of the customer and the attacker. However, with the knowledge of the market, i.e., the cost parameters of the customer c_u and the attacker c_a, the insurer aims to minimize the average effective loss of the customer while maximizing his operating profit.

Recall Corollary 7.2, the average effective loss of the customer at saddle-point is $\mathbb{E}(\xi) = (1 - s)\mathbb{E}(X) = (1 - s)R^* = (1 - s)\log\left(\frac{c_u}{c_a} + 1\right)$, which is monotonically decreasing on s. When the customer is under full coverage, the average loss with the payment T is $(1 - s)R^* + T\big|_{s=1} = T$. When the customer does not subscribe to an insurance, the average direct loss is R^*. Thus, the customer has no incentive to insure if the cost under fully coverage is higher than that under no insurance, i.e., $T > R^*$. Moreover, for $T \leq R^*$, the customer will choose to insure if the average loss under the given coverage level s is lower than under no insurance, i.e., $(1 - s)R^* + T \leq R^*$. Therefore, we arrive at the following conditions, where the annotation "-c" indicates them as relating to the customer:

Condition 1 (Individual Rationality (IR-c)) *The subscription fee must satisfy* $T \leq T_{max} := R^* = \log\left(\frac{c_u}{c_a} + 1\right)$, *so that the customer prefer to subscribe the insurance.*

Condition 2 (Incentive Compatibility (IC-c)) *For the subscription fee* $T \leq T_{max}$, *the customer will subscribe the insurance if the coverage level s satisfies* $s \geq s_0 = \frac{T}{R^*} = \frac{T}{\log\left(\frac{c_u}{c_a} + 1\right)}$.

The customer will enroll only when (IR-c) and (IC-c) constraints are satisfied. Note that when c_u is large and c_a is small, i.e., the saddle-point risk level R^* is high, T_{max} is large and $s_0(T)$ is small, i.e., when the cost of the customer to put local protections is large, and the cost of the attacker to conduct cyber-attack is small, the price of the subscription fee is large, but the minimum coverage is low. Note that s_0 is monotonically increasing on T, moreover, when $T = 0$, $s = 0$, i.e., the customer will accept any coverage level when there is no charge for the insurance premium.

When $T = T_{max}$, $s = 1$, i.e., the customer only accept a full coverage when the subscription fee is the maximum.

The insurer charges a subscription fee T from the customer, i.e., the insurer has a gross profit of T. However, the insurer also pays the customer an average amount of $sR^* = s \log \left(\frac{c_u}{c_a} + 1 \right)$ from Corollary 7.2. Thus, the average operating profit of the insurer is $T - sR^*$, which must be larger than or equal to 0 so that the insurer will provide the insurance. Thus, we have the following conditions for the insurer, again marked by the annotation "-i" hereafter:

Condition 3 (Individual Rationality (IR-i)) *The insurer will provide the insurance if* $T - sR^* = T - s \log \left(\frac{c_u}{c_a} + 1 \right) \geq 0$.

Recall Proposition 7.2, the insurer will provide the insurance when the customer is insurable, i.e., inequality (7.9) must be satisfied. Thus, we reach the following proposition that indicates the feasible coverage level.

Condition 4 (Feasibility (F-i)) *The coverage level s is feasible, i.e., the customer is insurable, when* $s > 1 - \frac{1}{\gamma \log\left(\frac{c_u}{c_a} + 1 \right)}$.

Conditions 3 and 4 indicate the individual rationality constraint (IR-i) and the feasibility constraint (F-i) of the insurer, respectively. With the (IR-c) and (IC-c) constraints for the customer and the (IR-i) and (F-i) constraints for the insurer, the insurer's objective to minimize the average effective loss of the customer and maximize the operating profit can be captured using the following optimization problem:

$$\min_{\{0 \leq s \leq 1, T \geq 0\}} J_i(s, T) := \gamma(1 - s) \log \left(\frac{c_u}{c_a} + 1 \right) + c_s (s \log(\frac{c_u}{c_a} + 1) - T)$$
$$\text{s.t. (IR-}c\text{), (IC-}c\text{), (IR-}i\text{), (F-}i\text{).}$$

(7.11)

Note that the first term of the objective function is the average effective loss of the customer under the coverage s, as the insurer also aims to reduce the loss of the customer from the attacker. Minimizing the second term of the objective function captures the insurer's objective of making profit. Note that parameter c_s indicates the trade-off of a safer customer and a larger profit of the insurer.

Furthermore, the solution of problem (7.11) and the corresponding equilibrium from Definition 7.2 yields an equilibrium for the bi-level game that can be defined as

Definition 7.3 Let AS_i be the action set for the insurer, $AS_u(s)$ and $AS_a(s)$ be the action sets for the customer and the attacker as dependent on the insurance coverage level s. The strategy tuple $(p_c^*, p_a^*, \{s^*, T^*\})$ is called an equilibrium of the bi-level game defined by the triple $\Gamma :=$ $(\{User, Attacker, Insurer\}, \{AS_u(s), AS_a(s), AS_i\}, \{K, -K, J_i\})$, if $\{s^*, T^*\}$ solves problem (7.11) with the equilibrium objective function J_i^*, and the strategy pair (p_c^*, p_a^*) is the saddle-point value of the zero-sum game defined in Definition 7.2 with the objective function K^* under the insurance policy $\{s^*, T^*\}$.

Note that the insurer's Problem (7.11) is a linear programming problem as the objective function and all the constraints are linear in s and T. Instead of solving this problem, we first observe that (IR-i) and (IC-c) together indicate that the insurance policy s and T must satisfy

$$T = sR^* = s \log \left(\frac{c_u}{c_a} + 1 \right). \tag{7.12}$$

Corollary 7.3 *Equality (7.12) indicates the following observations:*

(i) *Zero Operating Profit Principle: The insurer's operating profit is always 0, as* $T - sR^* = 0.$

(ii) *Linear Insurance Policy Principle: The insure can only provide the insurance policy s and T that satisfies (7.12), so that the customer subscribes to the insurance and the insurer provides the insurance.*

With (7.12), the linear insurance policy indicates that the ratio of the subscription and the coverage level only depends on the saddle-point risk R^*, which is determined by the costs seen in Remark 7.2. It provides a guidance for designing the insurance policy.

With (7.12), the optimal insurance for the insurer can be summarized using the following proposition:

Proposition 7.4 *The optimal insurance policy for the insurer is*

$$s^* = 1, \quad T^* = T_{\max} = \log \left(\frac{c_u}{c_a} + 1 \right). \tag{7.13}$$

Proposition 7.4 shows that a full coverage level and a maximum subscription fee are the optimal insurance policy of the insurer. Together with Proposition 7.1, we have the following proposition of the equilibrium of the bi-level game.

Proposition 7.5 *The bi-level game admits a unique equilibrium* $(p_c^*, p_a^*, \{s^*, T^*\}) = (0, 0, \{1, \log \left(\frac{c_u}{c_a} + 1 \right)\})$. *At the equilibrium, the insurer provides a full coverage for the customer and charges a maximum subscription fee from the customer. The customer and attacker have no incentives to take actions at the equilibrium as the cost would be too high. The equilibrium also demonstrates that cyber insurance will effectively mitigate the loss.*

The analysis of the bi-level structure of the game informs the optimal insurance policies to transfer risks from one node to the insurer. The framework can also be extended to a scheme over interdependent infrastructures. The cyber risks at one node can propagate to other nodes when there are cyber, physical, human, and social interdependencies. The insurance of one node can play a important role in well-being of the entire system. We can anticipate that the insurance problem should take into account network effects. This research can also be further extended to investigate the impact of dynamic evolutions of the risks on the mechanism of insurance and the protection behaviors of the agents.

7.4 Disappointment Rates

Extending the framework to the perspective of the customer, one question is to set the amount of insurance coverage level, or insured amount. Naturally, a person would buy insurance for the *expected loss*, and following a systematic method, one could use optimization and game theory to take actions towards optimizing this expectation. In principle, it thus seems simple and obvious to:

1. Use game theory to optimize actions for loss minimization,
2. and buy insurance for the expected loss occurring in the worst case (e.g., after playing a security strategy if we know nothing, or equilibrium, if we know something about the opponent's incentives).

This simple approach, however, may fail in practice.

A subjective problem of maximizing expectations is the potential subjective misunderstanding of the statistical concept of expectation that may incorrectly be taken as an outcome that "should occur most of the time" (it is "expected"). It is obvious that such thinking is flawed, as even the expected number of points on a normal dice is 3.5, but the dice can only come up with integer points $1, 2, \ldots, 6$. In playing a game with real-valued rewards (for simplicity since we can now take $\trianglelefteq$ as the familiar $\leq$ ordering on $\mathbb{R}$), for example, at an equilibrium strategy $\mathbf{x}^*$ that maximizes the payoffs expectable from it, the saddle-point value is nonetheless an *average*. This means that in many repetitions of the game, we will precisely get a reward below and above the average; only the long run average of these will converge to the expected value. Optimizing an expectation, however, does not automatically also optimize the fluctuation around it. Consider Fig. 7.2 as an illustration: there, two actions are shown with different expected outcomes marked by the vertical lines. Following a loss minimizing paradigm, one would clearly vote for action a_1 here, because action a_2 on average causes larger losses. However, the losses are in both cases *random*, and differently distributed, and in fact, taking action a_1 leaves a much larger chance of suffering very high damages, as opposed to action a_2, which on average incurs more loss, but will surely not admit as much damage as is possible by taking action a_1. Knowing this, a decision maker could reasonably argue for action a_2, which – seemingly – is a violation of expected utility maximization, but given the prospects and not only the average losses, deciding for a_2 is still not a case of bounded rationality.

In the concrete incident of the reward being less than the game's value, the decision maker or player in the game would be disappointed since it get less from the game than we expected. Conversely, we may be positively surprised to get more in

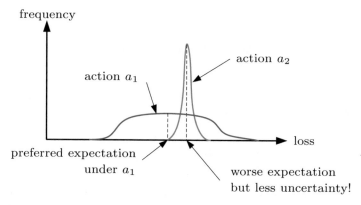

Fig. 7.2 Decision making disregarding uncertainty

the opposite case, but the subjective valuation of both does not need to be identical. Though mathematics considers only the long run average (by the law of large numbers) as $v = \lim_{n \to \infty} \frac{1}{n} \sum_{i=1}^{n} r_i$, so it does not matter if $r_i > v$ or $r_i < v$. But empirical evidence suggests that people do take either incident differently severe, as prospect theory describes with help of probability weighting functions (Figs. 5.1 and 5.2).

If an individual is not only striving for maximal average rewards but also seeks to avoid disappointment, then a further quantity that we seek to optimize is $\Pr(R > \mathbb{E}(R))$, when R is the random payoff obtained from the decision. Thinking in terms of effective losses, one could also set the limit as proportionally larger than $\mathbb{E}(R)$ or larger than the expectation by a fixed amount that reflects the maximum bearable effective loss. This would determine the coverage level.

None of this is in fact *not* covered by standard game theory, which maximizes $\mathbb{E}(R)$ only. The *theory of disappointment* [7, 14] centers decision making on the disappointment ratio $d = \Pr(R > \mathbb{E}(R))$ for a minimizing, equivalently $d = \Pr(R < \mathbb{E}(R))$ for a maximizing decision maker. It is easy to see that a conventional equilibrium calculation does not tell us anything about the shape of the random variable R, which we have to compute from the equilibrium. For a standard matrix game with real-valued entries, $\mathbf{A} \in \mathbb{R}^{n \times m}$, the expectation under the equilibrium strategy $(\mathbf{x}^*, \mathbf{y}^*)$ is directly computable as $v = (\mathbf{x}^*)^T \cdot \mathbf{A} \cdot \mathbf{y}^*$, and the ratio d can to be approximated by simulating the game and counting the cases where the random choice $(\mathbf{e}_i^*)^T \cdot \mathbf{A} \cdot \mathbf{e}_j$ is less or larger than v.

Example 7.1 (Disappointment at Equilibrium in a Matrix Game) Consider the zero-sum two-player game with payoff structure

$$\mathbf{A} = \begin{pmatrix} 3 & 2 \\ 1 & 6 \end{pmatrix}$$

whose equilibrium is $\mathbf{x}^* = (5/6, 1/6)$ and $\mathbf{y}^* = (2/3, 1/3)$ with the saddle-point $v = 8/3$.

What would be the disappointment rate that a *maximizing* first player would experience if both (the defender and the adversary) follow the equilibrium?

For the answer, observe that the payoff matrix $\mathbf{A}$, only those entries $a_{ij} < v \approx 2.667$ are relevant and count as 1; all others count as zero. This gives the new matrix of disappointment indicators being

$$\mathbf{D} = \begin{pmatrix} 0 & 1 \\ 1 & 0 \end{pmatrix}$$

Calling this matrix $\mathbf{D}$, its reward at $\mathbf{A}$'s equilibrium is $(\mathbf{x}^*)^T \cdot \mathbf{D} \cdot \mathbf{y}^* = 7/18 \approx 38.9\%$, which is the probability for a maximizer to get disappointed. That is, even though both players optimize their behavior, player 1 has a dramatically high chance of *not* as much as the equilibrium promised.   ◊

The disappointment rate is always linked to a goal and equilibrium in the game, and if there are multiple security goals, each can induce its own disappointment rate. Formally, let a game Γ be given, where player i is minimizing and follows the equilibrium $\mathbf{x}_i^*$, and her/his opponent play $\mathbf{x}_{-i}^*$. The disappointment on player i's utility u_i is defined as

$$d_{u_i}(a_i, \mathbf{a}_{-i}) := \begin{cases} 1, & \text{if } u(a_i, \mathbf{a}_{-i}) > \mathbb{E}_{(\mathbf{x}_i^*, \mathbf{x}_{-i}^*)}(u_i) \\ 0, & \text{otherwise.} \end{cases} \tag{7.14}$$

on all choices of pure strategies $a_i \in AS_i$ and $\mathbf{a}_{-i} \in AS_{-i}$. Apparently, this function already depends on the equilibrium, and in turn, including it in the game, makes the equilibrium conversely depend on the function d_{u_i}. More problematically, if defined as in (7.14), this is a *discontinuous* utility function, so that none of the so-far given existence results (Nash's Theorem 3.2 or Glicksberg's Theorem 3.3) apply any more. See Fig. 7.3 for an illustration.

In fact, the discontinuity can even remove all equilibria from the game, leaving it as one without optima in the sense we are looking for. Restoring the existence of equilibria then calls for more technical conditions imposed on the game, which we not further explore here [40].

An immediate way to escape the issue is to redefine the disappointment as a continuous function, e.g., by putting

$$d_{u_i}(a_i, \mathbf{a}_{-i}) := \max\left\{0, \lambda \cdot [u(a_i, \mathbf{a}_{-i}) - \mathbb{E}_{(\mathbf{x}_i^*, \mathbf{x}_{-i}^*)}(u_i)]\right\}, \tag{7.15}$$

for a parameter $\lambda > 0$ to penalize the amount of excess over the expected loss. This gives an amount proportional to the lot of "unexpected" damage, or zero otherwise (since the expected loss bound apparently was correct). This setting may be particularly interesting in its account to maybe accept losses that are slightly

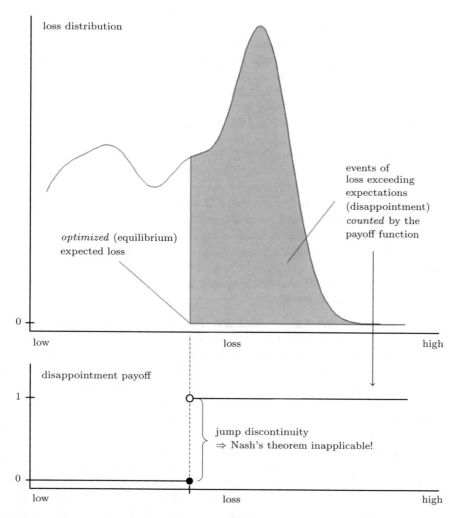

Fig. 7.3 Illustration of disappointment rate as utility

more than expected, but with the tolerance against such events to become less and less the more excess loss we suffer. By scaling the so-defined function by a positive factor, we can control the tolerance to be more or less.

In the form as (7.15), we are back with a game having all continuous payoff functions, and the existence of equilibria follows again by fixed point theorems (also underlying Nash's original result).

Technically, even simpler to handle are games where the payoffs are not real numbers but distributions as we described in Sect. 4.5. The disappointment associated with a random utility $X \sim F(a_i, \mathbf{a}_{-i})$ whose distribution F depends on the chosen actions is then given by $d_X := \Pr(X > \mathbb{E}(X))$, and easy to work out if the known distribution F is given (which it is not, if the game is using utilities taking values in $\mathbb{R}$ only). It can be shown that for finite (matrix) games, this is a continuous function without any further technical modifications like (7.15). The reason is actually not in the different game concept as such, but rather in the way of modeling it: while a conventional game is expressed with real-valued utility functions, crisp payoff values contain not enough information to derive the disappointment rate from this data. On the contrary, modeling the payoffs as distributions establishes enough data to (i) define a crisp payoff, and (ii) compute a disappointment rate from it, both of which can easily go into a conventional equilibrium calculation for the resulting Multi-Objective Game (MOG). The algorithmic details are very simple, yet we postpone their presentation until Chap. 12 for the sake of having introduced helpful some software tools until then.

Distribution-valued games as introduced in Sect. 4.5 have the appeal of allowing a direct computation of d from the saddle-point as a random variable V, since in this setting, V comes as a distribution function from which the area d to the left or right of the first moment is directly computable by integration. This process is even automatable in software, as we show in Sect. 12.6.1.

Games with payoffs defined from $\mathbb{R}$ *cannot* naturally include disappointment as a goal to optimize for two reasons:

1. If the utility only comes as a number, such values do not contain the information about the fluctuation around the mean, and this lost information hence cannot be included in the optimization. Doing so requires an enriched model, expressing the utility as a whole distribution, rather than only as a real number. In that case, however, it is perhaps cheaper and more efficient to use stochastic orders and distribution-valued games as in Sect. 4.5 directly. Alternatively, one could include the variance as a second goal to optimize, making the game multi-objective (see Sect. 3.5.1).
2. The expression $\Pr(R > \mathbb{E}(R))$ is in fact the expectation of an indicator variable, and as such seemingly fits into the game-theoretic framework of optimizing an expected utility. Involving an indicator function, however, makes the resulting goal functional *discontinuous*, so that neither Nash's theorem, nor any result building upon it guarantee the existence of an equilibrium. In fact, games without equilibria are known [38], and the absence of equilibria is explicitly due to discontinuities in the payoff functions. Fixing this requires a smooth approximation to the indicator, some proposed examples are found in the literature [40].

Pessimism and optimism have both been recognized as linked to the shape of how individuals rate different outcomes of potential actions, expressible by utility weighting functions with different curvature (convex for a risk seeker, concave for a risk avoider , or linear for risk neutrality). If we assume u as twice-differentiable, then the quantity $\lambda_u(x) := -u''(x)/u'(x)$ is the *Arrow-Pratt coefficient of absolute risk aversion* [13, pg.23]. Its definition partly roots in Theorem 5.1, since the derivations "cancel out" the shifting and scaling (the variables a and b in the Theorem), so that we can think of an individual with utility function u being more risk averse than another individual with utility function v if and only if $\lambda_u(x) < \lambda_v(x)$ for all x. Note that this applies to *decision problems* again, and not to game problems, where the utility would be stochastic and depend on other player's actions. In such random outcomes, risk aversion or risk seeking can be linked to the skewness of the utility distributions [10, 42] and also the weight of the distributions's tails (with heaviness of the tails allowing extreme events to occur more often).

In light of this, let us reconsider the case of distribution valued games under the $\preceq$-ordering (of Definition 3.1) again: though the definition of $\preceq$ does not specifically depend on the third moment or any curvature, it does account for the mass in the distributions tails, and thus somewhat "controls" parts of the disappointment ratio (though not being designed for that purpose). A second element of pessimism is induced by the (not necessarily realistic) assumption of a zero-sum competition. While mainly adopted for matters of theoretical convenience and ease of modeling, it may be overly pessimistic to ascribe exactly the opposite of the defender's incentives to the adversary. However, this modeling appeals to the concept of disappointment aversion, and as such can be preferable over the attempt to create a more accurate adversarial model (see Sect. 6.4 that may still fail to predict the adversary's behavior (and hence cause disappointment). In essence, the zero-sum assumption is perhaps unrealistic in some cases, but it is a provable worst-case assumption and as such can never disappoint (only lead to positive surprise if the actual damage suffered is less than what the defender was prepared for).

Reacting on disappointments is in many cases a myopic behavior, which in turn is occasionally also coined an element of bounded rationality. However, only so if we define rationality as utility maximizing. A myopic attitude underneath decision making may also root in the simple thinking of why bother with what could happen in farer future if this cannot be reliably predicted anyway? Games are designed to optimize the average over the upcoming eternity, so, why should an individual care for a long run average revenue if the game cannot run forever in practice? Such thinking is particularly justified in security where the ability to take damage is naturally bounded from below. Even if the game guarantees some acceptable (desired) average reward, this assurance does not extend to saving the defender from suffering arbitrarily large damage. Thus, a myopic element is one of survival for the defender, since it cannot allow for the (cumulative) damage to eventually "destroy" the asset (kill the defender) before the game along further repetitions compensates these losses by larger rewards to get closer to the equilibrium payoff (reached in the theoretical limit).

References

1. Acemoglu D, Malekian A, Ozdaglar A (2013) Network security and contagion. Technical report, National Bureau of Economic Research
2. Altman E, Avrachenkov K, Garnaev A (2007) A jamming game in wireless networks with transmission cost. In: Network control and optimization. Springer, pp 1–12

3. Anderson R, Moore T (2006) The economics of information security. Science 314(5799): 610–613
4. Axelsson S (2000) Intrusion detection systems: a survey and taxonomy. Technical report, Technical report Chalmers University of Technology, Goteborg
5. Balakrishnan K (1996) Exponential distribution: theory, methods and applications. CRC Press, Boca Raton
6. Bolot J, Lelarge M (2009) Cyber insurance as an incentivefor internet security. In: Managing information risk and the economics of security. Springer, pp 269–290
7. Chauveau T (2012) Subjective risk and disappointment. Documents de travail du centre d'economie de la sorbonne, Université Panthéon-Sorbonne (Paris 1), Centre d'Economie de la Sorbonne. https://EconPapers.repec.org/RePEc:mse:cesdoc:12063
8. Chen J, Touati C, Zhu Q (2017) A dynamic game analysis and design of infrastructure network protection and recovery. ACM SIGMETRICS Perform Eval Rev 45(2):128
9. Christoffersen P, Pelletier D (2004) Backtesting value-at-risk: a duration-based approach. J Financ Economet 2(1):84–108
10. Eichner T, Wagener A Increases in skewness and three-moment preferences. Mathematical Social Sciences **61**(2), 109–113 (2011). doi:10.1016/j.mathsocsci.2010.11.004
11. Farhang S, Manshaei MH, Esfahani MN, Zhu Q A dynamic bayesian security game framework for strategic defense mechanism design. In: Decision and Game Theory for Security, pp. 319–328. Springer (2014)
12. Finkelstein M (2008) Failure rate modelling for reliability and risk. Springer Science & Business Media, London
13. Gintis H (2014) The bounds of reason: game theory and the unification of the behavioral sciences, revised edn. Princeton University Press, Princeton/Oxfordshire. http://www.jstor.org/stable/10.2307/j.ctt13x0svm
14. Gul F (1991) A theory of disappointment aversion. Econometrica 59(3):667. https://doi.org/10.2307/2938223
15. Hölmstrom B (1979) Moral hazard and observability. Bell J Econ 10:74–91
16. Holmstrom B (1982) Moral hazard in teams. Bell J Econ 13:324–340
17. Horák K, Zhu Q, Bošanský B (2017) Manipulating adversary?s belief: a dynamic game approach to deception by design for proactive network security. In: International conference on decision and game theory for security. Springer, pp 273–294
18. Huang L, Chen J, Zhu Q (2017) A large-scale Markov game approach to dynamic protection of interdependent infrastructure networks. In: International conference on decision and game theory for security. Springer, pp 357–376
19. Jajodia S, Ghosh AK, Swarup V, Wang C, Wang XS (2011) Moving target defense: creating asymmetric uncertainty for cyber threats, vol 54. Springer Science & Business Media, New York
20. Jhaveri RH, Patel SJ, Jinwala DC (2012) Dos attacks in mobile ad hoc networks: a survey. In: Advanced computing & communication technologies (ACCT), 2012 second international conference on. IEEE, pp 535–541
21. Kelly FP, Maulloo AK, Tan DK (1998) Rate control for communication networks: shadow prices, proportional fairness and stability. J Oper Res Soc 49:237–252
22. Kesan J, Majuca R, Yurcik W (2005) Cyberinsurance as a market-based solution to the problem of cybersecurity: a case study. In: Proceedings of WEIS
23. Lelarge M, Bolot J (2008) A local mean field analysis of security investments in networks. In: Proceedings of the 3rd international workshop on economics of networked systems. ACM, pp 25–30
24. Manshaei MH, Zhu Q, Alpcan T, Bacşar T, Hubaux JP (2013) Game theory meets network security and privacy. ACM Comput Surv (CSUR) 45(3):25
25. Miao F, Zhu Q, Pajic M, Pappas GJ (2018) A hybrid stochastic game for secure control of cyber-physical systems. Automatica 93:55–63
26. Minkova LD (2010) Insurance risk theory. Lecture notes, TEMPUS Project SEE doctoral studies in mathematical sciences

27. Miura-Ko R, Yolken B, Mitchell J, Bambos N (2008) Security decision-making among interdependent organizations. In: Computer security foundations symposium, CSF'08. IEEE 21st. IEEE, pp 66–80
28. Pal R, Golubchik L, Psounis K, Hui P (2014) Will cyber-insurance improve network security? a market analysis. In: INFOCOM, 2014 proceedings IEEE. IEEE, pp 235–243
29. Pawlick J, Zhu Q (2015) Deception by design: evidence-based signaling games for network defense. arXiv preprint arXiv:1503.05458
30. Pawlick J, Zhu Q (2016) A Stackelberg game perspective on the conflict between machine learning and data obfuscation. In: Information forensics and security (WIFS), 2016 IEEE international workshop on. IEEE, pp 1–6. http://ieeexplore.ieee.org/abstract/document/7823893/
31. Pawlick J, Zhu Q (2017) A mean-field Stackelberg game approach for obfuscation adoption in empirical risk minimization. arXiv preprint arXiv:1706.02693. https://arxiv.org/abs/1706.02693
32. Pawlick J, Colbert E, Zhu Q (2017) A game-theoretic taxonomy and survey of defensive deception for cybersecurity and privacy. arXiv preprint arXiv:1712.05441
33. Pawlick J, Colbert E, Zhu Q (2018) Modeling and analysis of leaky deception using signaling games with evidence. IEEE Trans Inf Forensics Sec 14(7):1871–1886
34. Peltzman S (1975) The effects of automobile safety regulation. J Polit Econ 83:677–725
35. Raiyn J et al (2014) A survey of cyber attack detection strategies. Int J Secur Appl 8(1):247–256
36. Rass S, König S (2018) Password security as a game of entropies. Entropy 20(5):312. https://doi.org/10.3390/e20050312
37. Rass S, Alshawish A, Abid MA, Schauer S, Zhu Q, De Meer H (2017) Physical intrusion games-optimizing surveillance by simulation and game theory. IEEE Access 5:8394–8407
38. Sion M, Wolfe P (1957) On a game without a value. Princeton University Press, pp 299–306. http://www.jstor.org/stable/j.ctt1b9x26z.20
39. Tague P, Poovendran R (2008) Modeling node capture attacks in wireless sensor networks. In: Communication, control, and computing, 2008 46th annual allerton conference on. IEEE, pp 1221–1224
40. Wachter J, Rass S, König S, Schauer S (2018) Disappointment-aversion in security games. In: International conference on decision and game theory for security. Springer, pp 314–325
41. Wang W, Zhu Q (2017) On the detection of adversarial attacks against deep neural networks. In: Proceedings of the 2017 workshop on automated decision making for active cyber defense. ACM, pp 27–30
42. Wenner F (2002) Determination of risk aversion and moment-preferences: a comparison of econometric models. Ph.D. thesis, Universität St.Gallen
43. Xu Z, Zhu Q (2015) A cyber-physical game framework for secure and resilient multi-agent autonomous systems. In: Decision and control (CDC), 2015 IEEE 54th annual conference on. IEEE, pp 5156–5161
44. Xu Z, Zhu Q (2016) Cross-layer secure cyber-physical control system design for networked 3D printers. In: American control conference (ACC). IEEE, pp 1191–1196. http://ieeexplore.ieee.org/abstract/document/7525079/
45. Xu Z, Zhu Q (2017) A game-theoretic approach to secure control of communication-based train control systems under jamming attacks. In: Proceedings of the 1st international workshop on safe control of connected and autonomous vehicles. ACM, pp 27–34. http://dl.acm.org/citation.cfm?id=3055381
46. Xu Z, Zhu Q (2017) Secure and practical output feedback control for cloud-enabled cyber-physical systems. In: Communications and network security (CNS), 2017 IEEE conference on. IEEE, pp 416–420
47. Yuan Y, Zhu Q, Sun F, Wang Q, Basar T (2013) Resilient control of cyber-physical systems against denial-of-service attacks. In: Resilient control systems (ISRCS), 2013 6th international symposium on. IEEE, pp 54–59
48. Zhang R, Zhu Q (2015) Secure and resilient distributed machine learning under adversarial environments. In: 2015 18th international conference on information fusion (fusion). IEEE, pp 644–651

49. Zhang R, Zhu Q (2017) A game-theoretic defense against data poisoning attacks in distributed support vector machines. In: Decision and control (CDC), 2017 IEEE 56th annual conference on. IEEE, pp 4582–4587

50. Zhang T, Zhu Q (2017) Strategic defense against deceptive civilian GPS spoofing of unmanned aerial vehicles. In: International conference on decision and game theory for security. Springer, pp 213–233

51. Zhang R, Zhu Q (2018) A game-theoretic approach to design secure and resilient distributed support vector machines. IEEE Trans Neural Netw Learn Syst 29:5512–5527

52. Zhang R, Zhu Q (2019) Flipin: a game-theoretic cyber insurance framework for incentive-compatible cyber risk management of internet of things. IEEE Trans Inf Forensics Secur 15:2026–2041

53. Zhang R, Zhu Q, Hayel Y (2017) A bi-level game approach to attack-aware cyber insurance of computer networks. IEEE J Sel Areas Commun 35(3):779–794

54. Zhu Q, Rass S (2018) On multi-phase and multi-stage game-theoretic modeling of advanced persistent threats. IEEE Access 6:13958–13971

55. Zhu Q, Fung C, Boutaba R, Başar T (2012) Guidex: a game-theoretic incentive-based mechanism for intrusion detection networks. Sel Areas Commun IEEE J 30(11):2220–2230

56. Zhu Q, Clark A, Poovendran R, Basar T (2013) Deployment and exploitation of deceptive honeybots in social networks. In: Decision and control (CDC), 2013 IEEE 52nd annual conference on. IEEE, pp 212–219

57. Zhuang J, Bier VM, Alagoz O (2010) Modeling secrecy and deception in a multiple-period attacker–defender signaling game. Eur J Oper Res 203(2):409–418

Chapter 8
Patrolling and Surveillance Games

*Under observation, we act less free, which means we effectively
are less free.*

E. Snowden

Abstract Patrolling and surveillance games both deal with a chasing-evading
situation of an adversary trying to escape detection by either a mobile defender
(patrolling) or a fixed defender (surveillance). Both kinds of games are played
on graphs as abstract models of an infrastructure, and we review a variety of
closed-form solutions for optimal patrolling in different classes of graph topologies.
Applications include patrolling along lines (borders, pipelines, or similar), harbors
(tree-structured graphs), and large geographic areas in general (planar graphs and
maps). For surveillance and patrolling, we give hints on how to estimate the
necessary resources, and how to include imperfectness and uncertainty, related to the
detection capabilities, but also the chances of the adversary escaping the view of the
patroller or surveillance. In complex terrain, we will discuss the use of simulation
and empirical games (over real-valued and stochastic orders).

8.1 The General Setting

Whenever a certain terrain defined by physical boundaries (museum, airport, city
district, warehouse, or similar) needs protection, patrolling games may be a model to
consider. They assume a mobile defender obliged with the continuous surveillance
of the area, but who is (generally) unable to rely on sensor information or has
otherwise no means of seeing remote areas beyond its proximity. If remote sensing
is doable, we are getting into *surveillance games* that we postpone until Sect. 8.8,
not the least so, since it changes the nature of the game from a pursuit-evasion style
for patrolling, into the different matter of optimizing sensor placement.

In a patrolling game, the defender usually acts alone and on its own. This is in
somewhat contrast to practical situations of security guards, which often work in

© Springer Nature Switzerland AG 2020
S. Rass et al., *Cyber-Security in Critical Infrastructures*, Advanced Sciences
and Technologies for Security Applications,
https://doi.org/10.1007/978-3-030-46908-5_8

teams, have access to surveillance equipment and are in continuous contact with each other. However, towards a simplification of the models, and without loss of too much generality, we may think of a collaborating team of defenders as a single entity, replacing the multitude of persons by the ability to check several locations at the same time. This amounts into a change of the area's topology, but leaves the underlying ideas and results about patrolling games unchanged.

Most (but not all) patrolling game models discussed hereafter are discrete, and start by dividing the terrain into a finite set of zones. This division can adhere to natural circumstances, such as floor layouts in buildings, or geographic regions separated by borders, rivers, mountains, but is generally unconstrained and up to our own choice. Commonly, some zones define regions of particular interest or importance (e.g., offices), while others are of less or secondary interest (e.g., hallways). As a systematic aid for the division, we may ask for how many zones are meaningful, and there is a famous theorem about art galleries that may help here.

8.2   The Art Gallery Theorem

This elegant result in combinatorics and graph theory is concerned with the following situation: given a museum with precious paintings on the walls, whose floor plan constitutes a polygon with n edges (walls), answer the following question: *how many stationary guards (each with a 360 degrees field of view) are necessary to keep all paintings (walls) in vision and protected?*

Computing the *minimal* number of guards required is NP-hard in general, but there is a simple upper bound to this number given by the following theorem:

Theorem 8.1 (Chvátal's Watchmen Theorem [11])   *If S is a polygon with n vertices, then there is a set T of at most n/3 points of S such that for any point p of S there is a point q of T with the line p—q lying entirely in S.*

The points in T are exactly the surveillance watchmen, or cameras, while the mentioned line is the line of sight from the camera placed at point q to the wall (painting) at position p. Chvátal's original proof was shortened by S. Fisk [15] into a five line argument of unbeatable elegance: treat the floor layout as a planar graph G, and add an artificial vertex v_0 that is connected to all vertices in G, making up the planar graph G'. By the four-coloring theorem [8], we can color G' using four colours, which, excluding the color for v_0 leaves G as three-colorable. Now, let T be the smallest set of vertices that all have the same color (there are only three of them to choose from). Then, $|T| \leq n/3$ necessarily. Moreover, every point p in S lies in some triangle (because every planar graph can be triangulated), and every triangle in S has a point $p \in T$ on it. Since the triangles are convex, there must be a line of sight from q to p, thus proving the theorem. Figure 8.1 depicts the proof idea.

Remark 8.1   Though the art gallery theorem allows for an architecture of arbitrary shape, it nonetheless assumes no "holes" in the layout, such as a patio that would be

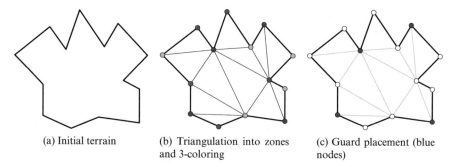

(a) Initial terrain   (b) Triangulation into zones   (c) Guard placement (blue
                             and 3-coloring   nodes)

Fig. 8.1 Art gallery theorem illustration

drawn as an inner polygon on its own (see Fig. 8.2a for an example, where the inner gray area could be a mountain, for instance). A deeper look at the argument above reveals that it implicitly uses Jordan's curve theorem to assure that the interior and the exterior of the polynomial are both connected regions, so that (i) the connection to the outer "artificial" node retains a planar graph, and (ii) the triangulation is possible without troubles. As long as the museum's layout is drawable by a single polygon, the theorem applies. Otherwise, we refer to more refined versions thereof, collected in [21].

8.3 From Art Galleries to Patrolling Games

The art gallery theorem is helpful for both, patrolling and surveillance games, since:

- it helps with the definition of zones by referring to triangulation of the area,
- and it also upper-limits the number of surveillance cameras (or guards) in terms of the area's boundary geometry: no more than $n/3$ cameras with 360 degrees panoramic view would be required for n walls to the outside.

Once the terrain is divided into zones, or locations, edges may express reachability of one zone from another. Figure 8.2 shows some examples: a natural habitat (Fig. 8.2a) may have a mountain in the middle, so that zones are reachable from each other if they are adjacent. Other terrains may be different, such as buildings (Fig. 8.2b) or harbours (Fig. 8.2c), define the edges in the graph in terms of physical reachability over doors, docks, or similar. For the building, but also for any other terrain, we can color the graph nodes to indicate the criticality of a zone, using a simple scale like green/yellow/red to denote normal, medium and high security areas.

The output of this modeling defines a graph that is the "playground" for the actual patrolling game. The use of a game-theoretic treatment lies in the insight of

Fig. 8.2 Example of dividing
different terrains into zones

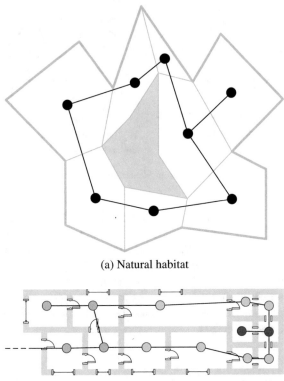

(a) Natural habitat

(b) Building, with color codes indicating zone criticality

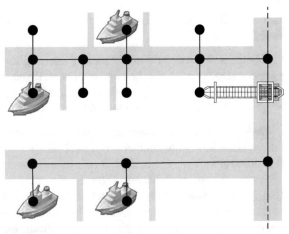

(c) Harbour

how the patroller should randomize its patrols to prevent being predictable by the attacker. However, the humble maximum entropy heuristic [16] that prescribes to uniformly at random visit each location $1, 2, \ldots, n$ with the same likelihood $1/n$ is also not recommended [5]. Game theory helps here to find an optimal randomization strategy.

In the following, let the playground for the patrolling game be a graph $G = (V, E)$ with $n = |V|$ vertices that correspond to zones that need regular checkups, and edges between zones i, j, whenever zone j is reachable from i. Depending on the topology (geography) or other constraints, it may be convenient to let G be directed (e.g., if a zone is reachable from another, but the direct way back is not possible), or to weigh the edges, for example, by assigning distances or other costs to them. We will make use of these modifications where appropriate.

8.4   A Simple Matrix Game Model

In the simplest setting, the strategies for both, the patroller and the attacker, are just visiting the spots or zones that correspond to the nodes in the graph (cf. Example 3.5). Both players pick their location i, j to visit (either for checking or for causing damage), and the game rewards the players accordingly with

$$u_1(i, j) = -u_2(i, j) = \begin{cases} 0, & \text{if } i = j, \text{ i.e., the guard catches the intruder at position } i; \\ 1, & \text{if } i \neq j, \text{ i.e., the intruder went uncaught,} \end{cases}$$

which defines the $(n \times n)$-payoff matrix $\mathbf{A}$ for a loss-minimizing defender as player 1. This is a diagonal matrix, and games with such payoffs admit closed form solutions that we will describe in more detail in Sect. 12.3.3. The particular game defined above has the saddle-point value is $\mathrm{val}(\mathbf{A}) = 1/n$. In a slight generalization, one could redefine the diagonal elements with the security clearance level of different zones, corresponding to the potential loss of business value or the amount of damage expected in zone i when the adversary is not discovered there.

Off-diagonal nonzero elements may be inserted if there are chances that the defender being in a zone will notice (or otherwise check) a nearby zone directly or a remote zone by using surveillance aids (including fellow guards patrolling at the same time).

In any case, this modeling is very simple and appealing for its computational feasibility, since the size of the graph, i.e., the number of zones, directly determines the size of the game.

However, this model oversimplifies practical matters, since it may prescribe practically cumbersome strategies: imagine the guard being instructed to check far-away rooms repeatedly, leaving aside locations along the way, which may intuitively appear reasonable to check first. Nevertheless, strictly following the randomization would mean, for example, checking the first floor (of a building), then, e.g., the

27th floor, and then go back to check the 3rd floor again, and so on. Such a naive randomization may just not be feasible in practice.

A more informed model [25] adds the cost for changing strategies over repetitions of the game: in the graph, let us label edges with the distances that the guard has to overcome when going from location i to location j. It is a straightforward matter to run a Floyd-Warshall shortest-path algorithm [12] on the so-obtained weighted graph, to obtain a distance matrix $\mathbf{S} = (s_{ij})_{i,j \in V}$, giving the shortest travel distance s_{ij} from zone i to zone j. Adding this cost as a secondary minimization goal in addition to the aim of catching the intruder makes the game multi-objective, and the techniques and solution concepts from Chap. 3 apply. We will postpone an algorithmic solution to the problem until Sect. 12.5; specifically, the equilibrium is obtained from solving problem (12.6).

8.5   Graph Traversal Games

The original term "patrolling game" has been introduced in the literature to describe a class of games being played on a graph, yet adding an element of time to the consideration (which was not present in the previous section, other than for the game being repeated at its natural periodicity, e.g., a work-day, shifts, or similar) [5].

Fix a discrete time horizon $T \in \mathbb{N}$ and consider the game played over the finite period $\mathscr{T} = \{0, 1, \ldots, T - 1\}$. We may think of $\mathscr{T}$ as defining a shift for the patroller, and the definition of its length is duty of some authority, obliged to pick T with respect to effectiveness in terms of cost criteria or other constraints. Later, we will give results supporting this choice.

As before, the attacker randomly picks a zone to visit (and cause damage, steal sensitive data, do poaching, etc.) over a time-span of m time units starting at time τ, thus being in zone i for the interval $I = \{\tau, \tau + 1, \ldots, \tau + m - 1\}$. The pair (i, I) then fully describes this attack strategy.

The patroller (defender), in turn takes routes through the graph, visiting a zone in each time unit. The patrollers strategy is thus a *walk* $w : \mathscr{T} \to V$, for which $w(t)$ gives the current location of the patroller at time $t = 0, 1, \ldots, T - 1$.

Remark 8.2 The terms "walk" and "path" are often used synonymously, but some authors distinguish the two as a walk admitting cycles in graph (meaning that a location may be visited along a walk more than once), while a path visits each location only once; in a way, it is the most direct (not necessarily shortest) route between two locations. We shall not make much use of this distinction in the following, and will use the terms walk and path widely synonymous.

The intruder is caught if the patroller visits the attacker's location during the time when it is active, i.e., if $i \in w(I)$ for the defender's strategy w and attacker's strategy (i, I). Correspondingly, the game is 0-1-valued in the sense of paying

$$u_1(w, (i, I)) = -u_2(w, (i, I)) = \begin{cases} 1, & \text{if } i \in w(I); \\ 0, & \text{otherwise.} \end{cases}$$

with the long-run average over u_1 again being the likelihood for the attacker to get caught. For repetitions, it is useful to think of $\mathscr{T}$ to define a *period* of the game, i.e., a single round of the game runs from time $t = 0$ until time $t = T - 1$, and the game starts afresh at time $t = T$ and runs until time $t = 2T - 1$, and so on. For this to be meaningful, we need to constrain the walks to start and finish at the same location. This actually simplifies matters of analysis, as it lets us assume that attacks take place equiprobable in any time interval. Nevertheless, repetition is no necessity, and the literature distinguishes one-shot from periodic variants, with different results about either type. Hereafter, we let Γ^o denote the one-shot variant, where the game is played over only a single period $\mathscr{T}$, and we let Γ^p be the periodic version just described. For both variants, we add the parameters V, T and m, to express properties of the game that depend on these values.

It is straightforward to set up another matrix game in this setting, if we fix a set of walks (routes), and consider all possible combinations of location i and interval I of adversarial activity there. The number of such strategies will become considerably large, so that methods for strategy reduction appear in order and will receive attention later.

Games defined in the framework above admit a set of closed form results about their saddle-point values, i.e., chances to catch the intruder. Among the most important facts [5] are the following:

1. $\mathrm{val}(\Gamma(V, T, m))$ is nondecreasing in m and does not increase if edges are added to the graph, irrespectively of whether the game is one-shot or periodic. This is due to the fact that catching a "faster" attacker (using a smaller m) is less likely than catching a slower one. Also, adding new strategies to the game (a new edge to use for a walk) cannot worsen the situation of the defending patroller.
2. $\mathrm{val}(\Gamma^p(V, T, m)) \leq \mathrm{val}(\Gamma^o(V, T, m))$, since the walks in the periodic game are constrained (see above) over those in the one-shot version.
3. If two adjacent zones i, j are merged into one new zone k, where the zone k is connected to all previous neighbors of i and j, then the value of the game on the so-simplified graph becomes larger or equal to the original value (irrespectively of whether the game is repeated or one-shot). Informally and intuitively, this is because the patroller is now checking two zones i, j at once by visiting location k, provided that the optimal route in the original is played "accordingly" on the smaller graph.
4. $\frac{1}{n} \leq \mathrm{val}(\Gamma(V, T, m)) \leq \frac{m}{n}$, for both, $\Gamma = \Gamma^o$ and $\Gamma = \Gamma^p$ and any configuration (V, T, m). Equality at both ends of the range is achievable by uniformly random strategies:

   - If the defender uniformly at random picks a location and waits for the attacker, then the minimum value bound is attained.
   - If the attacker uniformly at random picks a location to attack for a fixed interval I, then the maximum bound to the value is attained.

5. Regarding the shift length T, the following can be useful to make model design choices:

   - Extensions are not doing better, i.e., val($\Gamma^o(V, T + 1, m)$) $\leq$ val($\Gamma^o(V, T, m)$). This is because the attacker retains at least the same strategies in the shorter game, but gets more chances to act due to the longer period. This *does not* hold for periodic games, as counterexamples show [5].
   - Longer shifts can be better of the game is repeated: val($\Gamma^p(V, k \cdot T, m)$) $\geq$ val($\Gamma^p(V, T, m)$) for all $k = 1, 2, \ldots$, simply because the patroller can repeat any optimal T-periodic walk in $\Gamma^p(V, T, m)$ for k times in val($\Gamma^p(V, k \cdot T, m)$) for at least the same success rates as in the shorter game. However, there are more strategies available to the patroller in the game with period kT, so the value may even increase.

6. Optimal generic strategies: If a patrol w is such that a node i appears on every sub-path of w of length m, then we call this strategy intercepting, since it will eventually intercept every attack on nodes that it contains. A set of such intercepting strategies such that every node in V is contained in an intercepting strategy is called a covering set. A *covering strategy* is a uniformly random pick from a minimal (in terms of cardinality) such covering set C of patrols.

   Likewise, for the attacker, if a patrol w is such that it cannot at the same time intercept two nodes i and j within the time-span m, we call these two locations *independent*. This occurs if either location i and j are so far from each other that it takes longer than m to traverse from one to the other (in the one-shot case), or $T \leq 2d(i, j)$, since the patroller will return to its origin after finishing a round and thus not reach the other location in time. Suppose the attacker knows a maximal (in terms of cardinality) set D of independent nodes, then it can pick a location uniformly at random from D and attack there during an a priori fixed attack interval m. We call this the attacker's *independent strategy*.

   These two generic behavior patterns lead to tight bounds for the saddle-point value of the game, which are

$$\frac{1}{|C|} \leq \text{val}(\Gamma(V, T, m)) \leq \frac{1}{|D|}. \tag{8.1}$$

The defender can play the covering strategy to attain the lower bound. The attacker can play the independent strategy to attain the upper bound.

Certain classes of graphs admit sharper assertions:

Hamiltonian graphs: These are graphs that have a spanning cycle without self-intersections, i.e., there is a circular walk that visits each zone exactly once. A simple example is that of a cycle C_n with n nodes, such as could model the outer *fence* of a region, *military basis*, or similar.
If the graph G is Hamiltonian, then

$$\text{val}(\Gamma^o(V, T, n)) = \frac{m}{n}, \quad \text{and} \quad \text{val}(\Gamma^p(V, T, n)) \leq \frac{m}{n}.$$

Deciding whether or not a graph is Hamiltonian is generally NP-complete [17], but some sufficient conditions that imply the existence of a Hamiltonian cycle (though not giving any hint on how to find it), are easy to check:

- The complete graph is Hamiltonian (trivially, but interestingly, since this is the model where every location is reachable from every other location, such as is the case on *water areas*).
- Dirac's theorem: if every among the n vertices in G has a degree $\geq n/2$, then G is Hamiltonian.
- If any two non-adjacent nodes u, v have degrees that satisfy $\deg(u) + \deg(v) \geq |V|$, then G is Hamiltonian.

Bipartite graphs: A graph $G = (V, E)$ is bipartite, if its vertex set can be partitioned into two sets V_1, V_2 such that edges all run between V_1 and V_2 or vice versa, but no edge lies entirely within V_1 or within V_2.
Important special cases of such graphs include trees, which can naturally arise for areas like *harbors* (cf. Fig. 8.2c), *airports* and many others.
If the graph is bipartite, with $a = |V_1| \leq b = |V_2|$ (without loss of generality), then

$$\text{val}(\Gamma^o(V, T, m)) \leq \frac{m}{2b} \quad \text{and} \quad \text{val}(\Gamma^p(V, T, m)) \leq \frac{m}{2b}.$$

Both bounds are tight and attainable if the bipartite graph is complete in the one-shot case, or if it is complete and $T = 2kb$ for some integer k in the periodic case.

Line Graphs: These are simply linear arrangements of nodes, i.e., each node has at most two neighbors.
Important applications include protection of a country's *border area*, a *pipeline*, or similar. Though such a terrain may not necessarily correspond to a straight line, the area under protection is often divisible into a sequence of adjacent zones, yielding a line graph according to the procedure above.
If G is a line and $n \leq m + 1$, then

$$\text{val}(\Gamma^o(V, T, m)) = \frac{m}{2(n-1)} \quad \text{and} \quad \text{val}(\Gamma^p(V, T, m)) \leq \frac{m}{2(n-1)}.$$

The latter bound is tight and is attained for $T = 2k(n-1)$ with some integer k.
Moreover, the bound (8.1) becomes explicit for line graphs [22], as

$$\left\lceil \frac{n}{\lfloor m/2 \rfloor + 1} \right\rceil^{-1} \leq \text{val}(\Gamma(V, T, m)) \leq \left\lfloor \frac{n+m-1}{m} \right\rfloor^{-1},$$

which again applies to both, one-shot and periodic games, where tightness of both bounds holds for the aforementioned generic covering and independent strategies. It is interesting to note that the intuitive behavior of traversing the line (or parts of it) back and forth turns out as optimal under certain conditions on the length n of the line and the attack duration m [22]. For example, if $n \leq (m+2)/2$, then the patroller can intercept every possible attack by simply going back and forth between the end nodes. Likewise, if $n < m + 1$, then the attacker has an optimal strategy of attacking simultaneously at both ends of the line (*diametrical strategy*). Other configurations are slightly more complex and we refer to the literature (including [7]) for details omitted here for brevity.

Line graphs are of particular use: first, they admits a closed form analysis also in continuous time (see Sect. 8.7). Second, many general graph topologies are decomposable into line graphs (trivially, since a single edge constitutes a line graph already, and any more complex structure can be built from this). We will make use of this in the next section.

8.6 Strategy Reduction Techniques

8.6.1 Decomposition

This means dividing the graph into smaller parts (line, complete and hence Hamiltonian, or others, not necessarily equally sized or isomorphic), to which specialized results, e.g., closed formulae, could apply. Imagine that the adversary enters the system at any point, at which it plays the "local game" against the patroller. Each local game has its own playground being another part in the graph and embodies all the timing matters of pursuit and evasion between the guards and the intruder. This elevates our view to see the patrolling as a nested "game of games". On the global view, it is a simple diagonal game like in Sect. 8.4 (see Sect. 12.3.3), in which each payoff is now defined by the subgame played locally. Under this intuition, the following result is evident:

Theorem 8.2 ([5]) *Suppose the graph $G = (V, E)$ is decomposed into N smaller graphs $G_i = (V_i, E_i)$ for $i = 1, 2, \ldots, N$ such that $V = V_1 \cup V_2 \cup \ldots V_N$, and where two nodes in G_i are adjacent if they were adjacent in G. Then, the value $val(\Gamma(V, T, m))$ of the patrolling game, whether one-shot or periodic, satisfies*

$$val(\Gamma(V, T, m)) \geq \left(\sum_{i=1}^{N} \frac{1}{val(\Gamma(V_i, T, m))} \right)^{-1} .$$

This means that, in absence of exact results on the perhaps complex zone topology at hand, we can resort to a decomposition of the graph into smaller graphs of easy-to-analyze topology, about which closed-form statements can be made. Theorem 8.2 then allows to combine the individually optimal strategies for patrolling the subgraphs into a performance estimate for the original (complex) topology.

At the same time, decomposition constitutes a heuristic for *strategy reduction*, if the smaller graphs can be chosen with a smaller feasible set of strategies.

8.6.2   Contraction

The idea of contraction is, like for domination, to retain strategies only if they contribute something nontrivial to the game. The intuition is simple: why consider longer routes in the system, if those bear larger risks for the intruder to be detected (as they give the defender more time and space to catch it)? If coming over a shorter route is preferable over the longer route, i.e., the faster strategy dominates the slower strategy, we can safely exclude them from the game. An algorithm from [9] roughly proceeds as follows:

1. Compute the pairwise shortest paths between all zones, i.e., nodes in the graph (e.g., by using the Floyd-Warshall algorithm).
2. Construct a new graph by unifying all shortest paths into the output graph.

Obviously, this procedure retains shortest connections between all nodes. In addition, we may decide to remove some nodes if they can naturally be covered along the way of checking other nodes. For example, if three zones are in line, corresponding to a part v_0—v_1—v_2, so that v_1 is bypassed on the way from v_0 to v_2 or vice versa, we may be able to merge v_1 into one of the neighbors to simplify the model further.

As an example, consider the graph in Fig. 8.3: initially, all the edges are in the graph, but retaining only those edges that are on some shortest path (between any two locations), we end up with two edges being removed in the example graph.

Fig. 8.3 Strategy reduction by retaining shortest paths only

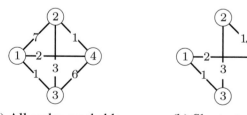

(a) All nodes reachable   (b) Shortest paths

8.6.3 Symmetrization

Less obvious is the possibility of treating certain short paths in the graph as
"equivalent", if we can reasonably assume that attacks occur equiprobable on them.
This technique is called *symmetrization* [5] and best illustrated by an example:
consider the graph in Fig. 8.4, where nodes 2 and 3 are symmetrically arranged
around the path from 1 to 4. Thus, patrolling through the nodes 2 and 3 and randomly
attacking nodes 2 and 3 is essentially no different to treating these two possibilities
as a single scenario, without changing the actual value of the game. More formally,
we can look for an automorphism on G. This is a permutation $\pi : V \to V$ that
maps a graph G into the graph $G' = (\pi(V), E')$ with $(u, v) \in E'$ if and only if
$(\pi^{-1}(u), \pi^{-1}(v)) \in E$, i.e., there is an edge in G' if and only if the same edge is in
G (likewise defined for undirected graphs). Intuitively, an automorphism thus only
"renames" the vertices but leaves all neighborhood relations (edges) unchanged.

Under such an automorphism π, we can call two patroller strategies, i.e., walks,
$w_1 \neq w_2$, *equivalent*, if w_1 maps to w_2 under π. For the example graph in Fig. 8.4,
an automorphism would be the mapping $\pi = \{1 \mapsto 1, 2 \mapsto 3, 3 \mapsto 2, 4 \mapsto 4,$
$5 \mapsto 5\}$, so that the walk $1 \to 2 \to 4$ under π maps to $1 \to 3 \to 4$ and vice versa,
thus making them equivalent. Likewise, an attacker possibly picking locations i, j
makes no difference in choosing one or the other, if $\pi(i) = j$ and $\pi(j) = i$. In that
case, we call the locations i and j again *equivalent*.

The crux of knowing a set $S = \{s_1, s_2, \ldots, s_d\}$ of equivalent strategies (for either
the patroller or the attacker) is that we can treat S as a *single* strategy, in which
we uniformly at random pick an element to play in the actual game. That is, the
game's setup and analysis has all equivalent strategies condensed into single entities,
while for the actual game play of such a (condensed) strategy S, the player would
uniformly pick a random element from S and play it. The choice of different such
sets S is, however, optimized by game theory. The resulting games typically have
less strategies and are thus easier to study, since the value of the game remains
unchanged upon symmetrization [1, 2].

Fig. 8.4 Kite graph

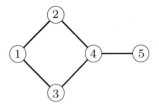

8.7 Further Variations of Patrolling Games

8.7.1 Patrolling in Continuous Time

In continuous time, the setup is as before, with the following changes [6]:

- The attacker can pick *any* position in the graph, not necessarily only vertices, but it can also be on any point on an edge.
- a patroller's strategy is a function $w : [0, \infty) \to V$ that satisfies $d(w(t), w(t')) \leq |t - t'|$ for all $t, t' \geq 0$, where d is the distance function on the graph, being a metric in the topological sense (see the glossary for a formal definition).
- The time horizon is infinite, so that periodic versions of the game merely appear if a walk is cyclic.

The remaining parts of the setting go unchanged, i.e., the attacker still picks a position x to attack during a time interval $I = [y, y+r] \subset [0, \infty)$ when the attacker hits us at time y for duration r. Many of the previous results hold in a similar form for continuous games, such as the bound (8.1) translates into

$$\frac{r}{\ell_{CPT}} \leq \text{val}(\Gamma) \leq \frac{r}{\ell_G},$$

in which ℓ_G is the total length of G, which is just the sum of all edge lengths, and ℓ_{CPT} is the length of a Chinese Postman Tour (CPT). This is a cycle in G that visits every node at least once, but, unlike in the Traveling Salesperson Problem or for a Hamilton cycle, is allowed to visit vertices more than just once. Still, we are interested only in the shortest such tour, but the relaxation lets us compute the shortest such tour polynomial time [14].

8.7.2 Accumulation Games

Some patrolling games are not concerned with catching an intruder itself, but with cleaning up his "footprints". For example, if the intruder leaves a certain amount of hazardous material in an area, which the defender may not clean up to the desired amount, then the game is lost for the patroller (likewise, if the patroller gets enough of the attacker's leftovers collected, the game ends in favour of the defender). These games enjoy many applications, such as for example, for keeping the *pollution* of a region under control.

Such models employ similar ideas and concepts as described above, and have been studied in [3, 20], with explicit optimal strategies for the attacker being known in various configurations of the game [4].

8.7.3   Covering Games

Similar to patrolling, yet different in the way the game is set up is the model [23]
related to guarding against terrorist attacks on airports. In a nutshell, the defender's
strategies are to guard (or not guard) certain locations, while the attacker's intention
is to hit the same locations or at least cause maximal damage. Each strategy (for both
players) is tailored to the particularities of the protection and the attack and damage
potential for the adversary. Eventually, it is again a game where the defender selects
an area to protect and the second player chooses an area to hit. It is generally a
nonzero-sum game, since the defender is additionally constrained to assign limited
resources to the defense (thus enforcing perhaps a reassignment at random from
time to time, if not everything can be covered continuously and at all times), and the
gain for the attacker is the damage, and not equal to the expenditures of the defender
(who has a gain from protecting its area, but also costs by doing so).

8.8   Surveillance Games

Surveillance extends patrolling by adding stationary equipment and other aids
to the protecting player. There is no apparent consensus in the literature on
how a surveillance game is typically defined, but the most common problem
summarized under the term concerns the optimal placement of sensors. The meaning
of "optimal" differs between applications, and includes metrics of field of view
coverage [10, 18, 19], localization error, rates of detection [24], or others. Generally,
the problem is thus one of multi-criteria optimization, in which cost constraints are
commonly included. In light of the latter, we may also seek to optimize the amount
of redundant sensor coverage (i.e., zones that are covered by multiple sensors,
watchmen, or other means at the same time). The example formalization given
below has this secondary goal included.

As before, the modeling starts with a division of the area into zones or locations
of interest, and introducing variables to measure the coverage of zones by the
surveillance equipment layout. The Art Gallery Theorem, and its relatives [21],
provide us with upper bounds to the number of sensors required. Surveillance games
strive for lower bounds and the additional information of where to optimally place
them. Note that the short proof of Theorem 8.1 is nonconstructive and silent about
where to place the watchmen; it exploits the mere existence of a (usually non-
unique) coloring to assert the claimed number of positions. Finding the coloring
itself is the computationally expensive part.

A simple yet instructive modeling is the following, based on [13]: let the zones
be enumerated from $1, \ldots, n$, and for each possible sensor *location* (not the sensor
itself, since we assume all of them to be identical), define an indicator variable

$$s_j := \begin{cases} 1, & \text{if the sensor is located at location } j; \\ 0, & \text{otherwise,} \end{cases}$$

where the variable j runs over all possible sensor locations, abstractly enumerated as $j = 1, 2, \ldots, m$, but augmented by background information on the real physical locations. This background information then determines the performance indicator and region cover set, which are

$$\lambda_{ij} := \begin{cases} 1, & \text{if zone } i \text{ would be covered by sensor at location } j; \\ 0, & \text{otherwise,} \end{cases}$$

$$N_i := \{ j : \lambda_{ij} = 1 \}$$

Overall, we are interested in a full, or at least maximal, coverage of the entire area, so we define a third indicator to tell whether a region is covered by *any* sensor: put $m_i := 1$ if and only if a region i is covered by some sensor, and set to zero otherwise. Likewise, we are free to define another indicator to account for multiple sensors covering the same area. Note that this kind of redundancy can be wanted, e.g., if the area is important and we care for reliability of the surveillance system, or unwanted, if cost is the primary concern, and the surveillance sensors are expensive or difficult to install or maintain. For simplicity, we leave the problem formulation with a single goal, leaving further goals as straightforward extensions to the above modeling.

Zones may have individual importance (values) for an attacker, which we reflect by assigning a *weight* α_i to the i-th zone. In absence of knowledge about the attacker's intentions or value that it assigns to the nodes, we may substitute our own security clearance or business value located in a zone or node as the value that it has for the attacker (worst-case assumption giving a zero-sum model).

The overall goal function for the optimization then comes to a sum of these scores, conditional on whether or not the zone is covered. Formally, the total score, measured in units of value for the zones, is then $\sum_{i \in \text{Zones}} \alpha_i \cdot m_i$.

Budget constraints can come in different forms, such as limits on the number of sensors, total cost to install sensors at different locations, and similar. If only the number of sensors is constrained by a maximum number s, but we do not care for cost, we add a constraint $\sum_j s_j \leq s$. If a sensor at location j has a cost c_j, and the total cost must not exceed a budget limit C, then the respective constraint is $\sum_j s_j \cdot c_j \leq C$. It goes without saying that the set of constraints is freely definable for the application at hand, and our choice is made only for illustration purposes.

The resulting problem is of mixed-integer kind, and stated as follows:

$$
\begin{aligned}
\text{maximize} \quad & z = \sum_i \alpha_i \cdot m_i \\
\text{subject to} \quad & \sum_{j \in N_i} s_j - m_i \geq 0 && \text{for all } i \text{ (full coverage);} \\
& \sum_j s_j \leq s; && \text{(limited number of sensors)} \\
& \sum_j s_j \cdot c_j \leq C; && \text{(limited installation budget)} \\
& s_j, m_i \in \{0, 1\} && \text{for all } i, j.
\end{aligned}
\qquad (8.2)
$$

Only the first and the last constraint are necessities, since the cost constraints can vary and depend on the context. The first constraint assures that the i-th zone is covered by a sensor, and the last constraint defines the variables as decision indicators. This prevents fractional assignments, like "0.5 sensors", which would not be meaningful.

The tricky part when implementing (8.2) is the definition of the set N_i, since this depends on the physical characteristics of the sensor (camera's field of view, motion detection's range and resolution, or similar). Further challenges and modifications of (8.2) include the following:

- Alternative goals: the definition of other relevant goals, such as location or motion detection errors, backup coverage (i.e., multiple sensors for the same location), up to matters of privacy infringement and user's convenience in presence of the surveillance system[24]. The latter two factors have no general mathematical model and must rely on empirical studies (interviews), besides being tightly constrained by legislation (e.g., the EU data privacy regulation or others).
- Imperfect performance: if a sensor is not fully reliable, then a straightforward modification to account for errors is letting some of the decision variables range over the unit interval $[0, 1]$ rather than $\{0, 1\}$. For example, $\lambda_{ij} \in [0, 1]$ can express a partial coverage of the i-th zone by a sensor, expressed as a (percentage) fraction of the zone's area. Likewise, redefining $m_i :=$ Pr(sensor detects adversarial activity in zone i) accounts for not fully reliable detection machinery. Though this leads to a partial relaxation of the mixed integer problem (which eases matters of solving it), this comes at the price of new challenges concerning reliable estimates for the now involved probabilities.

Even in an exact form like (8.2), closed form solutions to the optimization generally do not exist (or are at least unknown).

The above computational and modeling tractability matters merit the use of *simulation* for the surveillance problem. Here, we basically replace the goal function by a white-box simulator outputting the sought score(s), e.g., detection rates, motion errors, or others. Problem (8.2) models a *static* placement of sensors, but does not account for a rational adversary seeking to escape detection. A surveillance *game* extends the placement problem in this way:

Player 1 (the defender) is in charge of placing the sensors; a repeated game herein assumes that the sensory is mobile (thus, bringing the concept close to a patrolling game again, though the defender's strategy is no longer constrained to be a walk in a graph).

Player 2 (the attacker) plays "cops and robbers" with the defender, thereby bypassing locations that the defender would check (e.g., by placing sensors), and aiming at its own goal(s), which can include (but are not limited to):

- reaching a certain target zone: APTs, as being highly targeted attacks, typically fall into this category, since the goal is reaching a vulnerable spot to hit a maximum number of secondary targets in a single blow.

- causing damage in as many or as valuable zone(s) as possible: this is also relevant for APTs, insofar it models phases of information gathering to prepare for the planned attack.

A game can be useful to circumvent the NP-hardness of typical sensor placement problems, and be interesting to account for hypothesis on the adversary's behavior. This can be integrated by making the game Bayesian, with different adversaries playing against the defender. Modeling this is subject of independent lines of research (such as crime prediction or adversarial risk assessment; see Sect. 6.4), but we emphasize the following possibility: when a priori information about an attacker is available telling that some locations will be hit more likely than others, then we may convert the set of possible locations, i.e., the strategy space for the attacker, into an equally sized set of adversary *types*, each of which exclusively attacks the respective location. The a priori information on the likelihood of locations being under attack then naturally translates into the same likelihoods imposed on the attacker types, and a PBNE (see Definition 4.3) is the proper solution concept. We will come back to this in the context of the Cut-The-Rope game in Sect. 9.4.

In a simulation, we can let the attacker come over different a priori presumed routes, and also let it hit different targets with individual likelihoods (thus incorporating adversary behavior models in a different way), and test the performance of a proposed fixed sensor placement. This corresponds to a $(1 \times m)$-matrix game, in which the defender adopts a fixed layout of sensors, and the attacker takes one out of m possible routes. Transposing the payoff structure, we can also ask for the best among a given set of n sensor layouts against a fixed route of the attacker (playing an $(n \times 1)$-matrix game). If n sensor layouts have been fixed, and m routes for the attacker are presumed, then the game delivers an optimal randomized choice of sensor layout against the adversary.

In the full modeling with a matrix $\mathbf{A} = (a_{ij}) \in \mathbb{R}^{n \times m}$, only pure equilibrium strategies for the defender lend themselves to a physical layout of sensors. If the game has all equilibria in mixed strategies, then several options are open:

- approximate the optimal mixed strategy by a pure strategy available that is closest from above or below. Since the optimum must lie between the bounds that the minimax inequality (3.19) gives, whichever value in

$$\left\{ \max_{x=1,\dots,n} \min_{y=1,\dots,m} a_{ij}, \min_{x=1,\dots,n} \max_{y=1,\dots,m} a_{ij} \right\}$$

comes closer to the optimum can tell which action needs to be taken.

- treat the mixed strategy as if it would be pure, by a partial assignment of resources. For a mobile sensor (e.g., a mobile ID scanner) this could be a part-time installment somewhere, or for a team of several patrolling guards, a schedule that they may follow to check locations at random (similar to a patrolling game), or on-demand assignment of security guards that do not patrol, but rather protect their assigned location.

- adopt a greedy approach by taking the equilibrium probabilities as weights to prioritize pure actions. That is, we simply implement those actions with the highest probability in the equilibrium, to the extent that we can afford within the budget.

References

1. Alonso NZ, Terol PZ (1993) Some games of search on a lattice. Naval Res Logist 40(4):525–541. https://doi.org/10.1002/1520-6750(199306)40:4<525::AID-NAV3220400407>3.0.CO;2-B
2. Alpern S, Asic M (1985) The search value of a network. Networks 15(2):229–238. https://doi.org/10.1002/net.3230150208
3. Alpern S, Fokkink R (2014) Accumulation games on graphs. Networks 64(1):40–47. https://doi.org/10.1002/net.21555
4. Alpern S, Fokkink R, Kikuta K (2010) On Ruckle's conjecture on accumulation games. SIAM J Control Optim 48(8):5073–5083. https://doi.org/10.1137/080741926
5. Alpern S, Morton A, Papadaki K (2011) Patrolling games. Oper Res 59(5):1246–1257. https://doi.org/10.1287/opre.1110.0983
6. Alpern S, Lidbetter T, Morton A, Papadaki K (2016) Patrolling a pipeline. In: Zhu Q, Alpcan T, Panaousis E, Tambe M, Casey W (eds) Decision and game theory for security. Lecture notes in computer science. Springer International Publishing, Cham, pp 129–138
7. Alpern S, Lidbetter T, Papadaki K (2017) Periodic patrols on the line and other networks. ArXiv:1705.10399v1 [math.OC]
8. Appel K, Haken W (1989) Every planar map is four colorable, vol 98. American Mathematical Society, Providence. https://doi.org/10.1090/conm/098
9. Basak A, Fang F, Nguyen TH, Kiekintveld C (2016) Combining graph contraction and strategy generation for green security games. In: Zhu Q, Alpcan T, Panaousis E, Tambe M, Casey W (eds) Decision and game theory for security. Lecture notes in computer science. Springer International Publishing, Cham, pp 251–271
10. Bodor R, Drenner A, Schrater P, Papanikolopoulos N (2007) Optimal camera placement for automated surveillance tasks. J Intell Robot Syst 50(3):257–295. https://doi.org/10.1007/s10846-007-9164-7
11. Chvátal V (1975) A combinatorial theorem in plane geometry. J Combin Theory Ser B 18:39–41
12. Cormen TH, Leiserson CE, Rivest RL (1994) Introduction to algorithms. MIT Press, Cambridge
13. Debaque B, Jedidi R, Prevost D (2009) Optimal video camera network deployment to support security monitoring. In: 12th international conference on information fusion, 2009. IEEE, Piscataway, pp 1730–1736
14. Edmonds J, Johnson EL (1973) Matching, Euler tours and the Chinese postman. Math Program 5(1):88–124. https://doi.org/10.1007/BF01580113
15. Fisk S (1978) A short proof of Chvátal's Watchman theorem. J Combin Theory Ser B 24(3):374. https://doi.org/10.1016/0095-8956(78)90059-X
16. Fox CR, Bardolet D, Lieb D (2005) Partition dependence in decision analysis, resource allocation, and consumer choice. In: Zwick R, Rapoport A (eds) Experimental business research. Springer, Dordrecht/Berlin, pp 229–251. https://doi.org/10.1007/0-387-24244-9_10
17. Garey MR, Johnson DS (1979) Computers and intractability. Freeman, New York
18. Hörster E, Lienhart R (2006) On the optimal placement of multiple visual sensors. In: Aggarwal JK, Cucchiara R, Prati A (eds) Proceedings of the 4th ACM international workshop on video surveillance and sensor networks – VSSN'06, p 111. ACM Press, New York. https://doi.org/10.1145/1178782.1178800

19. Indu S, Chaudhury S, Mittal N, Bhattacharyya A (2009) Optimal sensor placement for surveillance of large spaces. In: 2009 third ACM/IEEE international conference on distributed smart cameras (ICDSC). IEEE, pp 1–8. https://doi.org/10.1109/ICDSC.2009.5289398
20. Kikuta K, Ruckle WH (2002) Continuous accumulation games on discrete locations. Naval Res Logist 49(1):60–77. https://doi.org/10.1002/nav.1048
21. O'Rourke J (1987) Art gallery theorems and algorithms. The international series of monographs on computer science, vol 3. Oxford University Press, New York
22. Papadaki K, Alpern S, Lidbetter T, Morton A (2016) Patrolling a border. Oper Res 64(6):1256–1269. https://doi.org/10.1287/opre.2016.1511
23. Pita J, Tambe M, Kiekintveld C, Cullen S, Steigerwald E (2011) GUARDS – innovative application of game theory for national airport security. In: IJCAI 2011, pp 2710–2715 . https://doi.org/10.5591/978-1-57735-516-8/IJCAI11-451
24. Rass S, Alshawish A, Abid MA, Schauer S, Zhu Q, de Meer H (2017) Physical intrusion games – optimizing surveillance by simulation and game theory. IEEE Access 5:8394–8407. https://doi.org/10.1109/ACCESS.2017.2693425
25. Rass S, König S, Schauer S (2017) On the cost of game playing: how to control the expenses in mixed strategies. In: Decision and game theory for security. Springer, Cham, Switzerland [S.l.], pp 494–505

Chapter 9
Optimal Inspection Plans

Truth is confirmed by inspection and delay; falsehood by haste and uncertainty.

Tacitus

Abstract In this chapter, we consider games for the computation of optimal strategies of how, how often, and when to inspect along a production line, or general industrial process. We review basic concepts of statistical tests, conducted whenever the defender chooses its action to "inspect", and to understand cheating strategies for the adversary trying to escape detection along the statistical test. This non-detection game is then embedded into an outer sequential game over several stages of inspection, accounting for limited resources and possibilities of the defender to check repeatedly. We also consider inspections as a defense pattern against advanced persistent threat (APT), with two models suitable for two distinct type of APTs: the FlipIt game is discussed as a model when the APT's goal is to gain longest possible control over an infrastructure, without wishing to damage or destroy it permanently. Complementary to this is the Cut-The-Rope game about defending against an APT whose goal is hitting a vital asset and to destroy or at least permanently damage a critical infrastructure.

9.1 Repeated, Independent Inspections

An inspection game typically runs between two players, which we hereafter call the *inspector*, player 1, and the *inspectee*, player 2. In many practical cases, the "inspectee" will not be a rationally acting entity, but rather some object of inspection that is only partially under an adversarial influence. For example, the inspector could be in charge of product quality control. The products under check themselves are clearly not acting against the intentions of the inspector, but an adversary seeking to damage the company's reputation may fiddle with the production line to cut down the production volume by decreasing quality or customer satisfaction due to low

© Springer Nature Switzerland AG 2020
S. Rass et al., *Cyber-Security in Critical Infrastructures*, Advanced Sciences and Technologies for Security Applications,
https://doi.org/10.1007/978-3-030-46908-5_9

quality that made it through the checks, because the (rational) adversary tricked the inspector to miss the deficiencies or flaws.

Our exposition here follows the models presented in [1], and we consider the object under inspection to define only the "playground" on which both, the honest inspector and the hostile inspectee act with opposite intentions as follows: The inspector seeks to detect illegal behavior, and suffers a loss of -1 for every violation of the inspectee that went undetected. For detecting a violation, the inspector gets the payoff $-a$ (with $a \in (0, 1)$), and the inspectee receives $-b$ with $b > 0$. If a false alarm is raised, then the inspector gets $-e$ and the inspectee receives $-h$ with $0 < h < b$. This (simple) model assumes that alarms are for both parties undesirable, whereas variations of it may declare a high rate of false alarms as an attack strategy on its own, driving the alarm response teams into *alert fatigue*. The change to the above setting is just a change of sign in the inspectee's payoff towards receiving $+h$ for a false alarm. We leave this change of numeric value without further discussion, except for stressing that the constraints and relations put on the values a, b, e and h may vary and depend on the application. For example, one could impose $e < a$ to express that a false alarm is less inconvenient than a detected violation that equivalently indicated a failure of the defenses violation (since the latter demonstrates that the safety precautions have already failed).

For the inspection itself, the inspector runs a (statistical) test on the objects or artefacts presented to him, maintaining an *alarm set* as a collection of events that raise an alarm. Generally, the inspector will decide upon alerting others by sampling either *attributes* or *variables* related to the alarm conditions:

- Attribute sampling means testing or estimating the *count* of items in a population that has certain characteristic features, based on a small selection of items (a matter of classical statistics, like inferring the number of smokers in a public community, based on a survey among a much smaller group).
- Variable sampling regards estimates of *values*, averages or other characteristics of some variable (on a continuous scale), to assess whether or not the performance lies within the limits told by the specification. For example, if the variable is the amount of electromagnetic radiation emitted by a device, the sampling aims at judging the whole batch regarding average performance within the (legal or technical) specification's limits.

Both techniques are closely related and have applications in accounting and auditing, environmental control, security checks at airports, crime control (by optimized patrolling), intrusion detection systems, and many others.

The alarm set defined by the inspector is generally a specification of conditions or circumstances that would raise an alarm. The form of these is generally unspecific and depends on the application.

Example 9.1 (possible definitions of alarm sets) Some, among many more possibilities to define alarm sets include:

- For standard statistical tests (e.g., quality checks): test statistics (e.g., F-tests of samples against reference groups), with the alarm set being defined as

$\{z > q_{1-\alpha}\}$, where z is the test statistic, and $q_{1-\alpha}$ is the respective $(1-\alpha)$-quantile of the test distribution. We will revisit the particular instance of likelihood ratio test in Example 9.2 later.

- For intrusion or malware detection, this could be a match of certain (known) virus patterns found within files, emails or similar. Formally, the representation would again be a set {virus pattern 1, virus pattern 2, ..., }, where the alarm is raised if the inspector comes across any of these patterns.
- For email filtering, the alarm set could be the event of sending more than a certain limit of emails per hour (e.g., any mail traffic above 100 mails per hour is suspicious activity).
- For tax fraud [11], one (among many more sophisticated indicators of fraud) regards the distribution of the first (leading) digit: those should adhere to the Benford distribution [2], which puts the mass $\Pr(\text{first digit} = d) = \log((d + 1)/d)$ for $d = 1, 2, \ldots, 9$. Any "significant" deviation from this distribution is a (weak) signal of manipulation in the data (not necessarily adversarial, but at least a reason to take a second look).

We remark that the test of Benford's law has much wider applicability than just tax fraud, and applies to many data series under the condition that these arise from complex interplay of at least two sources, and have neither a systematic assignment (like serial numbers, account numbers, psychological prices, etc.) nor been artificially modified (e.g., truncated, rounded, or similar).

- For credit rating [15], the alarm set could be the set $\{z > k\}$ for $0 < k < 1$ setting the threshold of the scoring z that typically comes out of a logistic regression model on the customer's attributes. It hence depends on the customer, and is calculated during an "inspection" of the customer to assess credit eligibility. The threshold value k is a constant and defines the desired minimal score for eligibility of a credit grant.

◇

As an inspection game is centered on the inspector performing a statistical test, whose outcome determines the revenues for both players, four basic cases are possible, which commonly come in arrangement of a *confusion matrix*:

In statistical test theory, false-negatives are called errors of the *first kind*, and false-positive events are errors of the *second kind*. Most statistical tests (in the classical sense of test theory) let the test designer fix an acceptance level α for errors of first kind, which in turn sets the bar on when to reject the test's hypothesis (in the example above and many others, the bar is the $(1 - \alpha)$-quantile of some test

Table 9.1 Confusion matrix – general structure

		Actual condition	
		Everything is okay	Illegal behavior
Condition decided	Everything is okay	True positive	False positive
Upon the test	Illegal behavior	False negative	True negative

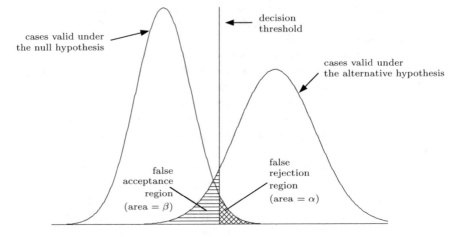

Fig. 9.1 Negative correlation between errors of first and second kind

distribution). It is basically the probability of raising an alarm though everything is actually okay, and typical values for α used in test theory are 5% or 1% (Table 9.1).

More subtle is the likelihood β for errors of the second kind. We typically have only implicit control over this parameter (neither can we choose or fix it a priori), and it negatively correlates with the value of α. Visually, one can think of the test separating "good" from "bad" cases, but both are randomly distributed around the respective null- and alternative hypothesis. The area to the left and right of the decision threshold is precisely the value of α and β that measures the performance of the test. Figure 9.1 illustrates how the two kinds of errors correlate when testing a null hypothesis against an alternative. Figure 9.1 visually explains the negative correlation by shifting the decision bar (vertical line) to the left or right, thus shrinking and enlarging the α and β areas complementary to one another.

Since the inspector can choose only one of the two error likelihoods, typically α, the inspectee can try maximizing the error of the second kind to escape detection. This error is governed by the alternate test hypothesis that the inspector can only assume, but which the attacker may control (see Fig. 9.1): the inspectee's strategy would be shifting the curves close to one another, in order to maximize the second kind error, i.e., the chances to go undetected with a violation.

The inspection *game* thus breaks down into two nested games:

1. Maximize the error of second kind to escape detection by the inspector, who only can choose an acceptable likelihood for false alarms. This is the *non-detection game*.
2. Maximize the actual payoff during the inspection itself, with the payoffs defined as above. This is the actual *inspection game*.

Towards a formal model, let us write δ for the statistical test, and let $\alpha(\delta)$ be the error of first kind in the instance of δ that the inspector defines. For a fixed test δ the strategy of the inspector is only the choice of an *alarm set* Z, generically taking the form $Z = \{\delta(\mathbf{x}) > k\}$, where $\mathbf{x}$ is the data that the inspector has, $\delta(\mathbf{x})$ is the test-statistic, and k is the threshold defined for the test. This generic formula includes, among others, all the examples given above.

For the inspectee, a strategy consists of a test cheating strategy ω (whose actual form depends on how the test works), and a probability q for acting illegal in this way in the actual inspection game. The value q is what the defender optimizes with help from game theory. Often, there may be infinitely many ways of fiddling with the test, but as we will later see (in Sect. 9.1.1), it is often more handy to stick with a finite set of possible adversarial actions. We let AS_2 be the set of all violation strategies (allowing for an infinitude or continuum, but working out examples using only finitely many).

The *non-detection game* is zero-sum and consists of the inspector choosing the best test among the available ones, to achieve the desired false alarm probability. Let us write $AS_1(\alpha)$ to mean the set of candidate tests with the chosen value α. Any (strategic) choice $\delta \in AS_1(\alpha)$ in turn defines the alarm set Z in the subsequent inspection game. The inspectee, in turn, tries to maximize the rate of undetected violations, β, by choosing ω from the set AS_2 of possible violation strategies, i.e., the non-detection game has the saddle-point value

$$\beta(\alpha) := \min_{\delta \in AS_1(\alpha)} \max_{\omega \in AS_2} \beta(\delta, \omega). \tag{9.1}$$

Again, graphically, the inspector's choice is between tests, which determine the shape of the distributions in Fig. 9.1. The shaping is done by choosing different tests, while the optimization of β is up to the adversary's choice of cheating strategy. Note that the inspection game is of Nash-type, while the inner non-detection sub-game (that is played whenever the inspectee chooses to cheat in the inspection game), is of Stackelberg-type, since the inspectee's cheating strategy follows from the test that the inspector applies. Opposed to this are the actions of the inspectee, trying to maximize the area β to the right of the decision bar, which depends on the value α that is known to both players.

In game theoretic thinking, the inspector wants to maximize the chances of detecting illegal behavior, by testing for the *normal* behavior formulated as some *null hypothesis* H_0. If the adversary acts illegally, then the *alternative hypothesis* H_1 is true, so that H_0 should be rejected. The probability for this to happen is known as the *power* of the test:

Definition 9.1 (Power of a statistical test) Let δ be a statistical test of some null hypothesis, and let H_1 be any alternative to H_0. The *power* of the test δ is the probability of rejecting H_0 if the alternative H_1 is true.

Example 9.2 will show an example calculation of the power. Rephrasing this definition in formal terms, with

- H_0 being the "baseline", in the sense of regular/expected/normal situation that the inspector wants to see, and
- H_1 being a potential adversarial influence by the inspectee,

then we have

$$\beta(\alpha) := 1 - (\text{power of the test against } H_1) = 1 - \Pr(\text{reject } H_0 | H_1 \text{ is true}),$$

where the dependence on α is herein implicit as it determines the testing procedure (more specifically the boundary between rejection and acceptance regions for the test statistic. It will follow in precise terms later as Theorem 9.1).

The power of a test is usually not exactly computable, unless all alternative hypotheses occur with known probabilities. Since those are exactly the strategies of the adversary deviating from the normal behavior being the null hypothesis H_0, one can simplify matters by assuming only a *finite* set of actions for the attacker, each of which results in a different alternative hypothesis H_1. For a *fixed* such hypothesis, we can work out the power (see Example 9.2), and concretely construct the most powerful test $\delta^* \in \Delta$ (in terms of optimal β). This is then the explicit solution to the inspector's minimization problem in (9.1).

To assure that (9.1) has a saddle-point, we must in addition impose the following standard assumption on β as a function of α:

$$\beta(0) = 1, \beta(1) = 0 \text{ and } \beta \text{ is convex and continuous at } 0. \tag{9.2}$$

9.1.1 Solving the Non-detection Game

For an a priori fixed α, the inspector can directly construct the optimal test for any given alternative hypothesis, i.e., adversarial action, H_1, against the expected (normal) behavior H_0.

For the test, let X be the inspected variable, which, under normal conditions, has the distribution $\Pr(X | H_0)$ with density $f_0(x)$. Likewise, under adversarial influence H_1 of the inspectee, X has a different distribution $\Pr(X | H_1)$, with density $f_1(x)$. Figure 9.1 illustrates these densities as the left and right bell-shaped curves.

The test statistic is the likelihood ratio

$$\Lambda(x) := \frac{f_1(x)}{f_0(x)}, \tag{9.3}$$

with the convention that $\Lambda := 1$ if the quotient takes a "0/0" form, and letting $\Lambda := \infty$, if the denominator is zero. These two special cases should either not occur, or correspond to situations where the test as such becomes trivial or deterministic. We will go into more details along Remarks 9.2 and 9.3. Let the auxiliary function

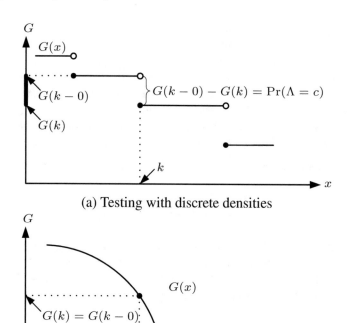

(a) Testing with discrete densities

(b) Testing with continuous densities

Fig. 9.2 Example plots of $\alpha(x)$

$G(\lambda) := \Pr(\Lambda(X) > \lambda | H_0)$ denote the likelihood for the quotient Λ to exceed a given value x under regular circumstances H_0.

The decision threshold (see Fig. 9.1) is any value k that lies within the bounds

$$\lim_{\varepsilon \to 0} G(k - \varepsilon) =: G(k - 0) \leq \alpha \leq G(k), \tag{9.4}$$

with the shorthand notation $G(k-0)$ meaning the limit towards k upon approaching from the left. To determine k, note that G is generally non-increasing and right-continuous. If f_0, f_1 are both continuous and $f_1 > 0$ everywhere, then k is the unique solution of the equation $G(k) = \alpha$, as in Fig. 9.2b. Otherwise, if Λ has jumps because f_0 and f_1 are discrete or categorical, *any* value on the vertical axis that lies "between the jump values" is admissible. Figure 9.2a marks an example such interval of possible choices for k with a thick line on the abscissa.

Remark 9.1 (Practical approximation of the decision threshold) A difficult step is usually the determination of the function G, which amounts to computing the probability of $\Lambda(X)$ when X is distributed assuming the null hypothesis. If a sufficient lot of "training data" regarding the normal situation (without any adversarial influence) is available, then we can approximate k empirically: let the data be a set of samples from X under H_0 enumerated as $x_1, x_2, \ldots, x_N$. We can construct an empirical distribution estimate $\hat{F}_\Lambda$ from the transformed data $\Lambda(x_i)$ for $i = 1, 2, \ldots, N$. The sought k is obtained from the quantile function of $\hat{F}_\Lambda$. This process is widely supported by software, such as R [12], which has the built-in functions `ecdf` to construct $\hat{F}_\Lambda$ directly from the precomputed values $\Lambda(x_1), \ldots, \Lambda(x_n)$, and the generic function `quantile` to compute k from.

With k being determined (numerically), the test's decision is as follows:

$$\begin{aligned} &\text{reject } H_0 \text{ if } \Lambda(x) > k, \text{ and} \\ &\text{accept } H_0 \text{ if } \Lambda(x) < k. \end{aligned} \tag{9.5}$$

The case of equality $\Lambda(x) = k$ will occur with zero probability if f_0, f_1 are continuous distributions, so the inspector does not need to worry about this possibility. This can also happen for discrete or categorical densities f_0, f_1. Both cases correspond to the optimum being a "pure" strategy for the inspector in this game.

Otherwise, when the event $f_0(X) = k \cdot f_1(X)$ has a nonzero chance to occur, then the test can be inconclusive for the given data (since the choice of k is such that we could make it in favour of both possible outcomes, by choosing k near either end of the interval; cf. Fig. 9.2a). In that case, we must repeated the test on fresh data. The final decision of the test then depends on the total set of collected test outcomes. This corresponds to the situation where the inspector's optimal strategy is thus a *randomized strategy*. The test's decision variable can then be defined between as the value

$$(\alpha - G(k))/(1 - G(k)) \in (0, 1). \tag{9.6}$$

The admissibility of this testing prescription is captured by Theorem 9.1. We state the result in full detail (from [9]). In a slight abuse of notation, let us write δ to denote the (entire) statistical test, applied to the data x in possession of the inspector. The concrete value $\delta(x)$ will be the decision outcome of the test, including the inconclusive case with the fractional value given above in (9.6).

Theorem 9.1 (Neyman-Pearson Lemma [9, 10]) *Let f_0, f_1 be densities conditional on a null-hypothesis H_0 and alternative hypothesis H_1. Define the likelihood ratio $\Lambda(x)$ as in (9.3), using (9.6) for the case of equality. Then, the following holds:*

1. First kind error level assurance: *The so-defined statistical test δ with the constant k satisfies*

$$\mathbb{E}_{H_0}(\delta(X)) = \alpha. \tag{9.7}$$

2. Optimality: *If a test satisfies (9.7) and (9.5), then it is the* most powerful *(in the sense of Definition 9.1) in testing H_0 against H_1 among all tests at level α (probability of error of the first kind).*
3. Uniqueness: *If a test is most powerful at level α for testing H_0 against H_1, then for some k it satisfies (9.5) with probability 1 (relative to the measure induced by H_0 and H_1). It also satisfies (9.7), unless there exists a test at level $< \alpha$ and with power 1 (see Remark 9.2 below).*

Remark 9.2 Taking $k = \infty$ in the theorem means rejecting whenever $f_1(x|H_1) > 0$. This is the most powerful test for level α, and has power 1, equivalently $\beta = 0$, i.e., no error of the second kind. However, it is also most impractical in all cases where the distributions induced by H_0 and H_1 overlap, and the test will basically reject whenever there is even the slightest chance of the alternative hypothesis to hold; however, this may include cases where there is simply a "natural" randomness in the data of the inspector that *could* but *not necessarily* point out an adversarial activity.

However, if the adversary's actions H_1 are, such that under any possible H_1, the supports of the densities f_0 and f_1 are disjoint, then any action of the attacker will have an unquestionable effect in the system that the inspector will detect. Such cases are usually coverable by *rule-based* alerting systems, which simply test for conditions that should never occur in a normal system state. If this is applicable, then the non-detection game becomes trivial, since there is no way to hide for the attacker due to the existence of a (100% reliable) deterministic test. This is again consistent with Theorem 9.1, since it admits that there are better tests, but only and precisely in the situation just described.

Remark 9.3 Likewise as in Remark 9.2, we can reject whenever $f_0(x|H_0) = 0$, i.e., only when there is zero chance of H_0 to hold. This is most optimistic towards accepting H_0 by taking $k = 0$ in Theorem 9.1, and the most powerful test at level $\alpha = 0$, i.e., without error of first kind. However, it also maximizes the error of second kind, obviously, much in the interest of the adversarial inspectee.

Example 9.2 (adapted [9]) Suppose that a system is such that the inspector measures (and checks) a certain parameter X that should, under normal circumstances, have zero mean and a known variance σ^2. As an example, you may consider X to be the deviation from the normal supply voltage level of an energy provider's network. Under regular conditions, the value should be, e.g., 220 V, and the *deviation* X from this nominal voltage is $X \sim \mathcal{N}(0, \sigma^2)$. Assuming a hacking attack, suppose that the adversary managed to deactivate parts of the energy system to "overpower" it and to cause damage. In that case, there may be a positive deviation X with the same variance.

The non-detection game is then about the inspector testing for such events, while the attacker wants to be stealthy and escape the detection.

Now, we design the optimal test for the inspector. The null-hypothesis is H_0 : $\mu = \mathbb{E}(X) = 0$, and the alternative hypothesis is $H_1 : \mu = \mathbb{E}(X) = \xi$, for some ξ that the inspector fixes a priori. A possible choice could be the amount of voltage excess that the system could handle or that naturally could occur (e.g., putting $\xi \approx 3\sigma$). Whatever rationale is behind the choice of ξ, let us assume that the inspector alerts the incident response forces if he determines that the nominal voltage hits the upper acceptable limit.

The likelihood ratio is the quotient of two normal distributions, which is

$$\frac{f_{\mathcal{N}}(x|\mu = \xi, \sigma^2)}{f_{\mathcal{N}}(x|\mu = 0, \sigma^2)} = \frac{\exp(-\frac{1}{2\sigma^2}(x - \xi)^2)}{\exp(-\frac{1}{2\sigma^2}x^2)} = \exp\left(\frac{\xi \cdot x}{\sigma^2} - \frac{\xi^2}{2\sigma^2}\right)$$

Since the exponential function is strictly increasing and $\xi > 0$, we can resort to the simpler term $\exp\left(\frac{\xi \cdot x}{\sigma^2}\right)$ for the maximization to get the region where $f_1(x)/f_0(x) > k$. This is, up to multiplicative constants, again a Gaussian density, so that we can equivalently look for the value k' for which $\Pr(X > k') = \alpha$, where the probability distribution is Gaussian. This value is $k' = \sigma \cdot z_{1-\alpha}$, where $z_{1-\alpha}$ is the $(1 - \alpha)$-quantile of the standard normal distribution. Thus, the test rejects H_0 if and only if the likelihood ratio is $> \sigma \cdot z_{1-\alpha}$ for the chosen α.

What is the power of this test against the (specific) H_1? It is given by region of rejection for H_0 if H_1 is true, i.e., it is the integral over the Gaussian density under H_1 for the region $> k = \sigma \cdot z_{1-\alpha}$. Conversely, for the specific violation strategy $H_1 \in AS_2$, the chance for the inspector to accept the observation as "normal" (adopting H_0), is

$$\beta(\alpha) = \int_{-\infty}^{\sigma \cdot z_{1-\alpha}} f_{\mathcal{N}}(t|\mu = \xi, \sigma)dt.$$

Taking, for example, $\sigma = 1, \xi = 1$ and $\alpha = 0.05$, we get $z_{1-\alpha} \approx 1.644854$ and $\beta(0.05) \approx 0.740489$.

Note that $\beta > \alpha$, which is a general fact that holds for *every* so-constructed most powerful test, unless $H_0 = H_1$.   ◇

Remark 9.4 Practical statistical tests often deliver the so-called *p-value* as an auxiliary output of the test function. The common use of this value is to accept the null hypothesis whenever the *p*-value is larger than α. Formally, the *p*-value is the area under the curve to the left of the test statistic's value. It thus measures how strongly the data contradicts the null hypothesis, and small *p*-values indicate that the observed data is unlikely under the null hypothesis.

Let us now return to the solution of (9.1): The definition of Λ already shows why it is usually helpful to pin down the alternative hypothesis, and to this end, assume a *finite* set AS_2 of actions for the adversary. Since the inspectee maximizes over violation strategies ω in (9.1), each of these defines its own alternative hypothesis H_1 by shaping the according conditional distribution to maximize the false acceptance region of size β (see Fig. 9.1). Thus, the inspector needs to construct an optimal test against the worst such alternative hypothesis, which is doable if there are only finitely many violation strategies $\omega \in AS_2$. Given these violation possibilities, the inspector's task is the determination of the resulting conditional densities $f_1(x|H_1(\omega))$ when $H_1(\omega)$ is the alternative hypothesis resulting from the adversary taking action $\omega \in AS_2$. The process is much like the *consequence assessment* as part of the risk analysis in risk management (see Chap. 6) or the general distribution-valued games of Sect. 4.5.

Let $AS_2 = \{\omega_1, \omega_2, \ldots, \omega_m\}$ be an enumeration of (presumed) violation strategies, with corresponding densities $f_{1,j} = f_1(x|H_1(\omega_j))$ for the alternative hypotheses implied by action j. For each of these, let δ_j be the individually optimal test as constructed like above. Since the inspector is unaware of what action the inspectee took, the inspector will need to pick a test δ_i for $i \in \{1, \ldots, m\}$ at random, whose power against the inspectee's choice $\omega_j \in AS_2$, i.e., the alternative hypothesis $f_{1,j}$ is the value (cf. Fig. 9.1),

$$\beta_{ij} = \int_{-\infty}^{k_i} f_{1,j}(t)dt,$$

when k_i is the decision threshold for the i-th test. The values β_{ij} for $i, j = 1, \ldots, m$ form an $(m \times m)$-matrix that represents problem (9.1) as a zero-sum matrix game, whose saddle-point value is exactly $\beta(\alpha)$.

9.1.2 Solving the Inspection Game

By definition, the game is of leader-follower type, since the inspector follows a publicly known policy (not the least so to establish transparency as a confidence-building measure), and the inspectee can adapt to it.

Written as a two-person game, the inspector's action space concerns only the choice of $\alpha \in [0, 1]$, which determines the test by techniques like the Neyman-Pearson Lemma (see Sect. 9.1.1). The inspectee's choice is only whether or not to violate, since if the latter entails the best violation strategy adapted to the test that the inspector uses. Note that the game in extensive form, shown in Fig. 9.3 has an information set for the inspector, since this one does not know whether a legal action or violation occurred. This secret choice by the inspectee determines whether H_0 or H_1 is true. However, the test is reliable when it rejects H_0, i.e., indicates a violation.

The game in Fig. 9.3 generally has a unique mixed equilibrium: Let $q \in [0, 1]$ be the probability of a legal action, and $(1 - q)$ be the likelihood for a violation, and let q^* be the optimal choice. If $q^* = 0$, then the inspector would have chosen a test δ^* without false alarms, but this necessarily increases the inspectee's chances to go undetected (since it increases the errors of second kind). Conversely, if $q^* = 1$, then the inspector could maximize its payoff by always raising an alarm, irrespectively of what happened. However, if $h < b$ (as we assumed above), then the optimal choice of the inspectee would not be $q^* = 1$ any more. The uniqueness of the strategies is only up to equivalence of their effects, since all equilibria strategies deliver the same value as the game is zero-sum.

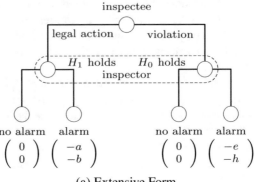

(a) Extensive Form

Inspectee \ Inspector	legal action (H_0)	violation (H_1)
$\alpha \in [0,1]$	$-h \cdot \alpha$ $-e \cdot \alpha$	$-b + (1+b) \cdot \beta(\alpha)$ $-a - (1-a) \cdot \beta(\alpha)$

(b) Normal form

Fig. 9.3 Leader-follower inspection game with perfect recall

Theorem 9.2 ([1]) *Assume that for any false alarm probability* α, *the non-detection game (9.1) has a value* $\beta(\alpha)$ *satisfying (9.2), where the optimal violation procedure* ω^* *does not depend on* α. *Then, the inspection game in Fig. 9.3 with parameters* a, b, e *and* h *has an equilibrium* (α^*, q^*), *where:*

- α^* *is the solution of the equation*

$$-h \cdot \alpha^* = -b + (1-b)\beta(\alpha^*),$$ (9.8)

where β *is given by (9.1) for the optimal test to a value* α *using the Neyman-Pearson Lemma (see Sect. 9.1.1).*
- *the value* q^* *is*

$$q^* = \Pr(\text{legal action}) = \frac{e}{e - (1-a)\beta'(\alpha^*)},$$

in which β' *is any sub-derivative of the function* β *at* α^*.

In response to α, *the inspectee would thus violate, whenever* $\alpha < \alpha^*$, *and act legally if* $\alpha \geq \alpha^*$.

To operationalize Theorem 9.2, one may follow these steps:

- Adopt assumptions on what the adversary could do to shape the distribution of the alternative hypothesis H_1 in the worst case for the inspector. Together with the distribution of how things should look like in the normal case (situation H_0), construct a likelihood ratio test statistic, using the Neyman-Pearson Lemma, i.e., Theorem 9.1.

  Upon input of a significance level α, implement an algorithm to compute the value β, resulting from the distribution of the likelihood quotient, which may need an analytical workout beforehand.
- Do a numeric (perhaps black-box) solution of (9.8), using the just implemented procedure to determine α^*. Note that this is the point where the hypothesis of the optimal violation strategy to *not* depend on α comes into play: if it were dependent on α, the two steps would become woven with each other, and we would not have a function to compute β on input α. By the above stage of identification of worst-case attack behavior, which is nothing else than fixing the optimal ω^* before setting any value for α, we assure this hypothesis of Theorem 9.2.

9.2   Sequential and Dependent Inspections

Some applications may restrict the inspector to perform a limited number of checks. As an example, consider a production line of n stages, along which at most $m < n$ inspections happen. Let $I(n, m)$ denote the equilibrium payoff for the inspector, and let $V(n, m)$ be the equilibrium payoff for the inspectee (violator). Suppose further that a legal action has zero payoff for both players (since the inspector did what was his job anyway, and the violator failed), and let a caught violation cause a loss of $-a$ for the inspector and $-b$ for the attacking inspectee.

Though the game spans several stages, it is 2×2 and as such has an explicit solution (given later in Sect. 12.3.2), attained if the inspectee is indifferent between legal action and violation. Thus, we come to a partially recursive equation for V and I,

$$V(n, m) = \frac{b \cdot V(n - 1, m) + V(n - 1, m - 1)}{V(n - 1, m - 1) + b + 1 - V(n - 1, m)} \tag{9.9}$$

with initial conditions given by

$$\left.\begin{array}{l} I(n, n) = V(n, n) = 0 \quad \text{for } n \geq 0; \\ I(n, 0) = -1, V(n, 0) = 1 \text{ for } n > 0. \end{array}\right\} \tag{9.10}$$

Theorem 9.3 *Assume that the recursive inspection game with payoffs as in Fig. 9.4 is such that after each stage, both players are informed about whether there has been a violation (i.e., both players know about the game state after a stage). Furthermore, let the initial conditions be (9.10). Then, the inspectee's equilibrium payoff is*

Fig. 9.4 Sequential
inspection game Example

Inspector \ Inspectee	legal action		violation
inspection	$I(n-1,m-1)$	$V(n-1,m-1)$	$-b$ $-a$
no inspection	$I(n-1,m)$	$V(n-1,m)$	1 -1

$$V(n,m) = \frac{\binom{n-1}{m}}{\sum_{i=0}^{m}\binom{n}{i}b^{m-i}},$$

and the inspector's payoff is

$$I(n,m) = -\frac{\binom{n-1}{m}}{\sum_{i=0}^{m}\binom{n}{i}(-a)^{m-i}}.$$

The equilibrium strategies are determined inductively (starting from $n, m = 0, 1, \ldots$, using the above quantities by substituting the values into the game matrix in Fig. 9.4, and computing the equilibrium strategies from the resulting (numeric) 2×2-matrices.

The applicability of Theorem 9.3 is sort of debatable by its hypothesis on the player's state to be common knowledge after each stage; this may not be accurately true in reality, since the inspector will typically not be informed about a violation that was missed. Changing the model accordingly to account for this lack of information renders the recursive structure generally invalid, and the game is usually not directly solvable any more. However, if the game is 2×2 or diagonal, one can obtain closed form expressions like (9.9) from the formulas given in Chap. 12 to get a partial recursion equation to define the equilibrium point, and numerically, i.e., iteratively, solve it starting from the initial conditions. In practice, this may already be enough to work with.

We devote the next Sect. 9.3 to games that accounts for such a *stealthy* inspectee.

9.3 Inspections Against Stealthy Takeover

Suppose that the inspector chooses random time instances to check and perhaps reset certain system components, thereby dropping out a potential intruder having conquered the component in the past (e.g., by installing a trojan backdoor or similar). Whenever the attacker becomes active, we assume that a component is captured, and (unknowingly) recaptured later by the inspector (since the attacker is stealthy). The game described and analyzed in the following carries the name `FlipIt` in the literature. It models the first type of APT as introduced in Sect. 1.3.

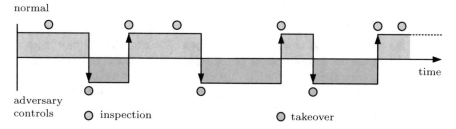

Fig. 9.5 Temporal takeovers against inspections over time

The game is about both players optimizing their strategies for inspection and takeover, where the goal function is like an indicator pointing up or down, depending on who is currently having control over the component.

Figure 9.5 illustrates the situation over time, with bullets marking points in time where inspections occur. Those "flip" the control indicator accordingly, and the overall revenue in the game depends on the time that either player has the component (or system) under its control, taking costs for moves into account. To start with, let $C_i(t) \in \{0, (-1)^{i-1}\}$ for $i = 1, 2$ be an indicator variable telling whether the i-th player has the control at time t; we let player 1 be the defender, and player 2 be the attacker. The *average* time of control for a player is then $\gamma_i(t) = \frac{1}{t} \int_0^t C_i(\tau) d\tau$ (this roughly corresponds to the average payoff obtained in a repeated game, discounted payoff in a stochastic game or continuous payoff functions in differential games). Apparently, the value $C_i(t)$ depends on the inspections and takeovers of both players, and we shall keep this dependence in mind implicitly, so as to ease notation by omitting the strategies for the moment.

Taking over a component is typically associated with costs; at least the defender may need to do a temporary shutdown for a patch, upgrade or reset of system parts. If, up to time t, the i-th player ($i \in \{1, 2\}$) took a total of $n_i(t)$ actions, each imposing a cost of $k_i \in \mathbb{R}$, then we can refine the game's payoff into the average time of control, minus the average efforts for it, formally being $u_i(t) := \gamma_i(t) - \frac{n_i(t) \cdot k_i}{t}$.

Remark 9.5 The practical challenge of putting this setting to work lies in the (reliable) estimation of the cost parameters k_1, k_2. The dependence of the equilibria, describing the optimal inspection and attack patterns below, depend on how these two quantities relate to each other. From the inspector's perspective, estimating k_1 may require careful efforts but is generally doable, yet the information to base an estimate of k_2 on is much sparser. One possible source of information are CVSS, since these typically involve assessments of the skill level, cost or general complexity entailed by a specific attack. This information may aid an expert estimate of the cost k_2.

For the actual game, let us assume that the inspection strategy is chosen once and forever at the beginning of the game, and the optimization is towards an infinite time horizon, by setting the player's revenue to

$$u_i := \liminf_{t \to \infty} u_i(t), \qquad (9.11)$$

where the actual value of u_i depends on the inspection and attack regime, both of which determine the temporal shape of the indicator function $C_i(t)$, as illustrated in Fig. 9.5. Having defined the utility function for both players, let us now write $u_i(s_1, s_2)$ to make the dependence on the strategies explicit, since the equilibria will depend on the choice of s_1, s_2, and are known for some concrete inspection regimes.

9.3.1 Periodic Inspections

In a *periodic inspection with random phase*, the defender inspects exactly every δ time units, with the first inspection occurring at a randomly chosen time $t_0 \sim \mathcal{U}(0, \delta)$. The random phase is thus the value t_0. For example, this regime would be played by an IT administrator conducting a weekly checkup (the time unit δ would be one week, accordingly), always on the same day of the week. The weekday is the random phase, and is chosen uniformly at random at the beginning.

This inspection strategy is widely deterministic, with the only uncertainty for the attacker being in the random phase. Consequently, we must assume that the attacker never knows or ever figures out when the inspections occur. Under this (practically strong) hypothesis, the game admits explicit equilibria. To distinguish the strategies from others, we write $periodic(\alpha)$ to denote a periodic strategy by either player. Then, we have the following equilibria [4]:

Let the game be as such that both, the defender and the attacker act periodically with rates $\alpha_i = 1/\delta_i$ and random phases $t_i \sim \mathcal{U}(0, \alpha_i)$ for $i \in \{1, 2\}$. Then, the game has the following expected revenues, where the averaging is over many game instances with random phases:

- If the defender plays at least as fast as the attacker, i.e., $\alpha_1 \geq \alpha_2$, then

$$u_1(periodic(\alpha_1), periodic(\alpha_2)) = 1 - \frac{\alpha_2}{2\alpha_1} - k_1\alpha_1;$$

$$u_2(periodic(\alpha_1), periodic(\alpha_2)) = \frac{\alpha_2}{2\alpha_1} - k_2\alpha_2.$$

- Otherwise, if the defender plays slower than attacker, i.e., $\alpha_1 < \alpha_2$, then

$$u_1(periodic(\alpha_1), periodic(\alpha_2)) = \frac{\alpha_1}{2\alpha_2} - k_1\alpha_1;$$

$$u_2(periodic(\alpha_1), periodic(\alpha_2)) = 1 - \frac{\alpha_1}{2\alpha_2} - k_2\alpha_2.$$

with the Nash equilibrium at the periods

$$\alpha_1^* = \frac{1}{2k_2}, \alpha_2^* = \frac{k_1}{2k_2^2} \quad \text{if } k_1 < k_2$$

$$\alpha_1^* = \alpha_2^* = \frac{1}{2k_1} \quad \text{if } k_1 = k_2$$

$$\alpha_1^* = \frac{k_2}{2k_1^2}, \alpha_2^* = \frac{1}{2k_1} \quad \text{if } k_1 > k_2.$$

Not surprisingly, the equilibria are determined by the cost of the moves for each player, and if the defender's and attacker's moves have roughly equal cost, then no player has a substantial advantage over the other. Conversely, if the defender plays much faster than the attacker, i.e., at rate $\alpha_1 > \frac{1}{2k_2}$, then the attacker's investment to reconquer exceeds the payoffs for having the control. Thus, the attacker's average gain would be negative, and the optimal (in the sense of economic) behavior in that setting is to not attack at all.

The (equilibrium) payoffs given above are averages over random phases, so to keep the performance promise, the defender needs to restart the strategy with a fresh random phase from time to time. Thus, the above results hold for a repeated game. Asymptotically, the payoffs as told above are attained when the defender rephases the game at times $T_j = j^2$. This has the appeal of growing intervals of "temporarily fixed" inspection intervals, yet comes at the price of not knowing when the convergence to the equilibrium kicks in (the convergence is by the law of large numbers, indicating that the variances of u_1, u_2 will ultimately vanish).

This suggests the idea of speeding up convergence by rephasing more often, which leads to another instance of the same game: let the inspections (and attacks) have random inter-occurrence times, which are *not* an integer multiple of some period. Interestingly, it can be shown that any such strategy is strongly dominated by a periodic strategy with the same rate, so this change of the game does not improve the situation for either player. This is practically good news for the defender, since a periodic inspection strategy is much easier to "live" than inspecting at random time instants.

9.3.2 Leading Defender and Following Attacker

Let us now drop the assumption that the attacker does not know when the inspection occurs. The leader-follower situation is thus such that the defender announces the inspection rate α_1, which the following attacker adapts itself to by picking $\alpha_2^* := \text{argmax}_x u_2(\alpha_1, x)$ as a best reply. The inspector presumes this in advance and will hence in first place pick α_1^* be the best reply to α_2^*. The resulting equilibrium is [4]:

- If $k_2 < (4 - \sqrt{12})k_1$, then the optimal rates are $\alpha_1^* = \frac{k_2}{8k_1^2}, \alpha_2^* = \frac{1}{4k_1}$, with maximal revenues for both players being $u_1(periodic(\alpha_1^*), periodic(\alpha_2^*)) = \frac{k_2}{8k_1}$ and $u_2(periodic(\alpha_1^*), periodic(\alpha_2^*)) = 1 - \frac{k_2}{2k_1}$.
- If $k \geq (4 - \sqrt{12})k_1$, then the optimal rates are $\alpha_1^* \searrow \frac{1}{2k_2}, \alpha_2^* = 0$, with revenues $u_1(periodic(\alpha_1^*), periodic(\alpha_2^*)) \nearrow 1 - \frac{k_1}{2k_2}$ and $u_2(periodic(\alpha_1^*), periodic(\alpha_2^*)) = 0$, where $\nearrow, \searrow$ denote convergence from below, respectively, above.

Thus, if the attacker's cost for a hostile takeover is larger than $\approx 53\%$ of the defender's cost to regain control, then the attack is simply not economic any more.

9.3.3   Inspections at Random Times

Now, let us assume that the attacker gets to know when the last inspection was done, and let the inspector choose the time of next inspection only based on the last inspection. That is, the inspector does not care about the past, but when the current inspection is finished, decides at random when to do the next inspection.

This convention implies that the times between two inspections must be *exponentially distributed*, since this is the only continuous distribution that is *memoryless*, meaning that the inter-occurrence times between two events are stochastically independent. Practically, this choice is appealing, since even if the attacker keeps record of the past inspections, this information is in no way helpful for determining the time of the next inspection. As in the leader-follower scenario just described, we assume that the attacker knows the average rate of inspections. Even if the defender does not announce this information explicitly, if the attacker notes the inspection (e.g., if the previously installed trojan just disappeared or stopped working due to a patch), then a statistical estimate the rate of inspections is easy.

The inspector's strategy is called *exponential at rate* λ, or equivalently, the number of inspections per time unit would follow a Poisson distribution (which is the only discrete memoryless distribution). Like as for periodic inspections, let us write $expon(\lambda)$ to denote the strategy to inspect at random times with $\mathscr{E}x(\lambda)$-distributed pauses (the number of inspections would thus be have a Poisson distribution). This strategy is generally superior over periodic inspections, at least against a periodically active adversary. Knowing the rate λ, the attacker's best action depends on the cost of conquer [4]:

- If $\lambda \geq \frac{1}{k_2}$, then the strongly dominant strategy (among the periodic ones) for the attacker is not playing at all.
- Otherwise, if $\lambda \leq \frac{1}{k_2}$, then the optimal attack period

$$\delta^* = \frac{1}{\alpha} \qquad (9.12)$$

is the unique solution of the equation $e^{-\lambda\delta^*}(1+\lambda\delta^*) = 1 - \lambda k_2$.

Under this optimal choice, the equilibria of the game are known [4]:

- If $k_1 \geq 0.854 k_2$, the inspector's optimal rate is $\lambda^* = \frac{1}{k_2} \cdot (1 - (1+z)e^{-z})$, with z being the unique solution to the equation $k_1/k_2 = z^{-3}(e^z - 1 - z)$. The attacker's maximum payoff is attained at period $\delta^* = z/\lambda^*$, and the equilibrium payoffs are

$$u_1(expon(\lambda^*), periodic(1/\delta^*)) = 1 - \frac{1 - e^{-z}}{z} - \frac{k_1}{k_2}[1 - (1+z)e^{-z}] \geq 1 - \frac{k_1}{k_2}$$

$$u_2(expon(\lambda^*), periodic(1/\delta^*)) = e^{-z}$$

- If $k_1 < 0.854 k_2$, then the inspector should play at rate $\lambda = 1/k_2$ to receive the payoff $u_2 = 1 - k_1/k_2$, and which forces the attacker to drop out.

This result naturally raises the question of other strategies for the attacker than periodic ones (even though these were recognized to dominate random inter-attack times, if the attacker does not know when the inspections occurred). A *greedy* attacker could seek to optimize u_2 over the time period when the inspector is inactive. Generally, if the inspection occurs at time steps t_i and t_{i+1}, then the greedy attacker would try taking over soon after t_i, but must expect the next inspection at a random time in future (since t_{i+1} is yet unknown). So, the question is when to take action for the attacker, since if the inspection will soon follow the takeover, the investment k_1 for this move could be more than the benefit of having control for a short time. Conversely, conquering too late will obviously reduce the revenue for the attacker. Thus, the attacker's problem would be optimizing the time of attack, based on an estimate $\hat{f}_0$ that is the probability density of the inspector's next move at time t_{i+1}.

It turns out that under exponential inspections, the optimum for the attacker is just the value δ^* told by (9.12), which generally dominates the periodic attack strategy. For other inspection regimes, such as periodic or variants thereof, a greedy attack strategy can often outperform the attacker's optima as given so far. Thus, letting the attacker be greedy generally invalidates the equilibria given above, and calls the defender for the computation of best replies explicitly in this situation.

9.4 Inspections Against Stealthy Intrusions

While `FlipIt` laid the focus on minimizing the time of being hiddenly under the attacker's control, inspections may alternatively or additionally aim at keeping the attacker away from the critical system parts at all (expressed as a proverb: *the devil takes the hindmost*; our point is that indeed *only* the hindmost is taken). This corresponds to the second type of APT introduced in Sect. 1.3, where the aim is killing the victim and not maximizing the time of control over it. The game described in the following has the name Cut-The-Rope [13].

Remark 9.6 The situation is actually similar to what the human immune system does each day: the immune system cannot possibly keep pathogenes out of the body, but it mostly keeps the disease agents under sufficient control to prevent them from reaching vital parts of the body or grows into a dangerous infection. The intrusion games described here follow exactly the same intuition.

Shifting the focus away from the duration of annexion to the event of hitting the target itself, the gameplay changes into a tug war between the inspector and the attacker. In particular, and consistent with the assumption of unlimited budgets backing up the APT attacker, the defender has no hope of playing such that the attacker will drop out.

The battle field can then be a physical area like in the patrolling games of Chap. 8, or a cyberspace, if the intrusion occurs through the network channels. Conceptually, the game models are indifferent, only the way they are set up is distinct. We shall illustrate the matter using a cyber intrusion scenario in which the attacker digs its tunnel through the logical enterprise network, by establishing an initial point of contact (e.g., by social engineering techniques like phishing, virulent USB gimmicks, stolen passwords, or other means), and works its way through the network in alternating phases of spying out the technical systems and penetrating them by tailor-made exploits. This is the practically reported pattern that contemporary APTs would follow, and it is exactly the dynamic that defines the game itself [14]; see Chap. 1.

TVA is the process of (semi-automatic) identification of scenarios along which an intruder could sneak into a system. It builds up graphs whose nodes correspond to physical or logical points in the system that the adversary can overtake hop-by-hop until it reaches the goal point in the system (which may be a database with sensitive business information, or any valuable asset of an enterprise.)

Remark 9.7 In other contexts such as hacking conventions, we find similar competitions called "capture-the-flag". There, the goal is for a team to conquer an asset, and the winner is typically the fastest team. Cut-The-Rope is similar in spirit, though it has only one player seeking to "capture the asset", while the second (defending) player's duty is to keep the attacker away from it. The battlefield of a capture-the-flag game and Cut-The-Rope, however, are essentially the same.

Figure 9.6 shows a toy example from [5–7] that we will use for our exposition throughout this section. The attack graph, whether obtained from a TVA or by other means, is already the first out of three ingredients for the intrusion game setup. We stress that TVA is only a (convenient) possibility of constructing the playground for Cut-The-Rope, but any other method that delivers a systematic model of how the attacker can break in and conquer the victim system is equally applicable.

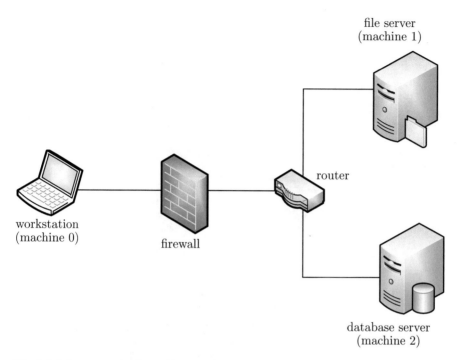

file server
(machine 1)

router

workstation
(machine 0)

firewall

database server
(machine 2)

Fig. 9.6 Infrastructure for the stealthy intrusion game; Example from [16]

9.4.1 Setting up the Game

9.4.1.1 Step 1: Preparing the Battlefield

Let an attack graph $G = (V, E)$ or attack tree (see Fig. 9.7 for an example) be obtained as an output artefact from a TVA or by other means. Calling the nodes in Fig. 9.6 0 (workstation), 1 (file server) and 2 (database server), let the attacker target machine 2. We assume that each machine runs certain services like File Transfer Protocol (FTP) $\mathsf{ftp}(i, j)$, Remote Shell (RSH) $\mathsf{rsh}(i, j)$ or Secure Shell (SSH) $\mathsf{ssh}(i, j)$ to denote the service on machine j being remotely available from machine i. The implementation of each service may be vulnerable to certain exploits like a buffer overflow (**bof**) or similar. The attack tree systematically enumerates the possibilities, and displays preconditions as rectangles, and (adversarial) actions as ellipses, both annotated by respective predicates as just described. For example, an FTP access exploit from machine i to a remote host being machine j is denoted as $\mathsf{ftp\_rhosts}(i, j)$. A buffer overflow exploit on the SSH implementation is denoted as $\mathsf{ssh\_bof}(i, j)$, when it starts from machine i towards machine j. The penetration thus proceeds by stepwise establishment of trust relations $\mathsf{trust}(i, j)$, meaning that machine i grants access from machine j, perhaps allowing execution privileges on machine j, written as the predicate $\mathsf{execute}(j)$.

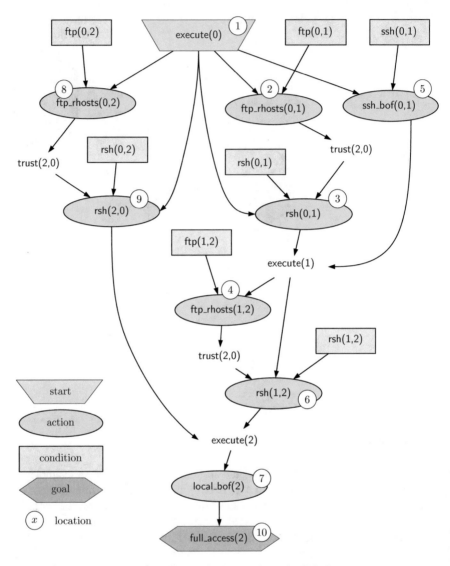

Fig. 9.7 Example Attack Graph $G = (V = \{1, 2, \ldots, 10 = v_0\}, E)$ [16]

The initial attacker's position is having full execution rights on the workstation, with the goal gaining full execution rights on the database server. Hereafter and for ease of referencing, let us assign numeric names to each node in the attack graph, displayed as circled numbers in Fig. 9.7.

From the tree, the inspector can successively enumerate all paths in the attack graph, starting from the entry node (here node ①) where the adversary can make the initial contact, up to the target node (here node ⑩). The number of paths is

Table 9.2 APT scenarios (adversary's action set AS_2, based on Fig. 9.7)

No.	Strategy description as attack path
1	execute(0) → ftp_rhosts(0,1) → rsh(0,1) → ftp_rhosts(1,2) → rsh(1,2) → local_bof(2) → full_access(2)
2	execute(0) → ftp_rhosts(0,1) → rsh(0,1) → rsh(1,2) → local_bof(2) → full_access(2)
3	execute(0) → ftp_rhosts(0,2) → rsh(0,2) → local_bof(2) → full_access(2)
4	execute(0) → rsh(0,1) → ftp_rhosts(1,2) → rsh(1,2) → local_bof(2) → full_access(2)
5	execute(0) → rsh(0,1) → rsh(1,2) → local_bof(2) → full_access(2)
6	execute(0) → rsh(0,2) → local_bof(2) → full_access(2)
7	execute(0) → sshd_bof(0,1) → ftp_rhosts(1,2) → rsh(0,1) → rsh(1,2) → local_bof(2) → full_access(2)
8	execute(0) → sshd_bof(0,1) → rsh(1,2) → local_bof(2) → full_access(2)

Table 9.3 Correspondence of attack trees/graphs and extensive form games

Extensive form game	Attack tree/graph
Start of the game	Root of the tree/graph
Stage of the gameplay	Node in the tree/graph
Allowed moves at each stage (for the adversary)	Possible exploits at each node
End of the game	Leaf node (attack target)
Strategies	Paths from the root to the leaf (= attack vectors)
Information sets	Uncertainty in the attacker's current position and move

generally exponential, so that strategy reduction heuristics (like in Sect. 8.6) are advisable; some of which are even part of automated tools for the vulnerability assessment [3].

Table 9.2 lists these for the example infrastructure displayed in Fig. 9.6. This list is already the action space for the attacking player 2, thus completing the first third of the model. The correspondence of extensive form games and attack trees, indeed, reaches much farther, as Table 9.3 summarizes.

9.4.1.2 Step 2: Inspection Strategies

We assume a periodic action of both players that does not necessarily entail them to take action simultaneously, but we let them become active in rounds. For the inspector, this corresponds to a periodic strategy in terms of Sect. 9.3; the same goes for the attacker, though we can allow multiple penetration moves to occur in between two inspections, by proper definition of the payoffs in step 3.

However, the action space for the inspector directly follows from the attacker's action space, i.e., by setting up a list of devices that should undergo a periodic checkup, *and* are feasible to inspect. This device list corresponds to a selection

$AS_1 \subseteq V$ of nodes in the attack graph $G = (V, E)$, since an inspection on a device manifests in G as the inspection of a certain node. It is important to keep in mind that an inspection hereby can mean various things, not only patching or installing some additional protection, but also a humble change of access credentials. In this context, we need to note that changing a password or other access credential is normally a simple matter, unlike taking off a whole machine for a complete re-installation, which disrupts the daily business and as such is typically avoided if possible. Hence, the latter is perhaps the more realistic setting, as it is (i) usually cheaper and faster to accomplish, and (ii) leaves the attack tree unchanged. Our game model in the following assumes a static such attack tree.

The number of strategies for the inspector is thus bounded by the number of nodes (e.g., exploits) in the attack tree. So, although the number of exploits may not be exponential, it can nonetheless be large, so that grouping exploits and vulnerabilities of similar kind together may be advisable to boil down a large number of spot checking points. For example, we may not distinguish different types of buffer overflow vulnerabilities, but generically call a node a "buffer overflow" (like ssh_bof in Fig. 9.7), or a remote shell exploit (not being too specific if this works using privilege escalation, theft of access credentials, or by other means). Keeping the attack tree generic in this sense also keeps it more static upon the defender's actions (under the hypothesis that no action of the inspector would clean all buffer overflows or remove all possibilities to acquire unintended permissions).

Based on Table 9.2, the inspector's strategies can be defined towards intercepting the known attack strategies. For example, if the inspector can spot check the "workstation (machine 0)", the "file server (machine 1)", or the "database server (machine 2)" in Fig. 9.6, the action space for the defender is essentially the entire set $AS_1 = V$ in the attack graph in Fig. 9.7.

This completes the second out of three ingredients for the simple stealthy intrusion game model.

9.4.1.3   Step 3: Payoff Assessment

The inspector's problem is finding the optimal strategy for spot checking, i.e., compute the optimal randomized choice $\mathbf{x} \in \Delta(AS_1)$, for which we will need a proper definition of payoffs. We adopt the worst-case perspective for the defender, and explicitly model the adversary's payoffs in the following.

Essentially, the attacker could be anywhere in G, and especially so since it can move at any point in time, while the defender may become active periodically (we will later generalize this towards letting the defender also act in continuous time, for example, if there is a 24/7 security response team in charge). In any case, the inspector must consider that the intruder could be anywhere in the network, and s/he can only hypothesize about certain likelihoods for the attacker to be somewhere after the inspection. Call $\Theta \subseteq V$ the set of possible locations, and let θ be a particular such location, for which the defender may presume a likelihood $\Pr(\theta)$. In absence of a reasonable guess for the distribution $(\Pr(\theta))_{\theta \in \Theta}$, the defender can put $\Pr(\theta) :=$

$1/|\Omega|$ as a non-informative choice of a uniformly random location. Each starting location then defines a distinct adversary type, making the game Bayesian.

We let the attacker take a random number, in fact any number, of attack steps between times when the defender becomes active. As in Sect. 9.3.1, let us assume a $periodic(1/\delta)$ defense, meaning that the defender becomes active every δ units of time. Then, allowing the attacker to take any random number of steps during idle times of the defender, the attacker will take a Poisson distributed number $N \sim Pois(\lambda)$ of steps, with λ being the *attack rate*. This value is the average number of attack steps accomplished per δ time units. It is assumed to be *common knowledge* (see Sect. 6.4.1), and a value that the defender needs to estimate a priori.

Further, we let the attacker be rational in choosing the path $\pi_\theta = (\theta, w_1, w_2, \ldots, v_0)$ according to being the shortest or "easiest" (e.g., offering perhaps not fewer but simpler exploits than an alternative) route towards v_0. Along the attack path π_θ, we assume that it leaves backdoors or other aids for a quick return later on, to further study and penetrate deeper into the system (to retain the stealthiness of the attack, to avoid causing too much noise to escape detection). Call $V(\pi_\theta) = \{\theta, w_1, w_2, \ldots, v_0\}$ the entirety of nodes on the attack path π_θ as defined before.

When the defender takes action, it picks a node to inspect, and cleans it from any malware that it finds. If, during this cleaning, one of the attacker's backdoors disappears (such as shown in Fig. 9.8a for a check at node ③), this sends the attacker back to the point right before where the defender has just spot-checked. Otherwise, if the defender was checking a node that the attacker has not passed yet or will not pass at all because s/he is on an entirely different route, nothing happens and the defender's action was without any effect (e.g., if the inspector checks at node ⑥ in Fig. 9.8a).

Example 9.3 (Success or failure of an inspection) If the attacker successfully installed a backdoor on the file server, and the inspector is currently on the workstation, then this is a miss, and the intruder will still be on the router. Likewise, if the intruder's malware is on the file server, and becomes dysfunctional upon a patch or software update done by the inspector, it sends the intruder back to an earlier point on the attack path. In either case, the inspector may not be aware of this success or failure, and can only update its hypothesis based on what would happen if the attacker were hit or missed by the inspection. However, the inspector is not successful in any of the following cases:

- Inspecting at a point that is not on the attack path chosen by the adversary
- Inspecting at a point that is on the attack path, but that the attacker has not yet reached (in that case, the defender was there too early)

◇

Remark 9.8 Our modeling assumes no lateral movement of the attacker. That is, if an attack path π gets intercepted by the inspector, the attacker is assumed to keep on moving along the same path π towards v_0. A real life attacker may nonetheless switch to another attack path π', but this is included in our modeling by assuming

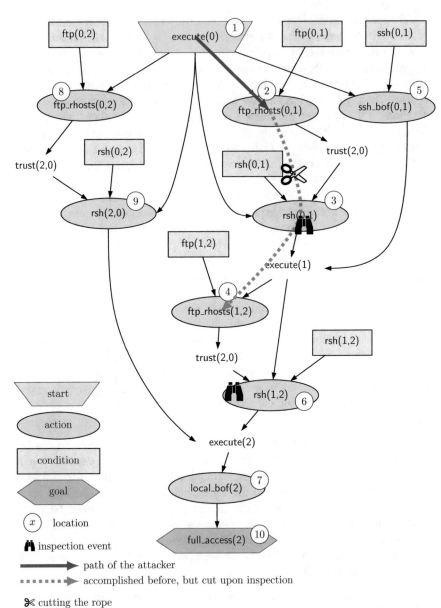

(a) Inspection at node ⑥ ("miss"), and inspection at node ③ ("rope cut")

Fig. 9.8 Gameplay of Cut-The-Rope

yet another adversary type that uses the path π' instead. Since the game is such that the defender acts against all possible adversary types, whether or not the real adversary moves laterally from the path π to the path π' makes no difference for the inspector's defense, since s/he is anyway taking both adversary types into account.

The chain of backdoors or other aids that let the attacker return over and over again is called the "rope", and the event of the defender disabling one of these backdoors to "cut" the way back leads to the game's name being Cut-The-Rope.

The utility for the adversary starting from node θ is then found by computing its location on the attack path π_θ, conditional on the spot checks as done by the defender. To this end, let $d_\pi(u, v)$ be the graph-theoretic distance counting the edges on the subsection from u to v on a path π in G, and recall that the adversary takes a Poissonian number of steps (we can drop the θ-index of π to ease our notation here without creating ambiguities). Then, the probability for the attacker to be at node v on the path π has the conditional distribution

$$\Pr(\text{adversary's location} = v | V(\pi)) = \frac{f_{Pois(\lambda)}(d_\pi(\theta, v))}{\Pr_{Pois(\lambda)}(V(\pi))}, \qquad (9.13)$$

in which $f_{Pois(\lambda)}(x) = \frac{\lambda^x}{x!}e^{-\lambda}$ is the density of the Poisson distribution, and

$$\Pr_{Pois(\lambda)}(V(\pi)) = \sum_{x \in V(\pi)} \Pr_{Pois(\lambda)}(d_\pi(\theta, x)) = \sum_{x \in V(\pi)} f_{Pois(\lambda)}(d_\pi(\theta, x)).$$

See Fig. 9.9a for an example relating to attack path no. 1 in Table 9.2.

Now, the inspector comes along and spot-checks and in the best case, cuts the rope behind the attacker. Let $c \in V$ be the inspected node, then the afterwards perhaps truncated attack path is (cf. Fig. 9.9b),

$$\pi|_c = \begin{cases} (\theta, w_1, w_2, \ldots, w_{i-1}), & \text{if } c = w_i \text{ for some } w_i \text{ on } \pi \\ (\theta, w_1, \ldots, v_0), & \text{otherwise.} \end{cases}$$

Cutting the rope then means conditioning the distribution of the adversary's location on the updated path $\pi|_c$. The formula is the same as (9.13), only with π replaced by $\pi|_c$ now. If the inspector takes random spot checks $c \sim \mathbf{x} \in \Delta(AS_1)$ on $AS_1 \subseteq V$ follows the defender's mixed spot checking strategy $\mathbf{x}$ (possibly degenerate if pure), and the set of paths π along which the attacker steps forward (at rate λ) is determined by the random starting position $\theta \sim \Theta$, the utility distribution for the attacker is now fully given as

$$U(\mathbf{x}, \theta, \lambda) = (\Pr(\text{adversary's location} = v | V(\pi|_c)))_{v \in V}. \qquad (9.14)$$

Figure 9.9b shows an example of a truncated distribution upon an inspection of a node on the path.

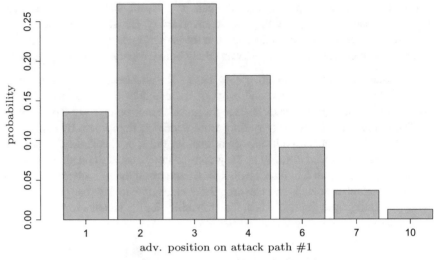

(a) Adversary's possible positions on attack path #1 before the inspection

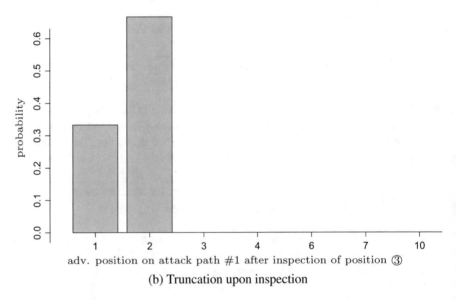

(b) Truncation upon inspection

Fig. 9.9 Payoff distributions in Cut-The-Rope

Expression (9.14) holds for the attacker starting from $\theta \in V$, and since the inspector is unaware of the attacker's location (stealthiness of the APT), s/he imposes its belief about the adversary's type, i.e., starting location, to define the ex ante payoff as

$$U'(\mathbf{x}, \lambda) = \sum_{\theta \in \Theta} \Pr_\Theta(\theta) \cdot U(\mathbf{x}, \theta, \lambda) \tag{9.15}$$

This is equivalent to writing $U(\mathbf{x}, \theta, \lambda)(v) = \Pr_\mathbf{x}(\text{adversary's location} = v|\theta, \lambda)$, since this is by construction the distribution of the attacker's location, *conditional* on the starting point θ. In this view, however, (9.15) is just the law of total probability, turning U' into the distribution of the attacker's location (by scalarization).

This distribution is $\preceq$-minimized by the defender, and likewise $\preceq$-maximized by the attacker. Since both, the attack graph and the defender's strategy sets are finite, the game is a mere matrix game, with distribution-valued payoffs, and as such falling into the category of games that we had in Sect. 4.5. Carrying out this optimization is a matter of implementing the steps so far, and in fact short enough to merit a full implementation as Listing 12.2, given in Chap. 12.

9.4.2 Inspections at Random Times

We can easily replace the assumption of a periodic inspection by an exponential strategy $expon(\lambda_D)$ for the inspector. Like in `FlipIt`, the inspector may become active in random intervals following an exponential distribution with rate parameter λ_D, and so does the attacker at rate λ. Both of these values refer to average counts of activity over the same unit of time.

The change to the model is trivial, as it merely means replacing the Poissonian $Pois(\lambda)$ distribution by a *geometric distribution* in (9.13) with parameter $\lambda_D/(\lambda + \lambda_D)$. Everything else remains unchanged and thus we won't got into further detail on this here.

9.4.3 Probabilistic Success on Spot Checks

Similarly, we can incorporate probabilities to capture the inspector's failure to clean a node from malware. Suppose that an inspection of node $c \in V$ is successful with probability p_c only. The according change to the model is again only in (9.13), which now becomes

$$\Pr(\text{adversary's location} = v) =$$
$$p_c \cdot \Pr(\text{adversary's location} = v|V(\pi|_c))$$

$$+ (1 - p_c) \cdot \Pr(\text{adversary's location} = v | V(\pi))$$

This is a simple mix of successful and unsuccessful cutting, but besides this formula to replace (9.13), nothing else changes.

9.4.4 Probabilistic Success on Exploits

Probabilistic success may also be a matter for the attacker, and with resilience values $q(e)$ that describe the probability to be successful on an exploit e, the likelihood to take d steps on the attack path π_θ is (assuming stochastic independence of exploits) given as

$$\Pr(d) = \prod_{k=1}^{d} q(e_k),$$

when the attack path is $\pi = (e_1, e_2, \ldots)$. Like as in the previous cases, we just replace the Poisson density f_{Pois} by $\Pr(d)$, and can leave everything else unchanged.

The challenge in this variation is the estimation of the exploit success probabilities $\Pr(e)$, and may receive practical aid from heuristic rules to derive $\Pr(e)$ from a CVSS score [8] or similar.

9.4.5 Security Strategy Computation

It turns out that a PBNE (Definition 4.3) fits as a solution concept here, endowing the inspector with an optimal defense against the presumed set of adversary types (with probabilities imposed thereon). Computationally, we can consider the inspector's problem as a one-against-all competition, where the defender simultaneously faces exactly one adversary of type θ for all types $\theta \in \Theta$. Then, condition (9.15) is just a scalarization of some multi-criteria security game, and the PBNE is equal to an MGSS (Definition 3.9) and vice versa. This lets us compute the PBNE by FP; see Fig. 12.3 and Chap. 12.

Cut-The-Rope is designed for ease of implementation and generalization, but unlike FlipIt does not lend itself to analytic studies as easily. It is more for practical matters of computing defenses, but has the unique feature of allowing different time axes (discrete and continuous) for the defender and the attacker. A generalization not explored but easy to bring in is an account for the costs of moving or spot checking, or similarly, measuring security in more than one dimension. Both generalizations cast the game from one to several goals (adding a quadratic cost term like described in Sect. 12.5, or adding further payoff distributions to measure further security goals relevant for risk management; see Chap. 6). In any case, the solution as implemented in Chap. 12 will remain applicable in this multidimensional setting, only requiring a change of how the game is constructed, but no change in how equilibria are computed.

References

1. Avenhaus R, von Stengel B, Zamir S (2002) Inspection games: 51. In: Aumann RJ, Hart S (eds) Handbook of game theory with economic applications, vol 3. Elsevier, Amsterdam, pp 1947–1987
2. Benford F (1938) The law of anomalous numbers. Proc Am Philos Soc 78(4):551–572
3. CyVision CAULDRON (2018) http://www.benvenisti.net/
4. Dijk M, Juels A, Oprea A, Rivest RL (2013) FlipIt: the game of stealthy takeover. J Cryptol 26(4):655–713. https://doi.org/10.1007/s00145-012-9134-5
5. Jajodia S, Ghosh AK, Subrahmanian VS, Swarup V, Wang C, Wang XS (eds) (2013) Moving target defense II – application of game theory and adversarial modeling. Advances in information security, vol 100. Springer. http://dx.doi.org/10.1007/978-1-4614-5416-8
6. Jajodia S, Ghosh AK, Swarup V, Wang C, Wang XS (eds) (2011) Moving target defense – creating asymmetric uncertainty for cyber threats. Advances in information security, vol 54. Springer. http://dx.doi.org/10.1007/978-1-4614-0977-9
7. Jajodia S, Noel S, O'Berry B (2005) Topological analysis of network attack vulnerability. In: Kumar V., Srivastava J., Lazarevic A. (eds) Managing Cyber Threats. Massive Computing, vol 5. Springer, Boston, MA
8. König S, Gouglidis A, Green B, Solar A (2018) Assessing the impact of malware attacks in utility networks, pp 335–351. Springer International Publishing, Cham. https://doi.org/10.1007/978-3-319-75268-6_14
9. Lehmann EL, Romano JP (2005) Testing statistical hypotheses. Springer texts in statistics, 3rd edn. Springer, New York
10. Neyman J, Pearson ES (1933) On the problem of the most efficient tests of statistical hypotheses. Philos Trans R Soc A Math Phys Eng Sci 231(694–706):289–337. https://doi.org/10.1098/rsta.1933.0009
11. Nigrini MJ (2002) The detection of income tax evasion through an analysis of digital distributions. Ph.D. thesis, University of Cincinnati
12. R Core Team: R (2018) A language and environment for statistical computing. R Foundation for Statistical Computing, Vienna. https://www.R-project.org/
13. Rass S, König S, Panaousis E (2019) Cut-The-Rope: a game of stealthy intrusion. In: Alpcan T, Vorobeychik Y, Baras JS, Dán G (eds) Decision and game theory for security. Springer International Publishing, Cham, pp 404–416
14. Rass S, Konig S, Schauer S (2017) Defending against advanced persistent threats using game-theory. PLoS ONE 12(1):e0168675. https://doi.org/10.1371/journal.pone.0168675
15. Siddiqi N (2017) Intelligent credit scoring: building and implementing better credit risk scorecards, 2nd edn. Wiley, Hoboken
16. Singhal A, Ou X (2011) Security risk analysis of enterprise networks using probabilistic attack graphs. Technical report, National Institute of Standards and Technology (NIST). NIST Interagency Report 7788

Chapter 10
Defense-in-Depth-Games

Without struggle, success has no value.

A. Lauritsen

Abstract In this chapter, we adopt a holistic cross-layer viewpoint towards a hierarchical structure of ICS and the attack models. The physical layer is comprised of devices, controllers and the plant whereas the cyber layer consists of routers, protocols, and security agents and manager. The physical layer controllers are often designed to be robust, adaptive, and reliable for physical disturbances or faults. With the possibility of malicious behavior from the network, it is also essential for us to design physical layer defense that take into account the disturbances and delay resulting from routing and network traffic as well as the unexpected failure of network devices due to cyber-attacks. On the other hand, the cyber security policies are often designed without consideration of control performances. To ensure the continuous operability of the control system, it is equally important for us to design security policies that provide maximum level of security enhancement but minimum level of system overhead on the networked system. The physical and cyber aspects of control systems should be viewed holistically for analysis and design.

10.1 The Need for Cross-Layer Security

The integration of IT infrastructure with ICS has created a closed network of systems embedded in the publicly accessible network. The integration brings many cost and performance benefits to the industry as well as arduous challenges of protecting the automation systems from security threats [6]. IT networks and automation systems often differ in their security objectives, security architecture and quality-of-service requirements. Hence, the conventional IT solutions to security cannot be directly applied to control systems. It is also imperative to take into account many control system specific properties when developing new network security solutions. In this chapter, we introduce a framework to assess the impact

© Springer Nature Switzerland AG 2020
S. Rass et al., *Cyber-Security in Critical Infrastructures*, Advanced Sciences and Technologies for Security Applications,
https://doi.org/10.1007/978-3-030-46908-5_10

of cyber-security policies on various control system performances and design an optimal security policy for a large-scale networked ICS.

Due to the total isolation of control systems from the external networks, control system security has historically been defined as the level of reliability of the system, with designs aimed at increasing the reliability and the robustness of the system, rather than considering the network security. Hence, merging a modern IT architecture with an isolated network that does not have built-in security countermeasure is a challenging task. From a mitigation perspective, simply deploying IT security technologies into a control system may not be a viable solution. Although modern control systems use the same underlying protocols that are used in IT and business networks, the very nature of control system functionality may make even proven security technologies inappropriate.

Sectors such as energy, transportation and chemical, have time-sensitive requirements. Hence, the latency and throughput issues with security strategies may introduce unacceptable delays and degrade acceptable system performance. The requirement of accessibility of control systems is a distinct factor that distinguishes it from its IT counterpart. Understanding the system tradeoff between security and system accessibility would enable us to learn the fundamental limitations and design principles for secure control systems. Security solutions at the physical and the cyber-layer of an integrated control system need to take into account the interaction between these two layers. Figure 10.1 illustrates the concept of defense in-depth as a nested structure. The target assets are safeguarded using multiple layers of protections including authorization and access control, firewalls, IDS, and DMZs, physical patrolling and surveillance. The figure also illustrate the interconnection between the physical and the cyber security of the system.

In this chapter, we adopt a holistic cross-layer viewpoint towards a hierarchical structure of ICS and the attack models. The physical layer is comprised of devices, controllers and the plant whereas the cyber-layer consists of routers, protocols, and security agents and manager. The physical layer controllers are often designed to be robust, adaptive, and reliable for physical disturbances or faults. With the possibility of malicious behavior from the network, it is also essential for us to design physical layer defense that take into account the disturbances and delay resulting from routing and network traffic as well as the unexpected failure of network devices due to cyber-attacks. On the other hand, the cyber-security policies are often designed without consideration of control performances. To ensure the continuous operability of the control system, it is equally important for us to design security policies that provide maximum level of security enhancement but minimum level of system overhead on the networked system. The physical and cyber-aspects of control systems should be viewed holistically for analysis and design.

10.2   Socio-Cyber-Physical Security Threats

With the emergence of cyber-physical systems in recent years and the increased interaction between physical infrastructures and people, on the one hand, as well as technology and computational components, on the other hand, a special focus has been laid on cyber-crimes and how to improve cyber-security in the contemporary complex workplace. In contrast, physical security has been relatively overlooked or improperly enforced in some places. Cyber security aims at protecting assets and resources in cyberspace by blocking unauthorized remote access to networks and assets. Still, the attackers have the opportunity to gain access by physically entering an organization's premises and connecting to the network, servers, or any other device available on site. Traditionally, physical security of enterprises has been mainly shaped by the castle (fortress) protection model, which aims at building a hard shell around a presumably trusted area encompassing different valuable assets varying from people, hardware and software to data and information resources. Therefore, various security controls have been deployed and mounted at the outer boundaries of the facility of interest establishing the so-called security perimeter. In general, the perimetric controls are designed and implemented with the *D5 strategy* in mind; i.e. Demarcation, Deter, Detect, Delay and Defend [9]. Demarcation refers to the process of creating boundaries around important assets. These boundaries should be visible to avoid innocent boundary crossings and to simplify identifying hostile intentions.

The goal of deterrence is to create an unattractive environment for potential adversaries. Security lighting, fences, monitoring points, and surveillance systems are, for example, effective deterrents since they are able to reduce an attacker's opportunity to commit his attack unobserved. However, deterrent efforts are not enough to keep adversaries out. Therefore, it is of vital importance for the perimeter to be able to detect unwanted activities and to delay potential perpetrators long enough to allow security forces or first responders to intercept and defend by denying access to the internal critical assets and resources. An enterprise's security perimeter can include common protecting components and mechanical barriers such as fences, walls, monitoring points, entrance gates or doors, vehicle barriers, security lighting, landscaping, video surveillance systems, alarm systems, guards, intrusion detection systems, among others. For example, closed circuit television cameras are increasingly used to provide features such as standard monitoring, recording, event detection and forensic analysis. They transmit video signals over dedicated coaxial cables or any common computer network to a specific video control system and to a limited number of monitors in a central monitoring station that is mostly serviced by a third party contractor. In addition to this, the physical access is sometimes limited to authorized users with valid credentials such as identity badges. Due to the fixed installation of these security solutions, they tend to remain static and inflexible in their operation.

Although all of these controls have been used to ensure that any contact with the protected assets is authorized, a breach of the physical security remains probable due to accidents, human errors or even targeted attacks. Attackers leverage the static nature of these mechanisms and the predictable placement of their devices to sneak into the facility. Moreover, the ongoing trend demanding enterprises to become more flexible and dynamic, as well as the increasing rate of collaboration and interconnection, make the solid security perimeter of these systems very porous. As a result, the perimeter is no more able to keep risk resources in the outside environment. The current tendency of an organization to extend beyond their conventional borders to reach other entities such as vendors, business partners, service providers or even customer, results in having different external entities within the system such as temporary workers, interns, independent contractors and subcontractors, or visitors. Even if the access to the industrial sensitive zones is tightly controlled at the borders, behind the borders the freedom of movement is almost entirely ensured for ordinary organization's personnel as well as for temporary ones. Therefore, potential adversaries can exploit the dynamic nature of the systems and lack of a proper resource management strategy to cause loss and damage.

As a result, the assumption that the inside does not have attack sources and it is outfitted with less protection is no more valid. Undetectability within the system complex will give the adversary a good opportunity to reconnaissance the target area, to gather some sensitive information, and to probably cover the tracks of ongoing attacks, too. Nowadays attackers will exploit different attack vectors trying repeatedly to adapt to the defender strategies [2]. Therefore, it is highly important to maintain situational awareness even within the system complex so that the potential intruders can remain detectable and the security managers are able to respond timely. Having dynamic and mobile surveillance systems (or strategies) will definitely increase the system robustness and increase the attack costs and complexity. This, in turn, will give the system's defenders the advantage to stay ahead of the attackers in the respective security game.

10.2.1   Cyber-Physical Threats

Cyber security is a broad term that comprises technical, organizational but also staff-related aspects, and as such cannot be covered by purely technical security notions. However, and conversely, organizational assumptions and those on the staff indeed are crucially underlying any technical precaution. The most important examples are security notions in cryptography, which effectively equal security to the secrecy of the underlying keys (we will discuss the formal notions in Chap. 11). Once a key is known, any security proof or other formal argument of security becomes void. This extends to many (if not all) other technical security precautions as well, since access credentials are to be protected, but at some point, the protection must ultimately rely on human behavior and awareness. This is essentially the point where attacks are mounted for the mere reason of efficiency: stealing an access credential or spying out a cryptographic key is in most cases much cheaper than running an expensive (and time-consuming) cryptanalysis.

The achievements of cryptography and related areas (see [3]) have made the attacker concentrate its efforts on the weakest element, which is the human; and this weakness is very well documented [8]. Although solutions to some vulnerabilities have been proposed [14], there remains much to be done. Essentially, social engineering is a matter of exploiting human's unawareness of a threat. Consequently, awareness training, spot checks, etc. appear to be only a few natural countermeasures, but none of which has a permanent effect. An effective protection against social engineering thus calls for a continuous information and training campaign, which costs money and time, and the question of an optimal investment in these measures directly relates security to economics. However, the tradeoff between security investment and the positive effects of it (the return-of-investment) also exists in the opposite direction on the attacker's side, which directly induces a non-formal yet practically effective understanding of security that we describe next.

10.2.2 Security Economics

Security is usually not about making an attack impossible, since it is fully sufficient to render an attack non-economic. That is, if the investment of the attacker exceeds the revenue received upon success, it is simply not meaningful to mount the attack. Consequently, our attack/defense model will consider costs incurred throughout the lifetime of the intrusion attempt, by measuring the "optimal" cost/benefit tradeoff in the defense. The resulting model, in light of this balance and conflict, is thus a two-player game between the defender seeking to maximize the attacker's costs, opposing an attacker attempting to minimize its investment in order to maximize the final payoff. This can be quantified by the damage caused, minus the investment that has been necessary, or, in a more sophisticated fashion, by considering a multi-objective optimization and treating investments and returns as separate goals (see Sects. 3.3 and 12.5).

10.3 Multi-layer Framework for Defense in Depth

The cyber-infrastructure serves as an interface between the controller and the physical plant. The control signals are sent through a security enhanced IT infrastructure such as wireless networks, the Internet and Local Area Networks (LANs). The security architecture of the IT infrastructure is designed to enable the security practice of defense in depth for control systems [13]. Figure 10.1 illustrates the concept of in-depth defensive as a combination of different game models, with an adversary playing different games per layer towards the target assets. The cascading countermeasures using a multitude of security devices and agents, ranging from physical protections to firewalls and access control, can offer the administrators more opportunities for information and resources control with the advent of potential threats. However, it also creates possible issues on the latency and the packet drop rate of communications between the controller and the plant.

Here, we describe a unifying security model for this cyber-physical scenario and investigate the mitigating strategies from the perspectives of control systems as well as of cyber-defenses. At the physical layer, we often aim to design robust or adaptive controllers that take into account the uncertainties and disturbances in the system to

physical security game,   cyber-security game
e.g., patrolling, surveillance, ...   e.g., inspections, ...

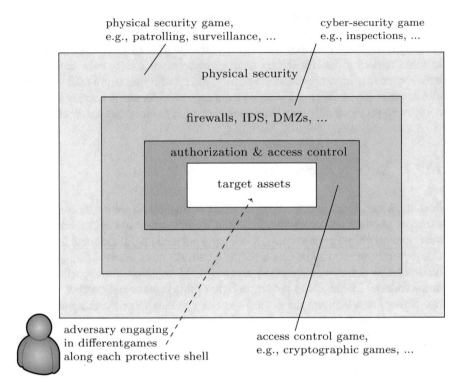

Fig. 10.1 Defense in-depth security mechanism to protect the target assets

enhance robustness and reliability of the system. At the cyber-level, we often employ
IT security solutions by deploying security devices and agents in the network. The
security designs at the physical and the cyber-layers usually follow different goals
without a unifying framework.

APTs are one of the emerging threats for ICS. Unlike the automated probes and
amateurish hackers, APTs have specific targets and conduct thorough research to
expose the system architecture, valuable assets, and even defense strategies so that
they can tailor their strategies and make the cryptography, firewalls, and intrusion
detection systems invalid. APTs are also deceptive and can stealthily stay in the
system for a long time, e.g., a replay attack which replays the prerecorded standard
sensory data to deceive the monitor during the attack. Various examples have been
reviewed in Sect. 1.4.

Defense in depth is a useful class of strategies to defend against such sophisticated attacks. The defense-in-depth employs a holistic approach to protect assets
across the cyber-physical layers of the system, taking into account interconnections
and interdependencies of these layers. The design of defense-in-depth strategies
provides efficient cross-layer protection that holistically deters attacks and reduces
cybersecurity risks.

To develop a formal design paradigm, we leverage game-theoretic methods to capture constraints on the defense, consequences of attacks, and attackers' incentives quantitatively. In particular, we introduce a multistage game framework to model the long-term interaction and stealthiness of APTs. We divide the entire life cycle of APTs into multiple stages where at each stage, both players take actions, make observations, and then move to the next stage. Section 1.3.2 outlined the details of this multi-stage nature of an APT that we call the kill chain [2]. The main idea of defense-in-depth games is to model each stage as its own distinct game, and letting the outcome of one game determine the initial strategic position of the player when the next game starts. This can mean having (or not having) certain strategies or having a payoff structure that depends on what happened in the previous stage of the chain of games. A defense-in-depth game model is thus a set of game models that are in a way nested, and with each game being chosen to best model the respective phase. In particular, this can entail structurally very different game models in each phase. We describe one possible set of models in the following, emphasizing, that the full spectrum of game theoretic models (as described throughout this book) may find application in different practical circumstances.

During the reconnaissance phase, the threat actor probes the system and obtains intelligence from open-source or inside information. The reconnaissance phase identifies vulnerable targets and increases the success of the initial compromise. After APTs obtain the private key and establish a foothold, they escalate privilege, propagate laterally in the cyber-network, and finally either access confidential information or inflict physical damages. The chain structure of APTs can be mapped into a multistage game, where each stage is described by a local interaction between an attacker and a defender whose outcome leads to the next stage of interactions. The goal of the APT is to reach the targeted physical assets while the defender aims to take actions at multiple stages of the kill chain to thwart the attack or reduce its impact.

One key challenge of the game-theoretic framework is to capture the stealthy and deceptive behaviors in the network. Let us use the multistage game of incomplete information [4] to model the information asymmetry between players. To this end, we introduce the notion of types to characterize the private information of the players in the game. For example, the type of users, legitimate or adversarial, is private to the users or the attackers themselves. On the other hand, the level configuration of the network, which can be low or high, is private to the network administrator, i.e., the defender. The defender does not know the users' type because of the stealthy nature of APTs, and she can also hide or obfuscate her type via defensive deception techniques such as the moving target defense and the honeypot deployment [1, 7]. Players' types determine their utilities and affect their behaviors. Thus, each player observes and analyzes behaviors of the other player at each stage to form a belief of the other's type. When observations are available during their persistent interactions, players continuously update their beliefs via the Bayesian rule (see Definition 4.3).

Both players act strategically according to their beliefs to maximize their utilities. The PBNE provides a useful prediction of their policies at every stage for each type since no players benefit from unilateral deviations at the equilibrium. The computation of PBNE is challenging due to the coupling between the forward belief update and the backward policy computation. We first formulate a mathematical programming problem to compute the equilibrium policy pair under a given belief for a one-stage Bayesian game. Then, we extend this approach to compute the equilibrium policy pair under a given sequence of beliefs for multistage Bayesian games by constructing a sequence of nested mathematical programming problems. Finally, we combine these programs with the Bayesian update and propose an efficient algorithm to compute the PBNE.

The proposed modeling and computational methods have been shown to be capable of hardening the security of a broad class of industrial SCADA systems. In [5], we have used Tennessee Eastman (TE) process as a case study and provides design guidelines for defense against APTs that can infiltrate into the cyber-network through spoofing emails, then escalate privileges, tamper the sensor reading, and decrease the operational efficiency of the TE process without triggering the alarm. The dynamic games approach offers a quantitative way to assess the risks and provides a systematic and computational mechanism to develop proactive and strategic defenses across multiple cyber-physical stages of the operations. First, we observe that defense at the final stage is usually too late to be effective when APTs have been well-prepared and ready to attack. We need to take precautions and proactive responses in the cyber-stages when the attack remains "under the radar" so that the attacker becomes less dominant when they reach the final stage. Second, the online learning capability of the defender plays an important role in detecting the adversarial deception and tilting the information asymmetry. It increases the probability of identifying the hidden information from the observable behaviors, threatens the stealthy attacker to take more conservative actions, and hence reduces the attack loss. Third, defensive deception techniques are shown effective to increase the uncertainties of the attacks and their learning costs and hence reduce the probability of successful attacks.

10.4   Multi-layer Games for Strategic Defense in Depth

Recent work on game-theoretic modeling of APTs [10–12] suggests decomposing a general attack into the following three phases as a simplified representation of the kill chain. The main observation here is that a physical intrusion can work along the same lines:

(1) Initial penetration (phase one): this phase is completed if the attacker has established in initial contact with the victim infrastructure; more precisely, a contact point within the system has been set up; typically, a piece of malware has been installed via a successful phishing email (see also Example 3.8), or similar. For a pure physical access attempt, a logical entry point can be used to

gather information about legitimate employees, so as, for example, to create a validly looking fake identity.

(2) Learning and Propagation (phase two): this phase comprises a possibly long period in which the attacker remains stealthy and gathers information about the victim's infrastructure. The goal is to penetrate the facilities as deep as possible, in order to get near neuralgic points where damage can be caused subsequently.

(3) Damaging (phase three): this is the time when the attacker becomes visibly active and attempts to cause damage by virtue of all previously acquired knowledge. In particular, this phase may not require physical presence if it is supported by a (parallel) cyber-attack having left secret installations of malware and backdoors in the victim's infrastructure. Each of these phases is, in the game-theoretic treatment of [12], modeled as a designated type of game; specifically:

Phase one is captured by a matrix game since there are only finitely many strategies (social engineering, compromising third party suppliers, etc.), to get into the infrastructure at first place. Also, no such strategy may be guaranteed to work, so repeating the attempts is usually necessary (phishing is a good example, since the success of it is due to the mass of emails, which corresponds to a repetition of the strategy).

Phase two is modeled as a sequential game, in which the attacker's goal is to complete a stage-by-stage path from the initial contact point (end of phase 1) up to the target asset (beginning of phase 3). In case of cyber-attacks, a "stage" may physically correspond to a certain sub-network or computer and the respective "next stage" is any subnet or computer that is "closer" to the target. The understanding of closeness herein depends on the domain context, and can be a certain distance in terms of network node hops, a certain level of privileges due to the position in the company's hierarchy, or –in the case of physical security that we consider hereafter– the number of rooms and security checks to pass until the target asset becomes accessible.

Phase three is again modeled as a matrix game, with the concrete set of possibilities (for the attacker and the defender) being dependent on the type of access that the attacker has gained. For example, if the adversary has full access to a central server, then the configuration or data within the server may be compromised. The respective defense strategy, in that case, is an integrity check of both, the configuration and data. Clearly, both actions have to be repeated, thus justifying the matrix game model here. For high-security applications, a physical separation between the high-security domain and the outer company intranet (having an internet connection) may be found. In such cases, strategies from social engineering, e.g., bring-your-own device, must be added to the phase three game. For example, the physical separation between networks can be overcome, if a legitimate person with access to the target can be tricked into plugging a virulent USB stick inside, so that the malware can jump over the physical boundary. Likewise, compromising the machinery used inside the high-security domain by compromising a third party supplier is another way of penetrating a physical separation. However, all of these

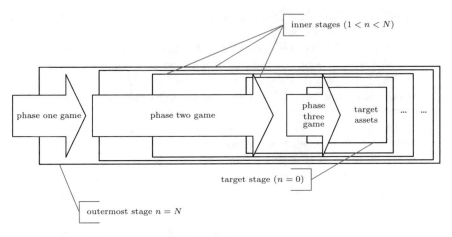

Fig. 10.2 Attack Phases in a game-theoretic view

strategies, depend on information learned in phase two, so the games are clearly connected to each other.

Figure 10.2 displays the modeling graphically and illustrates its relation to defense in depth. That is, defense in depth considers a number of protective shells around an asset, assuming that all of them have to be penetrated in sequence to get to the inner treasure. Our phase two game models exactly this view, with the games in phase one and three corresponding to the start and finish of the attack.

References

1. Boumkheld N, Panda S, Rass S, Panaousis E (2019) Honeypot type selection games for smart grid networks. In: Alpcan T, Vorobeychik Y, Baras JS, Dán G (eds) Decision and game theory for security. Springer International Publishing, Cham, pp 85–96
2. Chen P, Desmet L, Huygens C (2014) A study on advanced persistent threats. In: IFIP international conference on communications and multimedia security. Springer, pp 63–72
3. Goldreich O (2007) Foundations of cryptography: volume 1, basic tools. Cambridge University Press, Cambridge
4. Harsanyi JC (1967) Games with incomplete information played by "bayesian" players, I–III part I. the basic model. Manag Sci 14(3):159–182
5. Huang L, Zhu Q (2018) Analysis and computation of adaptive defense strategies against advanced persistent threats for cyber-physical systems. In: International conference on decision and game theory for security. Springer, pp 205–226
6. Kuipers D, Fabro M (2006) Control systems cyber security: defense in depth strategies. Technical report, Idaho National Laboratory (INL)
7. La QD, Quek TQS, Lee J (2016) A game theoretic model for enabling honeypots in IoT networks. In: 2016 IEEE international conference on communications (ICC). IEEE. https://doi.org/10.1109/icc.2016.7510833
8. Mitnick KD, Simon WL (2009) The art of intrusion: the real stories behind the exploits of hackers, intruders and deceivers. Wiley, New York

9. Murray J (2012) Securing critical infrastructure: perimeter protection strategy at key national security sites including transport hubs, power facilities, prisons and correctional centres. A senstar corporation white paper. Senstar Corporation, Ottawa, White Paper
10. Rass S, Alshawish A, Abid MA, Schauer S, Zhu Q, De Meer H (2017) Physical intrusion games-optimizing surveillance by simulation and game theory. IEEE Access 5:8394–8407
11. Rass S, König S, Panaousis E (2019) Cut-the-rope: a game of stealthy intrusion. In: Alpcan T, Vorobeychik Y, Baras JS, Dán G (eds) Decision and game theory for security. Springer International Publishing, Cham, pp 404–416
12. Rass S, Zhu Q (2016) Gadapt: a sequential game-theoretic framework for designing defense-in-depth strategies against advanced persistent threats. In: International conference on decision and game theory for security, pp 314–326. Springer, Springer International Publishing. http://link.springer.com/chapter/10.1007/978-3-319-47413-7_18
13. Stouffer K, Falco J (2009) Recommended practise: improving industrial control systems cybersecurity with defense-in-depth strategies. Department of Homeland Security, Control systems security program, national cyber security division
14. Suo X, Zhu Y, Owen GS (2005) Graphical passwords: a survey. In: 21st annual computer security applications conference (ACSAC'05). IEEE, p 10

Chapter 11
Cryptographic Games

Security is a process, not a product.

B. Schneier

Abstract The term "game" has substantially different meanings within the security area, depending on whether we speak about cryptographic security in particular, or system security in a more general setting that includes quantitative security with help of game theory. Game theory and cryptography are, however, of mutual value for each other, since game theory can help designing self-enforcing security of cryptographic protocols, and cryptography contributes invaluable mechanisms to implement games for security. This chapter introduces both ideas, being rational cryptography for the design of protocols that use rationality to incentivize players to follow faithfully, but also addresses the classical security goals like confidentiality, integrity, availability and authenticity by describing security games with quantitative and unconditional security guarantees. The chapter closes with a connection between network design for security and the P/NP question whose discovery is made with help from game theory.

11.1 Rational Cryptography

The combination of games and cryptography dates back to the days when public key cryptography has first seen formal notions of security. Initially, games were used to formalize notions of security and aid proofs in cryptography. However, this application of games differs from the newer area of rational cryptography in the particular use of games and its entailed understanding of security:

- In *classical cryptography*, a game is used to define a specific attack scenario and the resilience against this scenario in "almost" all cases. The game itself fixes how the attacker may interact with the honest party towards some security breach (i.e., disclosure of secrets, impersonation, etc.). In describing a form of interaction, the "game" is here similar to an algorithm, which is then further used

© Springer Nature Switzerland AG 2020
S. Rass et al., *Cyber-Security in Critical Infrastructures*, Advanced Sciences and Technologies for Security Applications,
https://doi.org/10.1007/978-3-030-46908-5_11

in complexity-theoretic reduction arguments. To illustrate how this is different to the game-theoretic treatments elsewhere in this book, we will exemplify below one particular game and one notion of security for a specific encryption function, and show how the game is used to prove security in the classical cryptographic realm.

- In *rational cryptography*, we seek to incentivize the attacker in a way that it is in its best interest not to attack at all. The game is here somewhat less restrictive in allowing the attacker to follow or deviate from a protocol. The goal lies in finding a design such that there is no incentive for the attacker to deviate from the cryptographic protocol, or in other words, to behave honestly. Thus, security would be *self-enforcing*.

Classical cryptography pursues strict worst-case assurances like claiming that the attacker will learn nothing significant in almost all but a negligible fraction of cases. Utility maximization arguments of rational cryptography are weaker in demanding only average case optimal behavior, thus allowing the attacker to succeed or fail in the intended security breaches alternatingly. It is then a matter of clever combinations of cryptographic primitives with game theoretic incentive design (utility functions) to keep the likelihood of a security breach within acceptable bounds. Examples related to very basic security notions like confidentiality or authenticity will be postponed until Sect. 11.2. For now, we start with the more "traditional" use of games in cryptography, to prepare the transition to modern concepts of rational cryptography, but also to shed some more light on what kind of security assurance cryptography can offer using game-based definitions.

Remark 11.1 We will intentionally deviate slightly from the notation used throughout the rest of this book, denoting cryptographic security games with the symbol **Game**, and letting algorithms use calligraphic letters. This is for the twofold reason of (i) emphasizing the conceptual difference to the mathematical games (denoted as Γ) discussed so far and later, and (ii) being more consistent with the cryptographic literature, where the security parameters that the games depend on play a central role. Hence, the "more complex" notation is important and justified, and for that reason also adopted in this chapter.

11.1.1 Game-Based Security and Negligible Functions

Much classical cryptography uses so-called "game-based security definitions". While leaving comprehensive accounts of this approach to textbooks on cryptology [15], we shall look at a game-based definition of confidentiality for probabilistic public-key encryption in the following example. This is to expose the difference in both, the way the game is defined as well as the way it is used for a security analysis.

Example 11.1 (Game-based Security by Indistinguishability against Chosen Plain Attacks (IND-CPA)) Given an encryption function, its security regarding confidentiality can informally be understood as follows: we say that an encryption function is *secure* if it produces "indistinguishable" ciphertexts. Herein, we say that two ciphertexts c_0, c_1 encapsulating distinct plaintexts $m_0 \neq m_1$ are *indistinguishable*, if no randomized algorithm that runs in polynomial time (in the length of the ciphertexts and the other related system parameters) can tell an encryption for m_0 apart from an encryption of m_1. In other words, the relation between two ciphertexts $\{c_0, c_1\}$ and the known inner plaintexts $\{m_0, m_1\}$ should remain effectively undecidable (in polynomial time and with a noticeable chance of correctness, where "noticeable" has a rigorous meaning given later). It can be shown that this requirement is equivalent to the inability to compute anything meaningful from a given (single) ciphertext; a property that we call *semantic security*.

A game-based definition of security in terms of indistinguishability frames the above decision problem into a non-cooperative interaction between two parties, where the *challenger* (player 1) asks the *attacker* (player 2) to decide which plaintext among a given set of two candidates (chosen by the attacker) has been encrypted. If the attacker answers correctly, it wins the game, otherwise, it looses (and the challenger would score accordingly). Thus, the game is implicitly zero-sum, though this is usually not stated or used anyhow.

The game takes three phases and is generally described as an experiment, similar to an extensive form game, in a notation like the following: As is common, we write pk, sk to mean public and private keys, as well as we let $\mathbb{G}_c$ and $\mathbb{G}_m$ denote the ciphertext and plaintext spaces (typically Abelian groups). The public-key encryption scheme comes as a triple of algorithms (G, E, D) where G is a probabilistic algorithm for key generation. This algorithm takes a security parameter $n \in \mathbb{N}$ to construct keys, plain- and ciphertext groups $\mathbb{G}_m$ and $\mathbb{G}_c$ with bitlengths, i.e., cardinalities, depending on n (practically, the choice of n would allude to recommendations published by related institutions, such as available at [14], among others). The symbols E_{pk}, D_{sk} denote the encryption function using key pk and decryption function using the key sk, respectively. Both mappings are between the plain- and ciphertext groups $\mathbb{G}_m$ and $\mathbb{G}_c$, in the respective order. The attacker, in turn, is modeled as a probabilistic Turing machine $\mathscr{A}$ that is assumed to run in polynomial time in its inputs.

Following the above outline, the *security game* is then played like shown in Fig. 11.1 for a *chosen plaintext attack*: it starts with the challenger (honest player) initializing the cryptographic mechanisms using the publicly known security parameter n. Then, the attacker (dishonest player 2) is allowed to ask for encryptions of plaintexts of its own choice (the number of which can be at most polynomial in n). In the second phase, the attacker chooses two plaintexts m_0, m_1 of the same length (to avoid a trivial win), and sends them to the challenger, claiming that it can still recognize m_0 and m_1 after being encrypted. The challenger picks on out of the two random, and sends back the ciphertext. The attacker wins, if it can correctly tell which $m_b \in \{m_0, m_1\}$ has been encrypted. In that case, the game rewards the attacker with 1, or rewards zero otherwise. This outcome is denoted as the random variable $\mathbf{Game}_{\text{IND-CPA}}^{\mathscr{A}^{E_{pk}(\cdot)}}(n)$. The superscript notation $\mathscr{A}^{E_{pk}(\cdot)}$ hereby

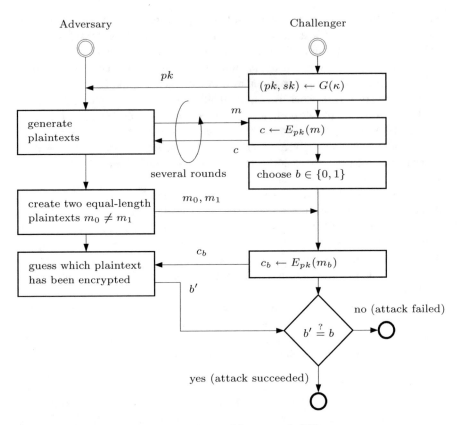

Fig. 11.1 Game-based security against chosen plaintext attacks [38]

denotes oracle access of $\mathscr{A}$ to the encryption function E_{pk}. Its dependency on the security parameter n is crucial in the following, since the game's analysis regards asymptotic properties of the sequence of random variables indexed by the security parameter n. However, despite Fig. 11.1 being interpretable as a game in extensive form, the analysis is *not* for saddle-point values, but rather seeks an upper bound to the game's worst-case outcome.

This game is essentially zero-sum in the sense that it assigns a utility to the attacker, usually called the attacker's *advantage*. For this example, it is defined to be $\mathbf{Adv}_{\mathrm{IND\text{-}CPA}}^{\mathscr{A}}(n) := \Pr[b^* = b]$, where the dependency on n is via $\mathbf{Game}_{\mathrm{IND\text{-}CPA}}^{\mathscr{A}^{E_{pk}(\cdot)}}(n)$, assuming that both players act optimal. $\diamond$

Based on the description given in Example 11.1, a game theoretic analysis would look for saddle-points to determine the simultaneously optimal behavior for both parties in the game. On the contrary, cryptography assumes that both parties a priori act optimal already, and tries to bound the game's outcome $\mathbf{Adv}_{\mathrm{IND\text{-}CPA}}^{\mathscr{A}}(n)$ asymptotically in n. Along this route, the game then becomes the object of a complexity-theoretic reduction argument that generically runs as follows, where different notions of security (e.g., against chosen ciphertext attacks, or others) mainly differ in the games under study:

1. Assume the existence of an attacker $\mathscr{A}$ who would gain a non-negligible utility $\mathbf{Adv}^{\mathscr{A}}_{\text{IND-CPA}}$ in $\mathbf{Game}^{\mathscr{A}^{E_{pk}(\cdot)}}_{\text{IND-CPA}}$ (aiming at a proof by contradiction). The exact meaning of "negligible" will be clarified later in Definition 11.1; essentially, it is an asymptotic condition based on the parameter n to the above game.

2. Construct another algorithm $\mathscr{A}'$ that invokes $\mathscr{A}$ as a subroutine, and show that $\mathscr{A}'$ could be used to solve some computationally intractable problem P that has been fixed a priori (typical examples are factorization of large integers, the computation of discrete logarithms, deciding upon quadratic residuosity among many others).

3. Since P is believed to be unsolvable in polynomial time (asymptotically as the sizes of the instances of P grow), but the polynomial-time algorithm $\mathscr{A}'$ could still do it, this contradiction refutes the existence of the attacker $\mathscr{A}$, thus proving the encryption to be secure using the cryptographic game.

On an abstract level, this is the basic structure that reductionist security proofs throughout cryptography follow (see [41] for numerable examples). This approach to security is successful for cryptographic purposes, but different from a game theoretic analysis in two respects: first, it does not look for optimal behavior, but rather rules out winning strategies (only). Second, the analysis looks for worst-case bounds to the utility values, rather than for average-case optimality as game theory would do.

This entails that some impossibility results found in cryptography need reconsideration under the game theoretic perspective [11], which is one benefit that game theory offers for cryptography. In order to properly study cryptographic competitions like the game in Fig. 11.1 from a game theoretic viewpoint, we need to first incorporate the computational notions into the game theoretic definitions. Primarily, this concerns the dependency of the game on the security parameter, that optimality conditions of game theory (such as for equilibria) should include. The basic elements of this formalization are summarized in Definition 11.1.

Definition 11.1 (Asymptotic Relations and Negligible Functions) Let $f, g :$ $\mathbb{N} \rightarrow \mathbb{R}$ be given. We write $f \leq_{\text{asymp}} g$ if an integer n_0 exists for which $f(n) \leq g(n)$ whenever $n \geq n_0$. The relation $\geq_{\text{asymp}}$ is defined mutatis mutandis. The function f is called *negligible*, if $f \leq_{\text{asymp}} 1/|p|$ for all polynomials p, and called *noticeable* if $g \geq_{\text{asymp}} 1/p_0$ for some polynomial p_0.

As with Landau-symbols, we write $\text{negl}(n)$ to anonymously denote a function that is negligible in n. With this, we write $f \approx_{\text{negl}} g$ if $|f - g|$ is negligible. Likewise, we write $f \leq_{\text{negl}} g$ if $f \leq_{\text{asymp}} g + \text{negl}$, and $f \geq_{\text{negl}} g$ if $f \geq_{\text{asymp}} g - \text{negl}$. The complement relation to $\leq_{\text{negl}}$ is written as $>_{\text{noticbl}}$ and means $f \geq g + \varepsilon$, where ε is a *noticeable* function.

Example 11.2 (negligible and non-negligible functions) It is easy to see that all constants are non-negligible, as well as are functions like $1/\log(n)$ or reciprocal polynomials such as n^{-3}. On the contrary, the function 2^{-n} would be negligible, as it decays faster than any polynomial.   ◇

The set of negligible functions is closed under addition and scalar multiplication, and as such behaves like a vector space; a property that is frequently used implicitly. The above reductionist security argument for Example 11.1 would then be formalized by writing $\mathbf{Adv}^{\mathscr{A}}_{\text{IND-CPA}}(n) \approx_{\text{negl}} 1/2$. This expresses that once the security parameter is sufficiently large, the attacker has no noticeably better chance of winning than it would have by a fair coin toss. To see this at work, Example 11.3 instantiates the generic argument for the game in Fig. 11.1.

Example 11.3 (Game-Based Security Analysis of ElGamal Encryption) ElGamal encryption [24] is an asymmetric scheme that works over a finite cyclic group $\mathbb{G}$ with generator $g \in \mathbb{G}$. The encryption under a public key $pk = g^s$ sends a plaintext $m \in \mathbb{N}$ to a ciphertext $(c_1, c_2) = (g^r, m \cdot pk^r)$, where r, s are random secret elements chosen from $\mathbb{G}$. Here, pk is the public key, and the secret decryption key would be $sk = s$. For completeness (yet not needed in the following), we note that the decryption would first construct the value $K = (c_1)^s = (g^r)^s = g^{sr} = (g^s)^r = pk^r$ and then recover $m = c_2 \cdot K^{-1}$ in $\mathbb{G}$.

Let us go back to Fig. 11.1, and assume that the attacker would have a "winning" strategy in the sense that it can act towards $\mathbf{Adv}^{\mathscr{A}}_{\text{IND-CPA}}(n) >_{\text{noticbl}} 1/2$. The problem P to base the security upon is the following: given a triple of values (a, b, c) in some finite cyclic group $\mathbb{G}$ generated by some element $g \in \mathbb{G}$, decide if there are integers x, y such that $(A, B, C) = (g^x, g^y, g^{xy})$. We call this decision the *Diffie-Hellman (DH) problem*, and its input is called a *DH triple*. It is widely believed that for properly chosen groups $\mathbb{G}$ that every polynomial time algorithm $\mathscr{A}$ taking (A, B, C) as input and running in a time that is at most polynomial in the order of the group $\mathbb{G}$, has a chance $\Pr(\mathscr{A}(A, B, C) = 1) \leq_{\text{negl}} 1/2$ of recognizing DH triples correctly among all possible triples (i.e., $\mathscr{A}(A, B, C)$ returns 1 if a DH triple is found, or zero to indicate a non-DH triple). This is the *DH assumption* and the computational intractability hypothesis upon which our security proof will rest (i.e., be conditional on assuming "only negligibly better chances to decide correctly that by an uninformed yet fair coin toss").

Assuming an attacker $\mathscr{A}$ that could win $\mathbf{Game}^{\mathscr{A}^{E_{pk}(\cdot)}}_{\text{IND-CPA}}(n)$, we can deduce that the DH problem is solvable with noticeable chances, i.e., $\Pr(\mathscr{A}(A, B, C) = 1) = 1/2 + \varepsilon >_{\text{noticbl}} 1/2$ whenever (A, B, C) is a DH triple. Here, $\varepsilon(n)$ is noticeable as a function of n. We construct a *distinguisher* algorithm D that works as follows: on input of a triple $(A = g^x, B = g^y, C = g^z)$, it starts $\mathbf{Game}^{\mathscr{A}^{E_{pk}(\cdot)}}_{\text{IND-CPA}}(n)$ with the attacker (i.e., algorithm) $\mathscr{A}$ taking (A, B, C) as input and outputting a guess for it to be a DH triple. In the challenge phase, upon receiving (m_0, m_1) in the IND-CPA game, we pick a random $\beta \in \{0, 1\}$ and respond with the ciphertext $(B, m_\beta \cdot C)$. Now, distinguish two cases:

1. if $z = xy$, then the response constructed by D is a properly distributed ElGamal ciphertext, and by our hypothesis on $\mathscr{A}$, the attacker will correctly recognize the inner plaintext with noticeable likelihood $\Pr[\mathscr{A}(g^x, g^y, g^{xy})] >_{\text{noticbl}} 1/2$.
2. if $z \neq xy$, then the response $(B, m_\beta \cdot C)$ is stochastically independent of β, and hence no algorithm, regardless of complexity, can correctly guess β with a chance $> 1/2$. Hence, $\Pr[\mathscr{A}(g^x, g^y, g^{xy})] = 1/2$.

By the law of total probability, we thus find that the algorithm D can use $\mathscr{A}$ to recognize DH triples with a likelihood of

$$\Pr(D(A, B, C) = 1) = \frac{1}{2}\Pr(D(g^x, g^y, g^{xy}) = 1) + \frac{1}{2}\Pr(D(g^x, g^y, g^z) = 1)$$

$$= \frac{1}{2} \cdot \left(\frac{1}{2} + \varepsilon\right) + \frac{1}{2} \cdot \frac{1}{2} = \frac{1}{2} + \varepsilon >_{\text{noticbl}} \frac{1}{2},$$

where ε is a noticeable function. However, the DH assumption states just the opposite for any algorithm running in polynomial time, including D in particular. This contradiction means that the attacker $\mathscr{A}$ cannot exist, conditional on the computational intractability of the DH problem.   ⋄

Remark 11.2 In being asymptotic, claims of security phrased in terms of negligibility are usually silent about "how large" the security parameter n in general should be for the asymptotic relation to hold. This lack of detail in many asymptotic security proofs has been subject of criticism and a subsequently quite emotional controversy [21].

11.1.2 Honesty and Rationality

Rational cryptography changes the perspective by not bounding but rather optimizing the attacker's advantage, being redefined as a proper utility function. To properly transfer the computational asymptotic relations used in the traditional cryptographic setting, it starts by redefining equilibria and optimality in terms of the relations given in Definition 11.1. This modification allows for a negligible amount of "violation" of optimality, which is necessary to account for the underlying cryptographic primitives and computational intractability hypothesis behind them. The point of negligibility in all game theoretic definitions in rational cryptography is making an allowance up to anything that is less than what a polynomially bounded attacker could achieve (since an exponentially fast attacker is neither realistic nor theoretically meaningful).

Rational cryptography in most instances is about one of the following things:

Protocol analysis (game theory):   here, we analyze a cryptographic protocol from a game-theoretic view, i.e., by asking for the best moves of each player per stage of the game (like in Fig. 11.1 or suitably adapted to the situation at hand). In many cases, this uses relaxed definitions of security that in turn allow for better complexities than achievable in classical schemes [4].

Protocol construction (mechanism design):   here, we look for cryptographic primitives to enforce some desired optimal rewards (in the defender's interest) in each stage of a (cryptographic) security mechanism. Game theoretic concepts are then used as a guidance for the design itself, whereas cryptography is used to enforce certain desired equilibria.

As far as it concerns the analysis, Example 11.3 has a clearly defined incentive in recognizing one out of two plaintexts. An implicit assumption is here the willingness of the players to engage in the game and to follow its rules. In other cases, this engagement cannot be safely assumed, and the players may be willing to interact with another party only if there is an incentive to do so; the players would act *rational*.

Likewise, the attacker would be rational in the sense of perhaps pursuing other goals than what the game defines. To illustrate this, we look at *secret sharing* as another popular cryptographic primitive.

Example 11.4 ((k, n)-secret sharing)   Secret sharing is one particularly rich and widely studied cryptographic concept studied in rational cryptography. In one of its basic forms [39], we have a secret $m \in \mathbb{F}$ from some, usually finite, field $\mathbb{F}$, and embody it in a randomly chosen polynomial $p(x) = m + a_1 x + a_2 x^2 + \ldots + a_{k-1} x^{k-1}$. The goal is storing m with a set of $n \geq k$ (potentially distrusted) parties $P_1, \ldots, P_n$, such that no party or coalition of parties can discover the secret m. The secret remains reconstructible only under a set of defined conditions. For the original scenarios of the polynomial, each party P_i receives the value $s_i := p(i)$, called a *share*. Clearly, it takes at least k or more samples from the polynomial p to uniquely recover it by interpolation. This already represents the above mentioned condition: we can distribute m over a set of n servers, assuring that no coalition of $\leq k - 1$ parties can recover it. However, with $\geq k$ data items $p(i_1), p(i_2), \ldots, p(i_k)$, the secret m can be recovered by interpolating through the data points and evaluating the resulting polynomial as $m = p(0)$. Hence, this is called a k-out-of-n secret sharing, or *threshold* scheme.   ◇

In case of secret sharing, the "game" could relate to various stages of the protocol, i.e., the initial construction and distribution of the shares, their use (e.g., for computations on the shared secret), or the reconstruction phase where the secret shall be recovered with help of (at least) k out of n players. This is the stage that we are interested in here, so let us take a closer look.

In a *reconstruction game*, let the players act rational by letting them freely decide with whom they want to exchange information and potentially open their privately owned shares. By construction, this is an n-person game for an (k, n)-secret sharing scheme. Example 11.4 assumes that the players just cooperate in the reconstruction, but why would they? From a fixed player's perspective, e.g. player P_1, it is not necessarily beneficial to open its own share, depending on how many others are willing to do the same: let k be the number of required collaborators among n players in total, and let t of them be present for the reconstruction. Then, the reconstruction game admits the following Nash equilibria [17]:

- For any t, k, n, it is a Nash equilibrium to refrain from revealing the share. Then, nobody learns anything, but in turn, also nobody has an information gain over, e.g. P_1, which it could count as a personal loss.
- If $t > k$, then it is an equilibrium for all players to reveal their shares, so that all players learn the same information and are thus "equally rewarded" again.

If $t = k$, then P_1 can gain even more by letting all others reveal their shares but keeping its own share secret. Then every other player has only $t - 1 < k$ shares, thus learns nothing, and only P_1 gets to know the secret. So, this configuration is not even a Nash equilibrium.

These considerations show that the incentive to interact with others can be individually different, and even be a matter of timing (i.e., if player P_1 were the last to reveal, it could just change its mind to remain the only one to reconstruct the secret). Thus, studying secret sharing from this more utility-driven perspective is nontrivial and paying, if a self-enforcing reconstruction protocol can be created to end up in the second of the aforementioned equilibria (in all cases).

From the above cases, we can derive local preferences for each player, which then induce utility functions as in Sect. 5:

- player P_i prefers outcomes in which it learns the secret, and
- if player P_i learns the secret, then it prefers outcomes in which as few other players as possible learn the secret as well.

These preferences can be formalized by indicator variables telling whether or not a player learns the secret, and adding them up to count the total number of players having the secret in the end. The utility implied by this preference relation is what rational secret sharing protocols can optimize by drawing from the whole set of possible protocol structures: it is up to each player to decide for itself what to do with the shares, e.g., broadcast them to everyone, or sending them individually to others over private channels, or act otherwise. Further, cryptographic protocols for consensus finding and commitments can be involved, in order to establish mutual trust relationships and to form trusted coalitions for the reconstructions before any share is actually revealed. The entirety of protocols arising from such considerations are summarized as *rational secret sharing* [19].

What a player can (and would) do of course depends on the goal, which so far has been learning the secret itself, but preventing others from learning it. In dropping the first goal, a player could merely mount a denial-of-service by blocking shares, or otherwise may fiddle with the reconstruction to cause an incorrect result. The scheme of Example 11.4 is particularly rich in structure here, as it enjoys a natural relation to Reed-Solomon codes for error correction [23]. If we let t be the number of actively malicious players in the reconstruction, e.g., we add strategies of blocking shares or sending incorrect values for the reconstruction, then the following facts about (k, n)-secret sharing hold:

1. without any error correction being applied to the set of shares before the reconstruction, a malicious player can send a modified share to let the other (honest) players reconstruct the value $m - c$ for any chosen value c instead of m [42].
2. up to $t < n/3$ shares can be corrupted (in the sense of actively modified by a malicious party P_i) but the secret m remains recoverable.
3. If $n/3 < t < k$ shares are modified, then an incorrect secret will be recovered without notice by the honest players, even though an attacker still did not learn the secret.
4. Up to $t < \min\{k, n/2\}$ shares can even be held back by the malicious parties. In that case, the adversarial players cannot learn anything about the secret, but the secret remains reconstructible by the honest players too.

Hence, the attacker can mount an active attack by modification (cases 2 and 3) or a denial-of-service (case 3); the list is further comprehended in [29].

Now, suppose the game modeling of the secret sharing reconstruction phase is such that the action sets of players include sending modified shares, blocking their own shares or those of others. Then, at least the above four outcome possibilities need to be added to the game dynamics, besides others that are possible due to the structure of the reconstruction process.

A rationally cryptographic analysis can then proceed by defining a utility function for each player, in which it assigns different rewards to each possible outcome (e.g., learning the secret, preventing others from learning it, allowing only the fewest possible number of others to get the secret, enforcing a changed output, etc.). Then, we can redefine the utility function as a weighted sum of revenues per security incident or protocol outcome. Given a list of security incidents indexed by an integer j, together with respective numeric rewards denoted as γ_j for the j-th such incident, the utility for an attacking player could be set to

$$u_{attacker} := \sum_j \Pr(\text{attacker causes security breach no. } j) \cdot \gamma_j, \qquad (11.1)$$

which would be usable for all players $i = 1, 2, \ldots, n$, besides the revenue for acting honestly (as was outlined above). In many cases, acting dishonestly can even yield to losses. Among such possibilities are putting the attacker into danger of becoming discovered (and hence excluded from the protocol), or making the corruption of parties so costly and difficult that the investment would exceed the gains from recovering the secret (e.g., if the secret is short-term relevant only, so that by the time when the attacker would have reconstructed it, it already became useless).

11.1.3 Rational Interactive Proofs

Leaving the secret sharing showcase example behind, we will now look at a more general setting of a cryptographic protocol between two parties. Protocols consist of a sequence of steps, each of which offers different ways of action to cause certain security breaches [11]. The point of game theory herein is pinning down a behavior that optimizes the outcome for all players simultaneously. This optimization treats players as Interactive Turing Machines (ITMs), whose utility is determined by how the ITM behaves, e.g., using a formula like (11.1). Since optimization over the set of all ITM is not practically achievable (in absence of the typically necessary algebraic and geometric structure), the optimization is always towards enforcing a desired equilibrium by using, respectively prescribing, proper cryptographic mechanisms to keep the probabilities in (11.1) under control. For example, if a security breach concerns a private share to leak out to another player, encryption would be added to the protocol (see Example 11.3). Likewise, if players agree to mutually open their shares but are unsure if the information being sent is correct, then cryptographic commitments can be brought into the game for verifiability [12], and so forth.

Since many cryptographic mechanisms come with security assurances that are only valid asymptotically, the respective asymptotic versions of optimality to define equilibria need to be used. Example 11.3 shows one possible reason for such negligible terms to arise from the intractability assumptions on which the security of the cryptographic primitive rests. Since we cannot hope to find the globally optimal ITM among the set of all possible ITMs, the cryptography to build the protocol is chosen to enforce a certain desired equilibrium, based on the utility-induced preferences, while the numeric rewards themselves may be of secondary interest only.

In any case, the analysis will here consider utilities derived from likelihoods of certain security incidents, which in turn follow from the properties of the involved cryptography. General assumptions made in the game theoretic analysis often include confidential and authentic channels. Both can in turn be realized on grounds of game-theoretic considerations, and we will devote the entire Sect. 11.2 to exemplify how this can be done, leveraging the secret sharing scheme of Example 11.4.

Like game theory concepts can be relaxed by cryptographic notions, cryptographic objects can also be rephrased in game theoretic terms. Let us showcase one such approach by taking an example of how an interactive proof in cryptography would translate into a rational proof in rational cryptography.

Example 11.5 (Rational Proofs; definition taken from [4]) Let f be a function that a prover P shall evaluate on behalf of a verifier V (e.g., P could be a cloud, and V could be its customer). A rational proof system for f is a pair of algorithms (P, V) that satisfy the following properties:

1. The honest prover always replies correctly: let the output of the interaction between P and V upon the input x be denoted as $(P, V)(x)$. Then, we demand $\forall x : \Pr((P, V)(x) = f(x)) = 1$.
2. A dishonest prover always gains less than an honest prover (rationality): Let $u((P, V)(x))$ be the rewards for an honest player, and likewise let $u((\tilde{P}, V)(x))$ be the reward for a dishonest player (upon the same input and the same verifier). Then, we demand an equilibrium condition like (3.17):

$$\forall \tilde{P} \; \forall x : \mathbb{E}(u((P, V)(x))) \geq_{\text{negl}} \mathbb{E}(u((\tilde{P}, V)(x))).$$

In both cases, the asymptotics are w.r.t. the bitlength of x. To avoid trivialities, an obvious additional requirement is V taking much less effort to engage in the interaction with P than if V would evaluate f itself (efficiency). Formally, it should run in polynomial time in the size of the smallest circuit to evaluate f.

The difference between a rational proof and a (cryptographically strong) interactive proof is the latter demanding to catch a cheating prover with a chance $\approx_{\text{negl}} 1$, as opposed to a rational proof that only seeks to catch a cheater with a small yet noticeable probability. In many cases, this allows for much better complexities (i.e., running times) than under stronger cryptographic requirements.   ◇

In both branches of rational cryptography (protocol design or security analysis), the underlying cryptography or the general security primitives are in the center of attention. Regardless of whether the application operates on the rational versions of secret sharing, secure function evaluation, or others, the existence of cryptographic primitives realizing the mechanism is assumed to be at the core of the analysis. Typically, the investigation takes these primitives to be idealistic and replaces them by their realistic counterparts in a second step. In this context, these assumptions concern unconditionally confidentiality, integrity, availability and authenticity of communication channels (CIA+), a game theoretic analysis can help constructing secure realistic such ingredients. It therefore pays to look deeper into the game theoretic analysis and design of communication infrastructures; not only for them being fundamental to more sophisticated applications of rational cryptography thereon, but also for being themselves critical infrastructures. We devote Sect. 11.2 to that matter, alluding to the aforementioned matters of protocol analysis and infrastructure design aided by game theory.

11.2   Communication Games

Computational intractability assumptions are a successful yet nonetheless unverified foundation of secure communication [21], and even claims of unconditionally secure communication such as are attributed to quantum key distribution networks (as one example) are, strictly speaking, not without hypotheses, but come with verified or at least verifiable assumptions to base their security upon for quantum cryptography [8, 40], or for transmissions over multiple paths [9, 27, 43]. Both, quantum key distribution and multi-path transmission are open to game theoretic analyses [30, 32], with the appeal that the game and its analysis are particularly simple in that case, with the techniques themselves enjoying much wider applicability [13, 20, 22, 25].

11.2.1 *Confidential Transmission Games*

The playground in the following is a communication network connecting multiple entities in a bidirectional fashion. Let us think of this network as a graph $G = (V, E)$ with nodes V and direct communication channels u—v between (physically adjacent) connected nodes $u, v \in V$.

An attack scenario in this context involves a corrupted set of nodes $U \subseteq V$, with explicit constraints imposed on either which specific subsets can be corrupted (making up a family of potentially hostile nodes), which is called an *adversary structure*, or how many nodes can be corrupted at most (as would be the case in *threshold cryptography*). In both cases, the adversary structure $\mathscr{A}$ is a subset of 2^V, and lists all sets that the attacker can simultaneously compromise.

Like in a person-in-the-middle attack, the hostile (i.e., corrupted) nodes are located at unknown positions between the honest sender and receiver in the network; the exact locations are only known to the attacker. Furthermore, we assume that the attacker's servant nodes collaborate in the sense that they either drain information that they give to the attacker, or upon its request behave arbitrarily different to what the sender and receiver presume (active attacks by not following any protocol any more). In the realm of rational cryptography, we may additionally assign adversarial efforts to conquer nodes (called a *costly corruption* [11]), but let us keep things simple here by leaving this generalization aside.

A multipath transmission based on threshold secret sharing then roughly follows these steps (cf. [9] for one provably optimal example scheme):

1. Put the message m through an (k, n)-secret sharing scheme (as in Example 11.4), yielding the shares $s_1, \ldots, s_n$.
2. Send each share over its own channel to the receiver, which is node-disjoint with other channels to the same destination. These channels may indeed run through dishonest nodes, but as long as less than k such channels are intercepted, the secret is perfectly protected by the sharing mechanism.
3. At the receiver's side, reconstruct the secret upon all incoming shares (some of which may went missing if the nodes were actively blocking, or were actively modified, or simply eavesdropped and sent elsewhere for reconstruction).

It is evident that for this protocol to work, the network topology needs to provide multiple paths between the sender and the receiver, from which a random subset of n paths can be chosen. Similarly obvious is the fact that the attacker will seek to intercept as many paths as it can, to maximize its chances to get k or more shares. This is the game being played, which lets:

- player 1 (the sender) choose the delivery paths,
- player 2 (the attacker) choosing the nodes to compromise.

Remark 11.3 The enforcement of multiple, and in this case non-intersecting routes is technically possible in various ways, including virtual LANs, source routing

(RFC 7855 [28], though this is not usually supported for security reasons), or more recently, preferred path routing [6].

The game's payoff is a 0/1-valued indicator variable, which, depending on the defined security goal, is determined by the properties of the underlying sharing scheme. Threshold secret sharing is a particularly illustrative example here, as its effects depend on what we ask for (continuing the remarks following Example 11.4):

Confidentiality: the message is safe as long as no more than k shares get into the attacker's hands. Security is here defined "unconditionally" in the sense of not using any intractability hypothesis, as follows: We call a transmission ε-*private* if the transcript of a transmission (i.e., the entirety of information that can be collected by eavesdropping) for distinct messages $m_1 \neq m_2$ has approximately the same distribution up to a statistical distance of less than 2ε. The statistical distance here refers to metrics on the distributions of the random variables representing the transcript information (typically strings or more complex compound hence multivariate random variables). The rationale is that small distances between the distributions imply that there is no significant information about a message contained, since the empty message (that has trivially no information in it) would essentially produce an almost identically distributed transcript (this is the same idea as behind the computational indistinguishability described in Examples 11.1 and 11.3).

Integrity: a (k, n)-secret sharing can detect and correct up to $n/3$ modifications. A manipulation of up to $n/2$ shares can be detected but no longer corrected (security against passive adversaries) [23].

Availability: Up to $\min\{k, n/2\}$ shares can be blocked overall (thus imposing a natural bound on k relative to n), without affecting the privacy of the message (at least), whereas we would have to choose k so small that sufficiently many shares come in for a correct reconstruction. Otherwise, the secret may still be safe upon a failure yet the receiver may as well be unable to recover it. In this context, the related security concept is called δ-*reliability*, meaning that the receiver with probability $\geq 1 - \delta$ terminates having recovered the correct message emitted by the receiver.

Protocol analysis [43] would ask for the optimal coincidence of ε-privacy and δ-reliability that we call (ε, δ)-security. The notable difference made by game theory is the reversal of quantifiers: while the graph topology usually determines the ε, δ for the security, game theory in a subsequent step can help tuning these values towards any desired security level. If this fails, we can go back and try to alter the network structure to get better ε, δ values. Game theory can help in this aspect as well, and we shall look into this shortly afterwards in Sect. 11.2.4. The optimum is of course $(0, 0)$-security, which we call *perfect security*. The reachability of this limit is characterized by a result of [2], based on the concept of subconnectivity against so-called k-active adversaries: The graph $G = (V, E)$ is said to be $\mathscr{A}^{(k)}(s, t)$-*subconnected*, if removing the nodes within the union of any k sets in $\mathscr{A} \subseteq 2^V$

from G leaves at least one s—t-path. An active attacker, who can corrupt up to k sets in $\mathscr{A}$ is said to be k-active.

Theorem 11.1 *A sender s can do Perfectly Secure Message Transmission (PSMT) with a receiver t in the network graph G if and only if the network is $\mathscr{A}^{(2)}(s,t)$-subconnected.*

The point of a game theoretic treatment is the insight that we can run the same scheme like we would play a repeated game with constant payoff matrix: instead of transmitting all shares in parallel, we can transmit them sequentially and take another random path in each round. Overall, we end up with the attacker (player 2) catching the share in some rounds and missing it in other rounds of the game. If the attacker notes 1 for a caught share and notes 0 for a miss, then the saddle-point value v of the resulting matrix game has a neat interpretation as being a bound to

$$\Pr(\text{message } m \text{ gets disclosed}) \le v.$$

The strategy sets for the players in this game are easy to define (see Fig. 11.2):

- For the sender and receiver (player 1): a set $AS_1 = \{\pi_1, \pi_2, \ldots\}$ of paths in G that can be chosen for the transmission (the entirety of which is exponentially many in general, but can be restricted to any feasible choice thereof).
- For the attacker (player 2): a family of node sets $AS_2 \subseteq 2^V$ that can be compromised, assuming that the hostile nodes may change but do not accumulate (i.e., the attacker must release some nodes before conquering new ones, which can be enforced by other optimized inspections [7]). For example, a threshold assumption would limit the cardinality $|X| \le t$ for every $X \in AS_2$ to a fixed number t of compromised nodes per round.
- In each repetition of the game, let $\pi \sim \mathbf{x} \in \Delta(AS_1)$ be a random choice of paths, and let $Y \sim \mathbf{y} \in \Delta(AS_2)$ be a randomly chosen set of compromised nodes.

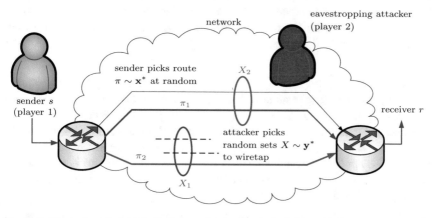

Fig. 11.2 Illustration of a Multipath Communication Game (Definition 11.2)

The attacker scores 1 if and only if $\pi \subseteq X$ (in a slight abuse of notation), i.e., the chosen paths all run through compromised nodes, and scores 0 otherwise (as implied by the cryptographic secret sharing). The sender, acting as player 1, minimizes the average of this indicator, being $\Pr_{(\mathbf{x},\mathbf{y})}(\pi \in X)$, over varying $\mathbf{x}$. Likewise, the attacker, acting as player 2, maximizes $\Pr_{(\mathbf{x},\mathbf{y})}(\pi \in X)$ by varying $\mathbf{y}$. Hence, the game is zero-sum.

If the sender plays its optimal strategy $\mathbf{x}^*$ from any equilibrium $(\mathbf{x}^*, \mathbf{y}^*) \in \Delta(AS_1) \times \Delta(AS_2)$, and with v being the saddle-point value of this 0-1-valued matrix game (recall that AS_1, AS_2 are both finite), this is nothing else than a security strategy, and as such satisfies

$$\Pr(\text{successful attack} \mid \mathbf{y}) \leq v \quad \text{for all} \quad \mathbf{y} \in \Delta(AS_2).$$

Furthermore, for every $\mathbf{x} \neq \mathbf{x}^*$, there is an attack strategy $\mathbf{y}_x \in \Delta(AS_2)$ for which $\Pr(\text{successful attack}|\mathbf{y}_x) > v$, i.e., the path choice $\mathbf{x}^*$ is optimal. This is nothing else than a moving target defense based on cryptography and optimized with game theory.

In this context, cryptography tells us that we cannot achieve *perfectly secure* communication unless the network graph has really strong connectivity (relative to the collection of distrusted nodes). However, the game theoretic analysis lets us reconsider problems in classical cryptography by resorting to a weaker notion of bounding the probability for a privacy breach rather than demanding it to be "zero up to a negligible error". In repeating the protocol as such (e.g., by a recursive application of it to send a single share only with chance $\leq v < 1$ of getting caught), we can even lower the bound v exponentially fast towards zero. This can be achieved by increasing the number of shares as well as the number of rounds and simply splitting the message initially into a sequence of bitstrings $r_1, r_2, \ldots, r_k$ whose bitwise XOR recovers $m = r_1 \oplus r_2 \oplus \ldots \oplus r_k$.

Remark 11.4 The mapping that sends a bitstring m to a set of n bitstrings r_i all of the same length as m and recovering m by the exclusive-or of all these strings is the special case of a n-out-of-n threshold secret sharing, since any out of the n strings r_i that is unknown to the attacker will perfectly conceal the message m similar as a one-time pad.

We send each r_i for $i = 1, \ldots, k$ over a multipath scheme (like the above), with the game theoretic assurance that the attacker can learn r_i with a likelihood $\leq v < 1$. If any of the r_i goes missing for the attacker, this piece acts like a one-time pad protecting m, thus the attacker learns nothing about m. Conversely, the chances to catch all r_i over k rounds is $\leq v^k \to 0$ as k increases. We call that an *arbitrarily confidential communication* [31, 33], since it can be made ε-private for any chosen $\varepsilon > 0$ and has an overhead that is polynomial in $\log(1/k)$. The precise setup is the following:

Definition 11.2 (Multipath Communication Game [31]) Let a graph $G = (V, E)$, an integer $k \geq 1$ and a pair of distinct nodes $s, t \in V$ be given. Assume that an s-t-communication runs over k paths chosen from the action set AS_1 for the sender, in the presence of an attacker described as an adversary structure $\mathscr{A} \subseteq 2^V$. This set of vulnerable nodes defines its action set $AS_2 = \mathscr{A}$ (the cardinality of AS_2 should be feasibly small). The quantitative security assurance (in the sense of Definition 3.9) of this s-t-communication is defined as $\rho(s, t) = 1 - \max_{\mathbf{x} \in \Delta(AS_1)} \min_{\mathbf{y} \in \Delta(AS_2)} \mathbf{x}^T \mathbf{A} \mathbf{y} = 1 - \text{val}(\mathbf{A})$, where $\mathbf{A} \in \{0, 1\}^{|AS_1| \times |AS_2|}$ models the zero-sum communication game with the payoffs as defined by

$$a_{ij} = \begin{cases} 1, & \text{if the } s\text{-}t\text{-transmission remained secret;} \\ 0, & \text{otherwise,} \end{cases}$$

and where the sender is a utility maximizer.

For general communication protocols, but in particular those that can be framed as a game in the sense of Definition 11.2, we can slightly weaken the notion of PSMT to be easier to achieve but retaining almost its full strength:

Definition 11.3 (Arbitrarily Secure Message Transmission (ASMT)) [31]) A communication protocol is called *arbitrarily secure*, if for every $\varepsilon > 0, \delta > 0$, we can run it to achieve ε-privacy and δ-reliability. We call the protocol *efficient*, if the overhead is polynomial in $\log \frac{1}{\varepsilon}$ and $\log \frac{1}{\delta}$.

The result making the above intuition rigorous is the following, whose proof is found in [36], and respective protocol is given in [31]. Note that the next result is *agnostic* of any underlying secret sharing or other means of channel encoding, and exclusively relies on how the communication game's payoff structure looks like (that in turn depends on the underlying transmission technique, e.g., secret sharing with transmission over multiple paths, or other). Furthermore, it does not depend on computational intractability hypotheses.

Theorem 11.2 *Let Alice and Bob set up their game matrix with binary entries $a_{ij} \in \{0, 1\}$, where $a_{ij} = 1$ if and only if a message can securely and correctly be delivered by choosing the i-th pure strategy, and the adversary uses his j-th pure strategy for attacking. Then $\rho(\mathbf{A}) \in [0, 1]$, and*

1. *If $\rho(\mathbf{A}) < 1$, then for any $\varepsilon > 0$ there is a protocol so that Alice and Bob can communicate with an eavesdropping probability of at most ε and a chance of at least $1 - \varepsilon$ to deliver the message correctly.*
2. *If $\rho(\mathbf{A}) = 1$, then the probability of the message being extracted and possibly modified by the adversary is 1.*

While appealing in its assurance that we can run arbitrarily secure communication by trading computational (hard to verify) intractability hypothesis by easily verifiable connectivity hypotheses on the network graph, this leaves the question of what to do if the vulnerability $\rho(\mathbf{A}) = 1$. This case can occur in either of the following circumstances:

- The network is such that it simply does not admit a sufficient number of paths to effectively do multipath transmission. In that case, we may try to extend the network towards stronger connectivity. We will discuss this possibility later in Sect. 11.2.4.

- There was an attack that the defender simply did not anticipate, e.g., a *zero-day exploit* that basically defeated all precautions available in the defender's action set. Handling zero-day exploits is a different story that whose discussion we postpone until Sect. 12.6.2.

In extending this idea further, robustness against routing can be taken into the game to extend the attacker's capabilities even towards redirecting the previously chosen paths to some extent [32].

11.2.2 Authentication Games

The goal of authenticity can also be cast into a multipath game between an honest sender-receiver pair $s, r \in V$ against an active adversary having corrupted a subset of nodes in the network $G = (V, E)$, excluding s and r to avoid trivialities. An *authentication game* [34] can mimic the way in which handwritten signatures would be verified in the paper-based world: there, a sample of a handwritten signature would be left with the verifying instance, to let a future signature be checked. The digital counterpart would either be an authentically available public key (to verify a digital signature), or, more simply, use common secrets known to the sender s and its neighbors, who may later verify a signature on behalf of s. In detail, we assume that the sender s of an authentic message m shares common distinct secrets with n of its neighbors $N_1, \ldots, N_n$ in the network graph G. Call these secrets $k_1, \ldots, k_n$, and write $MAC(m, k)$ for the Message Authentication Code (MAC) of m created under the secret key k. Further, let $Hash$ be a cryptographically secure hash function. To authenticate the message m, the sender attaches a set of MACs $\{a_i := MAC(Hash(m), k_i)\}_{i=1}^{n}$ to the message m. Upon reception of the message m' by the designated receiver r, it computes the hash value $h' = Hash(m')$ and asks each of s' neighbors to verify the given message-MAC pair (h, a_i) (see Fig. 11.3 for an illustration).

It is easy to see that if $m' = m$ went unmodified, then the MAC verification will come back positive and s' MAC on m is verified without revealing m to the verifying party, since it is covered as a hash code (this of course assumes that m has enough min-entropy to thwart guessing attacks based on repeated trial hashing). Conversely, if $m' \neq m$, then the MAC verifications will fail. The protocol's outcome can then be defined based on the condition under which r accepts m as authentic. For example, it could require all or at least one MAC verification to come back positive to the receiver for accepting m to be authentic. The "strength" of the authentication then of course depends on the condition of acceptance adopted by the receiver.

It is worth noting that for some instances of MACs based on universal hashing [5, 18, 26], the scheme offers unconditional security, meaning that no computational intractability assumptions are involved.

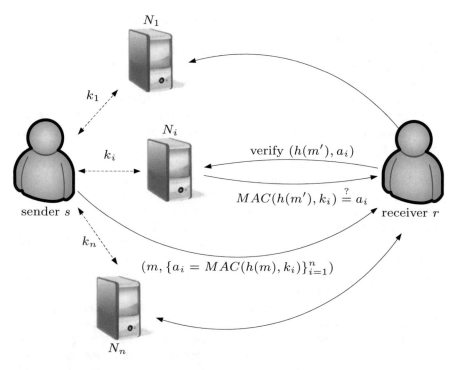

Fig. 11.3 Multipath authentication

Like for the previous multipath transmission, setting up the game is equally straightforward in terms of a matrix of indicator variables to quantify the protocol's outcome. From this matrix, a conventional equilibrium can be computed to deliver both, the best set of nodes or neighbors for the attacker to conquer (at random), and the best choice of neighbors to ask for the verification.

11.2.3 Practical Implementations

Note that for both schemes, the full set of paths through a network would in general be exponential and as such not feasible. Likewise, only few networks will have sufficient connectivity to allow for many different paths to choose from. Most likely, the topology will be such that there are "bottleneck" nodes or bridges, i.e., nodes or edges that must be passed by any path connecting s to r. However, the benefit of a game theoretic analysis lies in the game to naturally dig up such neuralgic spots, which can the be further secured by implementing additional security precautions, or at least adopt assumptions (as, for example, is done in the *trusted relay* model of most practical quantum key distribution networks). For the analysis, a feasibly small set of paths is already sufficient; if the game comes up with a winning strategy for the attacker, more paths (if they exist) or more security could be added. In fact, in case of a simple indicator-variable payoff matrix as we have above, adding only one strategy that avoids the attack in only one case can already be enough to run the protocol in an arbitrarily secure fashion again [35].

Similar concerns apply for the theoretically exponentially large set of subsets that an adversary may conquer. If we limit ourselves to saying that at most k nodes out of n are potentially hostile, this still leaves $O(n^k)$ possibilities, which is still infeasible even for moderately small k.

The more practical approach is letting the respective adversary structure (cf. Sect. 11.2.1) be defined by common properties of the nodes that would enable the same exploits. That is, if a set of machines run the same operating systems, presumably with the same vulnerabilities, then it can be assumed that an exploit will open up all machines of the same configuration. Likewise, if a certain set of machines is under administration of the same person, social engineering on this person may grant access to the entire set of machines, and so on. Each such exploit would then go into the action space for the attacker and define the set of hereby compromised nodes per round of the game.

Practical reports on prototype implementations of such schemes exist [33]: Fig. 11.4 shows screenshots from an experimental prototype that operationalizes Theorem 11.2 for confidential transmission that is authenticated by "playing" the game from Fig. 11.3. In the figure, the left window is the sender, the middle window shows what a person-in-the-middle would gain from eavesdropping, and the right window is the receiver's view.

In the experimental setup, the set of paths used for transmission is based on an enumeration of all paths in a software overlay network. Therein, the set of vulnerable nodes is determined by conditions on properties of the local computers on which the prototype software instances are running. The actual protocols run on the application layer. On lower layers, the paths could be realized by creating virtual LANs, each representing a specific connection, but the administrative burden to maintain all these for all pairs of nodes may quickly become cumbersome and impractical even for small network instances. In contrast, using an overlay network admits automated (self-configuring) paths and routing, which is almost a plug-and-play solution for confidential transmission with arbitrarily strong security, using game theory.

Returning to the discussion about rational cryptography, note that we here have an example of where game theory helps to design new protocols in light of theoretical limits (e.g., Theorem 11.1) imposed by the strength of cryptographic security concepts. In letting the attacker be successful in a small fraction of time, which is fixed and *noticeable* (as in the sense of Definition 11.1) but still controllable, we can run cryptographic schemes in a more powerful form than its standard use. This may be more than sufficient in many practical applications.

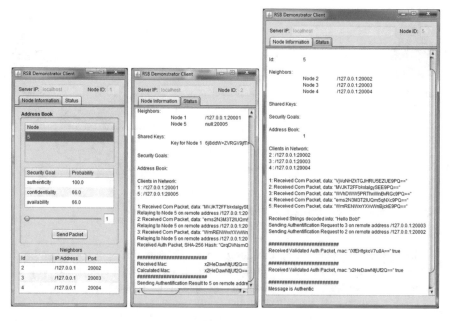

Fig. 11.4 Screenshots from an experimental multipath transmission prototype [33]

11.2.4   On Network Design Using Game Theory

Taking Theorem 11.1 as a starting point, we can go and look for network topology designs such that the communication game that we run in the network has optimal performance measured in game theoretic terms. Continuing the thoughts of Sect. 11.2.1, we can take the saddle-point value of the communication game as a goal function to optimize, i.e., we have a graph augmentation problem. This is done by looking at different edge-extensions to the network $G = (V, E)$, e.g., a subset $E^+ \subseteq V \times V$ with $E^+ \cap E = \emptyset$ (we would not add existing edges). Suppose further that each such new edge comes with a cost told by a function $c : V \times V \to \mathbb{R}$ with $c(e) = 0$ whenever $e \in E$ (existing edges do not cost anything). Then, we can formulate two versions of a network optimization problem, based on the communication game in Sect. 11.2.1, illustrated in Fig. 11.5.

Problem 1 (best security under limited budget) Minimize the saddle-point value v of the communication game played over the graph $G' = (V, E \cup E^+)$, within to a given budget limit $c(E^+) \leq L$. Here, we vary over $E^+ \subseteq (V \times V) \setminus E$ and put $c(E^+) := \sum_{e \in E^+} c(e)$.

Problem 2 (cheapest extension for desired security) Minimize the cost $c(E^+)$ towards lowering the attack chances below a given threshold (risk acceptance level) $L \geq v$. The symbols c, E^+ and v are the same as in the previous problem, i.e., the game is played over an extended network $G' = (V, E \cup E^+)$.

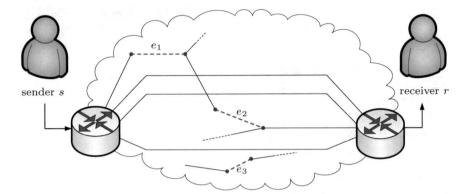

Fig. 11.5   Edge extensions to increase connectivity (dashed edges are in E^+)

Both problems relate to optimal graph extensions in terms of edges, and aim at increasing node connectivity. However, the game theoretic analysis reveals an interesting connection between efficient approximations to the above problems and the P/NP question [10], namely the following:

Theorem 11.3 ([31])   *The following three statements are equivalent:*

1. *There exists a polynomial time algorithm to find an approximate solution to Problem 1 (best security under limited budget).*
2. *There exists a polynomial time algorithm to find an approximate solution to Problem 2 (cheapest extension for desired security)*
3. *$P = NP$*

In this context, "approximate" characterizes an algorithm, which returns an answer to the optimization problem and the error relative to the true optimum is bounded by a constant.

Theorem 11.3 is somewhat devastating since it tells that security measured like in the communication game in Sect. 11.2.1 is computationally intractable to achieve unless the network is small or already has the necessary connectivity. On the bright side, however, the game theoretic analysis hereby provided a positive conclusion about the impact of the P/NP question on the possibility of confidential communication:

- If $P \neq NP$, then the known encryption algorithms (candidates are from the post quantum realm [3] such as McElice encryption) will most likely (though yet unverified [16]) continue in successfully protecting confidentiality against polynomially bounded adversaries. Thus, in this case we are (to nowadays public knowledge) safe by known public key encryption algorithms.
- Otherwise, if $P = NP$, then the design of communication networks that allow for arbitrarily secure communication against computationally unbounded attackers (implied by the use of secret sharing, which does not hinge on any computa-

tional intractability) becomes practically doable. Hence, we become safe again by building proper communication infrastructures, even though much of the classical intractability-based cryptography would be no longer secure.

11.3 Critical Remarks

Not only cryptography can benefit from game theory, also the opposite is naturally possible by faithful cryptographic implementations of equilibria. In fact, the validity of a game's prediction of behavior relies on the players following their best strategies (equilibria). The enforcement of these can be aided by cryptographic techniques [1].

The combination of game theory and cryptography is an interesting and fruitful field of research, yet a direct comparison of classical vs. rational cryptography appears opaque: the assurances of reductionist arguments are only as good as the underlying computational intractability assumptions are true, and hold asymptotically and hence not in every instance of the system. Likewise, the assurances of rational cryptography in turn are only as good as the utility definitions are accurate. If we assume actors in a cryptographic scheme to behave rational, then utilities that appear plausible from a cryptographic perspective, may still be subject to issues of bounded rationality (see Chap. 5), and as such be nonetheless inaccurate in reality.

Despite the potential of rational cryptography, it has not seen too many practical implementations, yet (at least by the time of writing this book); neither in its pure form nor as a foundation for more sophisticated matters of secure function evaluation. Perhaps a reason is that in many cases it appears simpler (and more "direct") to incentivize actors by contractual regulations, fees and penalties, rather than to count on the protocol involvement to be in the party's own natural interest. As for secure function evaluation, cryptographic solutions thereto compete with much simpler methods of just using virtualization technology, access control, good contract design and insurances, in order to cover all matters of misbehavior or technical failures. Moreover, most secure function evaluation frameworks apply transformations to a certain representation of a fixed function in order to securely evaluate it, and as such do not provide a universal computer that would be secure (any such computer built on encryption only would in simple instances be vulnerable to side-channel attacks [37] unless further security is imposed).

Simply imagine the options to be either (a) implementing a sophisticated, elegant yet individual cryptographic circuit to compute a fixed function or run a fixed algorithm, or (b) deploy your own hardened environment (operating system) with access control and sandboxing (in a cloud), so that you can compute any function or run any algorithm of your own choice. Which would you choose for an economic, simple to use, deploy and maintain?

References

1. Alwen J, Cachin C, Nielsen JB, Pereira O, Sadeghi AR, Schomakers B, Shelat A, Visconti I (2007) D.PROVI.7 summary report on rational cryptographic protocols (01 Jan 2007), Mar
2. Ashwin Kumar M, Goundan PR, Srinathan K, Pandu Rangan C (2002) On perfectly secure communication over arbitrary networks. In: PODC '02: proceedings of the twenty-first annual symposium on principles of distributed computing. ACM, New York, pp 193–202. https://doi.org/10.1145/571825.571858
3. Buchmann J, Ding J (eds) (2008) Post-quantum cryptography. Lecture notes in computer science 5299. Springer, Cham
4. Campanelli M, Gennaro R (2017) Efficient rational proofs for space bounded computations. In: Rass S, An B, Kiekintveld C, Fang F, Schauer S (eds) Decision and game theory for security, pp 53–73. Springer International Publishing, Cham
5. Carter JL, Wegman MN (1981) Universal classes of hashing functions. J Comput Syst Sci 22:265–279
6. Chunduri U, Clemm A, Li R (2018) Preferred path routing – a next-generation routing framework beyond segment routing. In: 2018 IEEE global communications conference (GLOBECOM), pp 1–7. IEEE, Abu Dhabi. https://doi.org/10.1109/GLOCOM.2018.8647410. https://ieeexplore.ieee.org/document/8647410/
7. Dijk M, Juels A, Oprea A, Rivest RL (2013) FlipIt: the game of stealthy takeover. J Cryptol 26(4):655–713. https://doi.org/10.1007/s00145-012-9134-5
8. Elliott C, Colvin A, Pearson D, Pikalo O, Schlafer J, Yeh H (2005) Current status of the DARPA quantum network. ArXiv:quant-ph/0503058v2
9. Fitzi M, Franklin MK, Garay J, Vardhan SH (2007) Towards optimal and efficient perfectly secure message transmission. In: TCC, LNCS, vol 4392. Springer, pp 311–322
10. Fortnow L (2009) The status of the P versus NP problem. Commun ACM 52(9):78. https://doi.org/10.1145/1562164.1562186. http://portal.acm.org/citation.cfm?doid=1562164.1562186
11. Garay J, Katz J, Maurer U, Tackmann B, Zikas V (2013) Rational protocol design: cryptography against incentive-driven adversaries. In: 2013 IEEE 54th annual symposium on foundations of computer science. IEEE, Piscataway, pp 648–657. https://doi.org/10.1109/FOCS.2013.75
12. Gennaro R, Rabin MO, Rabin T (1998) Simplified VSS and fast-track multiparty computations with applications to threshold cryptography. In: Proceedings of the seventeenth annual ACM symposium on principles of distributed computing. ACM, New York, pp 101–111. https://doi.org/10.1145/277697.277716
13. Ghazvini M, Movahedinia N, Jamshidi K, Szolnoki A (2013) GTXOP: a game theoretic approach for QoS provisioning using transmission opportunity tuning. PLoS ONE 8(5):e62925. https://doi.org/10.1371/journal.pone.0062925
14. Giry D (2018) BlueCrypt – cryptographic key length recommendation. http://www.keylength.com/. Retrieved 29 May 2018
15. Goldreich O (2003) Foundations of cryptography 1, 2. Cambridge University Press, Cambridge
16. Goldreich O, Goldwasser S (1998) On the possibility of basing cryptography on the assumption that P=NP (01 Jan 1998)
17. Gordon SD, Katz J (2006) Rational secret sharing, revisited. In: de Prisco R, Yung M (eds) Security and cryptography for networks. Lecture notes in computer science, vol 4116, pp 229–241. Springer-Verlag GmbH, Berlin/Heidelberg. https://doi.org/10.1007/11832072_16
18. Helleseth T, Johansson T (1996) Universal hash functions from exponential sums over finite fields and Galois rings. In: Advances in cryptology CRYPTO '96. Springer, Berlin, pp 31–44
19. Kawachi A, Okamoto Y, Tanaka K, Yasunaga K (2013) General constructions of rational secret sharing with expected constant-round reconstruction. Cryptology ePrint Archive, Report 2013/874. https://eprint.iacr.org/2013/874
20. Kim S (2014) Game theory applications in network design. Premier reference source. Information Science Reference/IGI Global, Hershey. http://dx.doi.org/10.4018/978-1-4666-6050-2

21. Koblitz N (2007) The uneasy relationship between mathematics and cryptography. Not AMS 54(8):972–979. Paper and follow-ups available at http://cacr.uwaterloo.ca/~ajmeneze/ anotherlook/ams.shtml. Retrieved on 29 May 2018
22. Kodialam M, Lakshman TV (2003) Detecting network intrusions via sampling: a game theoretic approach. In: IEEE INFOCOM, vol 3. San Francisco, pp 1880–1889
23. McElice RJ, Sarwate DV (1981) On sharing secrets and reed-solomon codes. Commun ACM 24(9):583–584
24. Menezes A, van Oorschot PC, Vanstone SA (1997) Handbook of applied cryptography. CRC Press LLC, Boca Raton
25. Michiardi P, Molva R Game theoretic analysis of security in mobile ad hoc networks (01 Jan 2002)
26. Nevelsteen W, Preneel B (1999) Software performance of universal hash functions. In: EUROCRYPT'99, LNCS 1592. Springer, pp 24–41
27. Patra A, Choudhury A, Rangan CP, Srinathan K (2010) Unconditionally reliable and secure message transmission in undirected synchronous networks: possibility, feasibility and optimality. Int J Appl Cryptol 2(2):159–197. https://doi.org/10.1504/IJACT.2010.038309
28. Previdi S, Horneffer M, Litkowski S, Filsfils C, Decraene B, Shakir R (2016) Source packet routing in networking (SPRING) problem statement and requirements. https://tools.ietf.org/ html/rfc7855
29. Rabin T, Ben-Or M (1989) Verifiable secret sharing and multiparty protocols with honest majority. In: Proceedings of the twenty-first annual ACM symposium on theory of computing, STOC '89. ACM, New York, pp 73–85
30. Rass S (2013) On game-theoretic network security provisioning. Springer J Netw Syst Manag 21(1):47–64. https://doi.org/10.1007/s10922-012-9229-1. http://www.springerlink. com/openurl.asp?genre=article&id=doi:10.1007/s10922-012-9229-1
31. Rass S (2014) Complexity of network design for private communication and the P-vs-NP question. Int J Adv Comput Sci Appl 5(2):148–157
32. Rass S, König S (2012) Turning quantum cryptography against itself: how to avoid indirect eavesdropping in quantum networks by passive and active adversaries. Int J Adv Syst Meas 5(1 & 2):22–33
33. Rass S, Rainer B, Vavti M, Göllner J, Peer A, Schauer S (2015) Secure communication over software-defined networks. Mob Netw Appl 20(1):105–110. https://doi.org/10.1007/s11036-015-0582-7
34. Rass S, Schartner P (2010) Multipath authentication without shared secrets and with applications in quantum networks. In: Proceedings of the international conference on security and management (SAM), vol 1. CSREA Press, pp 111–115
35. Rass S, Schartner P (2011) Information-leakage in hybrid randomized protocols. In: Lopez J, Samarati P (eds) Proceedings of the international conference on security and cryptography (SECRYPT). SciTePress – Science and Technology Publications, pp 134–143
36. Rass S, Schartner P (2011) A unified framework for the analysis of availability, reliability and security, with applications to quantum networks. IEEE Trans Syst Man Cybern Part C Appl Rev 41(1):107–119. https://doi.org/10.1109/TSMCC.2010.2050686
37. Rass S, Schartner P (2016) On the security of a universal cryptocomputer the chosen instruction attack. IEEE Access 1. https://doi.org/10.1109/ACCESS.2016.2622724
38. Rass S, Slamanig D (2013) Cryptography for security and privacy in cloud computing. Artech House, Boston
39. Shamir A (1979) How to share a secret. Commun ACM 22(11):612–613
40. Shor PW, Preskill J (2000) Simple proof of security of the BB84 quantum key distribution protocol. Phys Rev Lett 85:441–444
41. Shoup V (2004) Sequences of games: a tool for taming complexity in security proofs. Cryptology ePrint Archive, Report 2004/332. http://eprint.iacr.org/
42. Tompa M, Woll H (1988) How to share a secret with cheaters. J Cryptol 1(2):133–138
43. Wang Y, Desmedt Y (2008) Perfectly secure message transmission revisited. IEEE Trans Inf Theory 54(6):2582–2595

Chapter 12
Practicalities

Nothing is more practical than a good theory.

K. Lewin

Abstract This chapter discusses the use of data and data science to choose values for model parameters, and suggests a few methods and literature pointers to techniques that can be helpful to instantiate models. Furthermore, we review a set of selected software tools that help with the setup and equilibrium analysis of practical game theoretic models. We revisit various examples throughout the book in a tutorial-like step-by-step approach describing how game models can be analyzed. The focus is herein on openly and freely available software, parts of which is open source. Where applicable, we also give closed form solutions to certain classes of games, and generic transformations to make game theoretic problems solvable with help of optimization software. This shall equip practitioners with direct tools to use in practice, and with further literature pointers.

12.1   Data Science and Choice of Model Parameters

After having made a choice for a particular (game-theoretic) model, the practical question of how to choose the model parameters in many cases is left unanswered in the literature. An explicit exception is the choice of parameters for probability distributions, for which a rich theory on point estimation and confidence regions is available (see, e.g., [30] for a start).

For the challenge of choosing parameters in the game theoretic models covered in this book and more widely found in the literature, it is useful to distinguish some types of parameters in models, to which different methods of parameter estimation are applicable. Broadly, and non-exhaustively speaking, the variables that appeared throughout are either probabilities, risks and losses, or general (other) parameters. We dedicate a subsection in the following to each type.

© Springer Nature Switzerland AG 2020

S. Rass et al., *Cyber-Security in Critical Infrastructures*, Advanced Sciences and Technologies for Security Applications,
https://doi.org/10.1007/978-3-030-46908-5_12

12.1.1 Probability Parameters

These are typically values between 0 and 1, and in most cases refer to chances or odds for some event to occur or some condition to hold. Examples include *beliefs* in Bayesian models, probabilities for success or failure of certain attacks, errors in statistical tests (cf. Chap. 9), or just ratings on the scale [0, 1], such as, e.g., in the insurance context (see Chap. 7).

As far as statistical testing is concerned, these tests enjoy a rich theory on experimental design [9] to obtain practical guidance from. In absence of such help, machine learning techniques (supervised and unsupervised) can help with an automated estimation of parameters. As one example, *logistic regression* provides a handy method of estimating a parameter in the range [0, 1] from labeled empirical data. The process roughly proceeds as follows:

- Identify variables that may determine the parameter p in question; call these variables $x_1, \ldots, x_k$, allowing them to be individually numerical or categorical (both types admissible within the so identified set).
- Collect a number n of instances with concrete values for this variable set, and, for each concrete set of values $(x_{1,i}, \ldots, x_{k,i})$, add an indicator $p_i \in \{0, 1\}$ related to this configuration, and do so for all records $i = 1, 2, \ldots, n$. The value $p = 1$ may herein indicate the event or condition about which p speaks to hold, and $p = 0$ would correspond to the opposite.
- To the so-labeled data, fit a logistic regression model, to get a point estimate of the value p by means of maximum-likelihood estimation, together with detailed information on the significance of the identified variables that is useful to revise the estimate or improve the data quality, if necessary.

Example 12.1 (Credit Risk Estimation) The most popular example to illustrate the above procedure is the estimation of risk for credit loans. There, the variables $x_1, \ldots, x_k$ relate to customer's properties like income, marital status, owned assets (like a house, car, ...), and others. Based on a collection of such information from (real) customers, and the according labels $p_i = 0$ if the customer was denied the loan, or $p_i = 1$ if the customer was granted the credit, logistic regression can take the data from a new customer to compute a rating in the range [0, 1], which banks can take as a risk score to make a decision about a pending credit loan application. The process is in almost all steps automated and implemented in R with the function glm ("generalized linear model"); see the documentation of this function within R for extensive examples and literature pointers.   ◇

Remark 12.1 (Data science requires, not replaces, human intelligence) The term "big data" can quickly become overstressed, and having *many* records in a database does not mean that they are all complete or contain useful information. *Data preparation*, which includes cleansing, outlier elimination, treatment of missing values, etc., is in order before moving to any analysis of the data. More importantly, it is crucial to not fall victim to "easy" mistakes in a statistical data analysis, which include:

- Formulating hypothesis or tests *after* the data has been collected. Such post-hoc analysis is often prone to looking for patterns that one might suspect to have seen in the data. The proper approach is to first have the question and then collect the data that is useful to answer it (the converse method may lead to *data dredging*, which effectively means forging scientific results).
- Concluding about causality from correlation: correlation measures are a popular measure to discover seeming interdependence between variables in data, however, pure data does cannot tell anything about causal effects. Those can only be assessed upon a properly designed experiment, but never recognized in plain data.

Big data *can* make a lot of sense, but, equally easy, can also lead to lots of nonsense in careless hands [29].

12.1.2 Learning Parameters Over Time

If a parameter has been chosen, it may be worth updating it continuously or from time to time, to improve the model accuracy. Bayesian methods are a natural way of doing this, besides other methods from machine learning (e.g., Q-learning or reinforcement learning).

We already met a simple Bayesian learning rule in Example 3.2: this method takes a probability parameter p as β-distributed random variable, and uses the concrete value $p = \mathbb{E}(\mathscr{B}e(a, b))$ as the concrete value for the probability parameter. If new data comes in independently, we can use a Poissonian likelihood, and do a simple update by either increasing the parameter a or the parameter b by 1, just as we described in Example 3.2. It can be shown that this method eventually converges to the true value of p. Moreover, the method is generalizable to more than one parameter even. See [28] for algorithmic details on this method.

Generally, this learning strategy has the issue of potentially slow convergence and some "symmetry" in the way of how trust is gained or lost: while human reactions may be towards losing trust faster than gaining it, a β-reputation model becomes more and more stable in the long run. As time goes by and updates come in, the parameter value computed from the β-distribution will become more and more insensitive to future updates. Practically, it thus pays to "reset" the model from time to time, meaning to restart the Bayesian updating afresh from a new a priori distribution that approximates the current information but with the smallest parameters possible (to make the model more reactive to updates again). Reinforcement learning also addresses the issue by randomly deviating from what the model told, to explore alternative possibilities and perhaps find even better parameter choices in this way.

12.2   Risk and Loss Parameters

The phenomena discussed in Chap. 5 (among others in this context), like the Ellsberg paradox, the Saint Petersburg paradox, or more recent discoveries like reversed common ratio effects [3], can be taken as justifications to as why in security risk management, people should never be confronted with or asked about specifying probabilities. As the common ratio effect and the Allais paradox demonstrate, decision making *can* consider probabilities, but there is an apparent tendency to simplify matters in many cases, such as options being certain or alternatives with roughly equal odds. Utility theory is very much exact in comparing options even in these cases, as opposed to human reasoning that at some point switches the decision mechanism to a simpler version that disregards some aspects for the sake of simplicity. For this reason, security risk management typically recommends to use *qualitative scales* of probability that abstract from numbers at all, in order to avoid such unwanted simplifications. Example 6.1 also illustrates the issue.

Graphical methods of risk specification [33] avoid asking for explicit numbers, and only require people to mark a point or "region of possibilities" on a given categorical scale to specify a risk (for example, as Fig. 6.2 proposes or similar).

Figure 12.1 shows a screenshot of a research prototype software implementing this method [28, 33]: the program resemblers the risk matrix picture as an entry field to specify a likelihood and impact range on categories displayed at the borders. The categories themselves then define a grid, on which the user can draw a rectangular region to indicate the presumed locations on the impact and likelihood scales, as well as the uncertainty about the risk. The area of the rectangle then visualizes the uncertainty in the risk estimate as the product of uncertainties related to the impact and likelihood. To account for weighting and valuation effects (see Figs. 5.1 and 5.2), the grid itself is also resizable, so that categories can be made smaller or larger, according to the subjective opinions. We remark that graphical specifications also exist for probability parameters, such as transition matrices for Markov chains. See [28] for details. An empirical study on several related (and similar) methods is provided by [10].

12.2.1   General Parameters Derived from Statistical Models

Many stochastic models assume Gaussian distributions for error terms or other probabilistic parameters. This is in most cases a choice for technical and analytical convenience, but despite its wonderful theoretical properties, the Gaussian distribution should not be overly used, since there are very many practical circumstances in which random values have a distribution that is far from Gaussian.

Many domains, especially in the insurance context, have well-studied distribution models, and actuarial science [11, 13, 15, 16] offers lots of distribution models specifically to describe extreme phenomena for risk management.

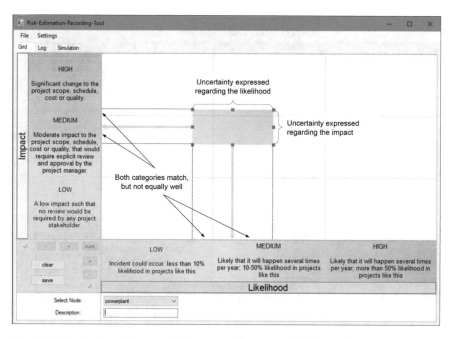

Fig. 12.1 Graphical specification of risks (research prototype [33])

In absence of domain knowledge, one may resort to non-parametric statistics [12] that works without explicit hypotheses about the shape of a distribution. One example are kernel density estimates that we will later in this chapter operationalize to model zero-day risks.

The method of computing a parameter (of any kind) from a distribution model behind it is generally applicable whenever data is available. Such data may come from experts directly (bearing in mind the impacts of human error in estimation, and phenomena as we discussed in Chap. 5), or from simulation models, such as those from Sect. 2.5, among others.

Generally, if parameters are to be estimated from data, the process to follow is the general data science workflow, summarized in Fig. 12.2. A game theoretic model, specifically parameters therein, are then obtained by transforming the data model, usually a probability distribution, into a representative value for the parameter in question. The simplest method is taking averages from the data, and if uncertainty shall be included, the statistical variance is a popular measure. The function "f" in Fig. 12.2 is then the simple arithmetic or expectation operator $\mathbb{E}$, w.r.t. the distribution that was modeled. More complex aggregation methods are, however, also admissible and applied (e.g., opinion pooling [7, 8]). Since aggregating a distribution model into one (or more) summary statistics is in most cases tied to information loss, a "lossless" use of the whole distribution is possible in the special case when the distribution is to define a payoff parameter, and the game uses stochastic orders. In that case, we can use the data model directly "as is" in the

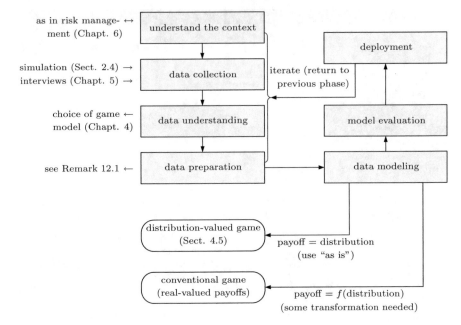

Fig. 12.2 Data Science Workflow (generic, adapted)

game model, and need no aggregation function f. See [24] for more details on this process.

12.3 Analytic Solutions for Special Games

12.3.1 Equilibria and Security Strategies

Let a two-player zero-sum game Γ be given by its payoff matrix $\mathbf{A} \in \mathbb{R}^{n\times m}$, implicitly fixing the player set $N = \{1, 2\}$ and the strategy spaces as labels to the rows and columns of the payoff matrix $\mathbf{A}$. The game's equilibrium and saddle-point value, which by Theorem 3.4 lead to security strategies as well, are computable from a linear optimization problem based on the payoff matrix $\mathbf{A}$.

Since the game is only between two players, the inner optimization in (3.22) is over the mixed strategies of the attacker, which is a convex set. The optimum on that set can only appear on some of the edge points, which makes the inner optimization easy, because $\min_{a_2\in\Delta(AS_2)} u_2(a, a_i) = \min_{a_2\in AS_2} u_2(a, a_2) = \min_{i=1,\dots,m} \mathbf{a}^T \mathbf{A} \mathbf{e}_i$, with $\mathbf{e}_i$ being the i-th unit vector. The outer optimization then boils down to a humble linear program, since all that the defender needs to do is determining the best mix of defense actions so as to maximize the revenue, which leads to the defender's problem:

$$
\left.
\begin{array}{ll}
& v \to \max_{\mathbf{x} \in \mathbb{R}^n} \\
\text{subject to} & v \le \mathbf{x}^T \mathbf{A} \mathbf{e}_i \quad \text{for } i = 1, \dots, m; \\
& \sum_{j=1}^{n} x_j = 1; \\
& x_j \ge 0, \quad \text{for } i = 1, \dots, n.
\end{array}
\right\}
\tag{12.1}
$$

The optimal solution is $v = \text{val}(\mathbf{A}) = \max_{\mathbf{x} \in \Delta(AS_1)} \min_{\mathbf{y} \in \Delta(AS_2)} \mathbf{x}^T \mathbf{A} \mathbf{y} = u_1(\mathbf{x}_0^*, \mathbf{y}_0^*)$, in the notation of equation (3.21). The value v obtained in this way is thus the lower bound (3.21) achieved by the security strategy $\mathbf{x}_0^*$ as being the optimal $\mathbf{x}$ obtained as output of (12.1). The validity of (3.21) then holds *conditional* on the exhaustiveness of the attacker's action set AS_2. Unless there are unexpected actions $\notin AS_2$ played by the attacker, v bounds the achievable revenue (or sufferable damage) irrespectively of how the attacker behaves within AS_2. If a new attack outside AS_2 was observed, we can extend the action space accordingly, and in some cases, even restore security against such unexpected actions (see Sect. 12.6.2.1).

12.3.2   2 × 2-Games

If the game is zero-sum, with only two actions for only two players, it admits a direct computation of the equilibrium from the game matrix

$$
\mathbf{A} = \begin{pmatrix} a & b \\ c & d \end{pmatrix}, \quad \text{with } a, b, c, d \in \mathbb{R}.
$$

If any of the two rows or columns dominates the other, then we have a pure strategy equilibrium, and are done. Otherwise, if the equilibrium is mixed, we can compute the saddle-point value as

$$
\text{val}(\mathbf{A}) = \frac{a \cdot d - b \cdot c}{(a + d) - (b + c)},
\tag{12.2}
$$

and with the optimal strategies $\mathbf{x}^* = (p, 1 - p), \mathbf{y} = (q, 1 - q)$ with

$$
p = \frac{d - c}{(a + d) - (b + c)} \quad \text{and} \quad q = \frac{d - b}{(a + d) - (b + c)}.
\tag{12.3}
$$

12.3.3   Diagonal Games

If the game is zero-sum between two players, and has a diagonal game matrix $\mathbf{A} = \text{diag}(a_{11}, \dots, a_{nn})$, we can also give a closed form expression for the saddle-point value as

$$\mathrm{val}(\mathbf{A}) = \left(\sum_{i=1}^{n} \frac{1}{a_{ii}} \right)^{-1} \tag{12.4}$$

and with the equilibrium strategy being the same for both players,

$$\mathbf{x}^* = \mathbf{y}^* = \left(\frac{v}{a_{ii}} \right)_{i=1}^{n} \tag{12.5}$$

12.3.4 Fictitious Play

An iterative method to compute an equilibrium is also suggested by another inspection of Example 3.5: by *simulating* the gameplay, optimizing the so-far average of (simulated) rounds of the game based on keeping records of what each player does, we can compute equilibria for more general classes of games, even. The resulting algorithm is shown in Fig. 12.3. It converges to an equilibrium for zero-sum games [31], all $2 \times n$-matrix games (even if not zero-sum [2, 19]), potential games, games with identical interests (i.e., when all players use the same utility function) [19], or games whose equilibria are found by systematical removal of dominated strategies [21] (by a continuation of the process started in Example 3.4).

FP has an exponential worst-case convergence rate [4]. It is, however, nonetheless interesting to note the combined use of minimax and Bayesian decisions: the algorithm starts from a minimax decision (in lack of better knowledge at the beginning), and continues with Bayesian decisions in each iteration, since it takes the decision optimal in consideration of so-far observed behavior of the other players.

Remark 12.2 The process has quite some similarity to learning parameters from β-distributions (see Example 3.2); indeed, the updating in step 3b is viewable as a Bayesian update on a β-prior.

FP is interesting for its technical simplicity, since it merely uses the ordering relation, either real ($\leq$) or stochastic ($\preceq$), on the payoff space, but nothing beyond this explicitly. Also, as an algorithm, its applicability extends widely beyond finite, two-player or zero-sum games.

A particularly useful application of FP is the computation of MGSS if the defender has multiple security goals, e.g., the protection of privacy (confidentiality), but also defending against a DoS (availability), or manipulation of processes (integrity, authenticity). The FP algorithm from Fig. 12.3 is generalizable to a one-against-all player game while retaining its convergence properties (cf. [26, 32]). We shall refrain from diving into deeper details, and refer the reader to the software support available, specifically the HyRiM package for the R statistical programming environment (see Sect. 12.4.3). Listing 12.1 shows an implementation of plain FP in R.

Given a game $\Gamma = (N, S, H)$, with finite set N and finite strategy sets AS_i for all players $i \in N$.
Initialization:

1. Let each player $i \in N$ maintain a vector $\mathbf{x}_i = (x_1, x_2, \dots, x_{|AS_i|})$, which we will iteratively update. In the k-th iteration, we will denote the current approximation of the equilibrium as $\mathbf{x}_i^{(k)}$.
2. Each player $i \in N$ determines the first action (0-th iteration) as a minimax decision, i.e., it starts with $\mathbf{x}_i^{(0)} \in \operatorname{argmax}_{a \in AS_i} \operatorname{argmin}_{\mathbf{a}_i \in AS_{-i}} u_i(a, \mathbf{a}_{-i})$.

Simulation of the game play:

1. Initialize the iteration counter $k \leftarrow 0$
2. Keep track of the total gains in variables t_i, starting from $t_i \leftarrow 0$ for all $i \in N$.
3. Loop over all players $i \in N$, and for the i-th player, do the following:

   a. Let $\mathbf{x}_{-i}^{(k)}$ be collection of all so-far recorded behavior of players $\neq i$. Choose the next action for player i to be optimal w.r.t. the expected utility based on the recorded behavior, i.e., choose
   $$a^{(k)} \in \operatorname*{argmax}_{a \in AS_i} \mathbb{E}_{(1/k) \cdot \mathbf{x}_{-i}^{(k)}}(u_i). \tag{12.6}$$

   b. Each player is told the action $a^{(k)}$, and accordingly updates the history of moves of player i into incrementing (by 1) the entry in $\mathbf{x}^{(k+1)}$ corresponding to action $a^{(k)}$. Keep record of the so-far total payoff per player by updating $t_i \leftarrow t_i + u_i(a^{(k)}, \mathbf{x}^{(k)})$.

4. Once this has been done for all players $i \in N$, increment the iteration counter $k \leftarrow k + 1$ and repeat the procedure from step 3. Stop the iteration, as soon as the average payoffs per round, i.e., $\left(\frac{1}{k} t_i\right)_{i \in N}$, has converged (in any norm applicable to that vector).

Fig. 12.3 FP for maximizing players in a finite game

12.4 Software Support

The GAMBIT software [18] and the GAME THEORY EXPLORER (GTE) both provide convenient analytics for games in normal and extensive form, and also come with a graphical user interface to specify the games. However, both are limited to games with scalar payoffs, i.e., admitting only a single security goal. Multi-goal security strategies for two-player games in normal form are subject of the HyRiM package [23] for R. Alas, the package cannot handle extensive form games as of version 1.0.4.

Our exposition in the following is not about showing that there is one system that can cover all needs, but rather pointing one (most) convenient software to deal with games of different type. For games in extensive form, GAMBIT is a convenient choice, as is the GTE, and we will use both to revisit some of the example games from previous chapters to recover the results given there. For empirical game theory and especially distribution valued games in normal form, the HyRiM package in R is a convenient choice, whereas it is fair to mention that the Comprehensive R Archive Network (CRAN) has lots of packages to offer for game theory (just to mention [6] as a non-exhaustive list of examples), all with their distinct strengths and useful for

Listing 12.1 Two-player zero-sum FP implemented in R [22]

```
1  #code for Fictious Play
2  #payoff matrix A with m rows, n columns, and iterations T must
       first be defined
3  #m and n must be at least 2 because of max and min functions
4  library(matrixStats)
5
6  T <- 1000
7  # A <- matrix(c(<your data>), ...)
8
9  m <- nrow(A)
10 n <- ncol(A)
11 x <- rep(0,m)
12 y <- rep(0,n)
13
14 # get starting points (arbitrarily, since convergence is
       assured)
15 rowMinima <- rowMins(A)
16 row <- which.max(rowMinima)
17 colMaxima <- colMaxs(A)
18 col <- which.min(colMaxima)
19 U <- A[,col]
20 y[col] <- y[col] + 1
21 V <- rep(0,n)
22
23 for(it_num in 1:T) {    # begin FP algorithm
24   # let player 1 move
25   Umax <-  max(U) # pick the best action
26   row <- which.max(U)
27   V <- V + A[row,] # accumulate payoffs
28   x[row] <- x[row] + 1 # record actions
29   # let player 2 move (exactly like player 1, only minimizing)
30   Vmin <- min(V)
31   col <- which.min(V)
32   U <- U + A[,col]
33   y[col] <- y[col] + 1
34 }
35
36 xeq <- x / T   # approximate equilibrium strategy for player 1
37 yeq <- y / T   # ...and for player 2
38 v <- t(xeq) %*% A %*% yeq # approximate the saddle point value
```

different purposes, extending up to cooperative game theory that we did not cover here.

Table 12.1 shows a comparative overview table including only the systems used for this book. The software used here all comes with free licenses, but as such also without warranties. Beyond these open-access available packages, general software for symbolic or numeric computations, such as MATLAB [17], OCTAVE [1], MATHEMATICA [14], in many cases offer sophisticated routines to compute equilibria by solving linear or nonlinear optimization problems. Since a

Table 12.1 Overview table of software used in this book

System	GAMBIT	GTE	R package HyRiM
Extensive form games	Yes	Yes	No
Normal form games	Yes	Yes	Yes
Distribution-valued games	No	No	Yes
Multi-objective and lexicographic orders	No	No	Yes

comprehensive account for this area would be a book on its own, we leave the pointer to this route here, but do not follow it any further.

The next three sections illustrate the three packages listed in Table 12.1 by revisiting the examples given earlier in the book, and show how to obtain the equilibria for them. The fact that we use a different package in each case does not mean that a package is incapable of handling the other two cases, unless Table 12.1 says so, so the reader is advised to choose the most convenient and most reliable package in each of her/his own cases. We start with GAMBIT to solve games in extensive form.

12.4.1 Solving Extensive-Form Games

GAMBIT is one of the earliest available implementations of an algorithmic toolbox to compute Nash- and other equilibria. It comes with a graphical interface to specify a game in extensive or normal form, and lets the user pick different algorithms to try when computing equilibria. Not all algorithms may be equally successful, so it is generally advisable to try more than one option offered by GAMBIT.

For illustration, let us start with a GAMBIT model of the phishing game from Fig. 3.2 in Chap. 3. Figure 12.4a shows the representation of the extensive form game as a graph, especially visualizing the information sets that appear as nodes connected by two dotted vertical lines. The nodes in the information set are labeled no sig. ("no signature") above and 1:1 below, indicating that these refer to the move "no sig." that is indistinguishable whether made by a phisher or an honest sender (appearing as labels next to the move), and belong to information set #1 that player 1 (the receiver) maintains (hence the label "1 : 1 = player:information-set-no."). Next to the leaf nodes and highlighted in the player's colors appear the utilities, as they have been set in Fig. 3.2, only that the game tree is plotted vertical there and drawn horizontally in GAMBIT. As a neat feature, we can ask GAMBIT to give the reduced normal form version of the game (where "reduced" means after elimination of dominated strategies), the result of which is shown in Fig. 12.4b.

The computation of equilibria is up to the user clicking the respective computation button in the toolbar (second from the right), which asks (see Fig. 12.5 for the dialogue) how many equilibria shall be sought, by which method, and on which representation of the game (extensive or normal form). The results then appear in

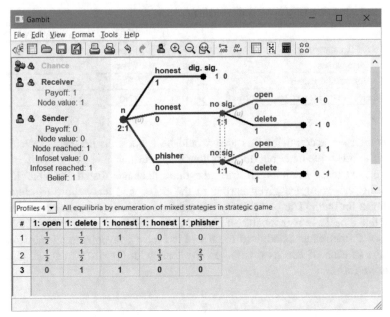

(a) Extensive form phishing game from Figure 3.2

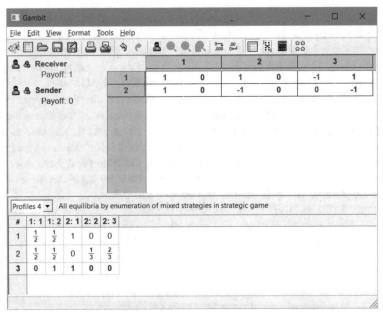

(b) Equivalent normal form representation

Fig. 12.4 Example of using GAMBIT to compute equilibria

Fig. 12.5 Equilibrium
computation dialogue in
GAMBIT

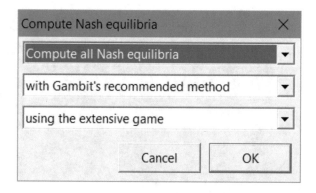

the lower part of the window, where this game admits apparently three equilibria,
one of which is even in pure strategies (the last one in the list).

Using the extensive form, we can also compute a Stackelberg equilibrium, such
as for the hide-and-seek game with, for example, a leading attacker in Fig. 4.3.
Figure 12.6 shows the model, equivalent normal form (cf. Fig. 4.4), and computed
equilibria.

GAMBIT is a convenient tool to study games "interactively", since the game is
easy to manipulate (e.g., change of strategies, modification of information sets,
altering of payoffs, etc.). The "Tools" menu offers also the computation of other
equilibria than Nash's, and what is particularly interesting, also an analysis for
dominated strategies. Doing this beforehand can provide interesting insights about
the model itself.

12.4.2   Solving Normal-Form Games

The GTE is an alternative to GAMBIT, offering roughly the same possibilities, but
doing so in a more wizard-like fashion. That is, to specify a game in extensive form,
the user follows a sequence of steps, starting with (i) the specification of the game
tree, (ii) specification of players, (iii) definition of information sets, (iv) entering the
moves (action sets), and (v) defining the payoffs (GAMBIT asks for the same inputs,
but lets a user take the steps them in any order; whether a systematic sequential
specification or a "more free" specification is more convenient is a subjective matter
of convenience for a user).

For larger strategic games, such as the 5×5 two-player nonzero-sum game from
Example 3.7, we can conveniently enter the payoff matrices for the two players
separately in the graphical user interface of the GTE, as shown in Fig. 12.7.

Upon solving for an equilibrium, it prints out (see Fig. 12.8) a detailed and
comprehensive description of equilibria obtained (of course, those match the ones
given in Example 3.7, which were found using GAMBIT).

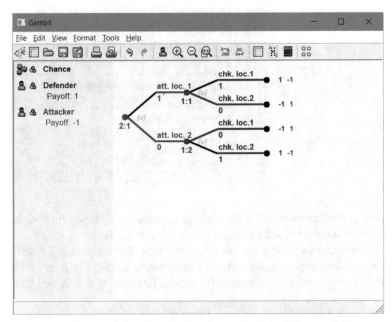

(a) Hide-and-Seek with leading attacker (Figure 4.3)

#	1: 11	1: 12	1: 21	1: 22	2: 1	2: 2
1	0	1	0	0	1	0
2	0	1	0	0	0	1

(b) Normal form representation

Fig. 12.6 Example of using GAMBIT to compute equilibria

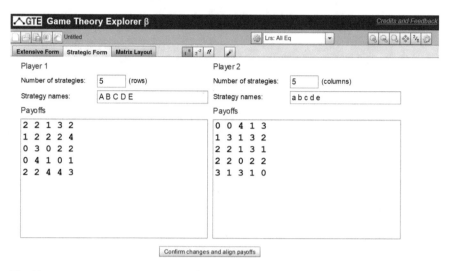

Fig. 12.7   Normal form game in the GTE

It is generally advisable to perhaps try different (both) systems on larger games or those with more than two players, since the algorithms implemented in GAMBIT or the GTE may perform different and may also individually fail for different game models.

Neither software can, however, solve games with multiple objectives or with uncertainty in the payoffs, which is that the HyRiM package for R can do.

12.4.3   Solving Distribution-Valued Games

Unlike the previous two systems, HyRiM is not a standalone software, but rather a package extension to the R language for statistics and data science. It is thus installable directly through the R environment from CRAN [6]. Similar to GAMBIT, the process to set up and analyze game puts is widely unconstrained in the order of completion of steps, but the overall steps are the same as for other systems. As a restriction, HyRiM does not support the construction of strategic form games (in R, the games package [5] can do this for example, but only for scalar and real-valued utility functions), and can only construct normal form games. Since the package is desigend for the computation of MGSS, it will only compute zero-sum equilibria. Internally, it applies FP (adapted and largely extended from the work of [36]), which makes the algorithm robust and applicable to very large games (perhaps with many dimensions), but at the price of delivering only approximate equilibria (higher accuracy takes more iterations).

Among the software discussed here and found elsewhere in the literature, HyRiM is the only implementation of games over stochastic orders and with

```
SUCCESS

Strategic form:

5 x 5 Payoff player 1

   a b c d e
A 2 2 1 3 2
B 1 2 2 2 4
C 0 3 0 2 2
D 0 4 1 0 1
E 2 2 4 4 3

5 x 5 Payoff player 2

   a b c d e
A 0 0 4 1 3
B 1 3 1 3 2
C 2 2 1 3 1
D 2 2 0 2 2
E 3 1 3 1 0


EE = Extreme Equilibrium, EP = Expected Payoffs

Rational:

EE 1 P1: (1) 0 0 0 1 0 EP=    2 P2: (1) 1/2 1/2 0    0    0 EP= 2
EE 2 P1: (1) 0 0 0 1 0 EP=    4 P2: (2)   0   1 0    0    0 EP= 2
EE 3 P1: (1) 0 0 0 1 0 EP=  8/3 P2: (3)   0 2/3 0 1/3    0 EP= 2
EE 4 P1: (1) 0 0 0 1 0 EP= 18/7 P2: (4)   0 4/7 0 1/7 2/7 EP= 2
EE 5 P1: (1) 0 0 0 1 0 EP=  9/4 P2: (5) 1/4 1/2 0  0 1/4 EP= 2
EE 6 P1: (1) 0 0 0 1 0 EP= 14/5 P2: (6)   0 3/5 0  0 2/5 EP= 2
EE 7 P1: (2) 0 0 0 0 1 EP=    2 P2: (7)   1   0 0    0    0 EP= 3
EE 8 P1: (2) 0 0 0 0 1 EP=    4 P2: (8)   0   0 1    0    0 EP= 3

Decimal:

EE 1 P1: (1) 0 0 0 1.0   0 EP=     2.0 P2: (1)  0.5      0.5     0         0       0 EP= 2.0
EE 2 P1: (1) 0 0 0 1.0   0 EP=     4.0 P2: (2)    0      1.0     0         0       0 EP= 2.0
EE 3 P1: (1) 0 0 0 1.0   0 EP= 2.66667 P2: (3)    0  0.66667     0 0.33333       0 EP= 2.0
EE 4 P1: (1) 0 0 0 1.0   0 EP= 2.57143 P2: (4)    0  0.57143     0 0.14286 0.28571 EP= 2.0
EE 5 P1: (1) 0 0 0 1.0   0 EP=    2.25 P2: (5) 0.25      0.5     0         0    0.25 EP= 2.0
EE 6 P1: (1) 0 0 0 1.0   0 EP=     2.8 P2: (6)    0      0.6     0         0     0.4 EP= 2.0
EE 7 P1: (2) 0 0 0   0 1.0 EP=     2.0 P2: (7)  1.0        0     0         0       0 EP= 3.0
EE 8 P1: (2) 0 0 0   0 1.0 EP=     4.0 P2: (8)    0        0   1.0         0       0 EP= 3.0

Connected component 1:
{1}  x  {1, 2, 3, 4, 5, 6}

Connected component 2:
{2}  x  {7, 8}
```

Fig. 12.8 Equilibrium computation report as delivered by the GTE

multiple objectives. Since it is based on the stochastic $\preceq$-ordering, it is as well the only package that can handle games using lexicographic orders by internally operationalizing Theorem 4.2. Like GAMBIT and GTE, it also supports finding an equilibrium for normal-form games with purely real-valued payoffs.

The general procedure to follow using the package is here given in an extended version that involves the collection of data about the competitive situation to define the utilities (including uncertainty). We include this step here to highlight that the embedding of the process in R, rather than invoking a standalone software for the game only, opens up all the possibilities to use data science techniques for the modeling.

We discuss the typical steps in the following, refraining from going too deep into the parameters or other settings taken by the respective functions (the R help with examples is the more extensive resource to consult here), but discuss what is generally to be done and what functions and commands could be helpful to this end.

1. Collect data about the situation that the game shall model. More specifically, gather empirical data and define a crisp utility values or payoff distributions. If the data is a set of observations (samples from a continuous outcome, or discrete and perhaps subjective risk assessments) called payoff_data, the construction of a loss distribution is straightforward by issuing

   ```
   > lossDistribution(payoff_data)
   ```

   perhaps adding the optional parameter discrete=TRUE to get a categorical distribution. If the payoff is a real number, and hence payoff_data is a list of numbers obtained by any means of data science, there is no need to call lossDistribution, as the next steps can take the number as direct input.

2. Construct the game: the package supports $n \times m$ matrix games with any number $d \geq 1$ of goals. The payoffs go into the game model as consecutive list of matrices $\mathbf{A}_i$ for all goals $i = 1, 2, \ldots, d$ that goes in as a list object in R with entries $(vec(\mathbf{A}_1), vec(\mathbf{A}_2), \ldots, vec(\mathbf{A}_d))$, where vec here denotes the "vectorization" of a matrix by rows or by columns (the choice is up to the user), which can be, for the example of a 3×3-matrix:

   - vectorization by rows:

   $$\mathbf{A} = \begin{pmatrix} a & b & c \\ d & e & f \\ g & h & i \end{pmatrix} \quad \text{becomes "flattened" into } vec(\mathbf{A}) = (a, b, c, d, e, f, g, h, i)$$

   - vectorization by columns:

   $$\mathbf{A} = \begin{pmatrix} a & b & c \\ d & e & f \\ g & h & i \end{pmatrix} \quad \text{becomes "flattened" into } vec(\mathbf{A}) = (a, d, g, b, e, h, c, f, i)$$

   The construction of the game is by a call to the function mosg ("multi-objective security game"):

   ```
   > G <- mosg(n, m, goals = d, losses = L, ...)
   ```

   in which L is a list object that consists of all payoff distributions (constructed by calls to lossDistribution) or a vector of real values if the game has payoffs in $\mathbb{R}$.

Depending on whether the payoff list is organized by rows or by columns, one can add the parameter byrow = TRUE or byrow = FALSE (meaning to take the list by rows or columns) to the call to mosg.
3. Compute an equilibrium, i.e., a multi-goal security strategy; the respective function is named by the according acronym mgss:

```
> eq <- mgss(G)
```

The number of iterations of FP that the function executes on the game object defaults to 1000, but can be set to any value by adding the respective parameter to the call. Other parameters for further fine-tuning are available, and explained in the manual for the package and function.

From here onwards where the equilibrium object has been constructed (we call it eq here), the analysis proceeds further "as usual" in R, i.e., the common generic functions for explorative analysis of data and objects (summary, plot, and others) are all provided by the package. This works like any other statistical data analysis, and thus left aside here.

Let us illustrate the use of the HyRiM package by providing a full implementation of Cut-The-Rope from Sect. 9.4.5, picking up on the equilibrium computation step explicitly and numerically.

Example 12.2 (Solving Cut-The-Rope with R/HyRiM)
Continuing the example running throughout Sect. 9.4, Listing 12.2 shows a full implementation of the game playground, specifically the paths in the attack graph (Fig. 9.6) in lines 2–12. Most of the code is about setting up the game data (action spaces with cardinalities n and m in lines 13–16, specification of attack rates and adversary types in lines 17–22, and constructing the game matrices in the loop from line 25 until 52). The actual application of the package is in lines 49 (where the loss distribution is constructed as a Poisson density), line 50 (where the payoffs are all collected in a linearly ordered list object), and lines 54 and 55, where the game is constructed (mosg) and solved (mgss) for a security strategy.

Running the code brings up a (9×8)-matrix game by an enumeration of 9 spot checking locations vs. 8 paths chosen by the attacker, with payoffs being discrete distributions over the location set V in the attack graph (variable V in the code). Figure 12.9 show parts of the resulting game matrix. The rows are indexed by the locations that the inspector can check, the columns correspond to the attack paths from Table 9.2. Each cell contains a distribution that is a truncated Poissonian (with $\lambda = 2$ for the example; line 17 in the code) and with its masses assigned to the locations as passed along each attack path. The vertical axis respectively shows the probabilities for the attacker to be at the respective position.

This game admits a PBNE in pure strategies advising the defender to periodically patch potential local buffer overflows at machine 2 (an optimal pure strategy being local_bof(2)), while the attacker is best advised to the attack path execute(0) → ftp_rhosts(0,2) → rsh(0,2) → local_bof(2) → full_access(2). This is intuitively not surprising, since all attack paths intersect at the node local_bof(2), making this

Listing 12.2 Full implementation of Cut-The-Rope in R

```r
1   library(HyRiM)
2   target <- 10   # this is the target node
3   # these are the routes (enumerated from the attack graph)
4   routes <- list(c( 1, 2, 3, 4, 6, 7, target),
5                  c( 1, 2, 3, 6, 7, target),
6                  c( 1, 8, 9, 7, target),
7                  c( 1, 3, 4, 6, 7, target),
8                  c( 1, 3, 6, 7, target),
9                  c( 1, 9, 7, target),
10                 c( 1, 5, 4, 3, 6, 7, target),
11                 c( 1, 5, 6, 7, target))
12  V <- unique(unlist(routes)) # get all nodes from all routes
13  as1 <- setdiff(V, target)   # the defender can check everywhere
14  as2 <- routes  # action space for the attacker
15  m <- length(as2)
16  n <- length(as1)
17  attackRate <- 2   # parameter lambda
18  # assume one adversary type per starting point
19  advTypes <- setdiff(V, target)
20  # no a priori information about the type
21  # i.e., a uniform prior
22  Theta <- rep(1/n, times = length(advTypes))
23
24  payoffMatrix <- list() # to take up the utility distributions
25  for(i in 1:n) { # run over all spots to inspect
26    for(j in 1:m) { # run over all attack paths
27      U <- rep(0, length(V))
28      path <- as2[[j]]
29      for(type in advTypes) {
30        # adversary moves only if it is on this path
31        if (type %in% path) {
32          route <- path[which(path == type):length(path)]
33          # let the adv. take a Poisson number of steps
34          pdfD <- dpois(x=0:(length(route)-1),lambda=attackRate)
35          pdfD <- pdfD / sum(pdfD)
36          cutPoint <- min(which(route == i),length(route)+1)-1
37          # truncate the distribution
38          payoffDistr <- pdfD[1:cutPoint]/sum(pdfD[1:cutPoint])
39          L <- rep(0, length(V))
40          L[route[1:cutPoint]] <- payoffDistr
41        }
42        else { # otherwise, the adversary doesn't move
43          L <- c(1, rep(0, length(V)-1))
44        }
45        # update the mix over all adversary types
46        U <- U + Theta[type] * L
47      }
48      # construct the loss distribution
49      ld <- lossDistribution(U, discrete=TRUE, dataType="pdf",
          supp=range(V), smoothing="always", bw = 0.2)
50      payoffMatrix <- append(payoffMatrix, list(ld))
51    }
52  }
53  # construct and solve the game
54  G <- mosg(n,m,1,losses = payoffMatrix, byrow = TRUE)
55  eq <- mgss(G)   # compute a multi-goal security strategy
56  print(eq)       # printout the equilibrium (optimal defense)
```

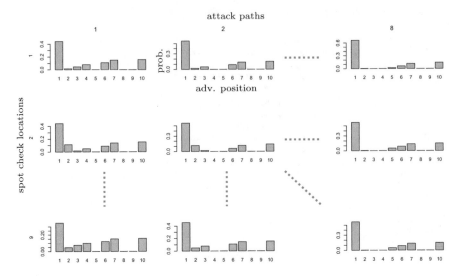

Fig. 12.9 Example game matrix for Cut-The-Rope

the system's "sweet spot" to guard. Under restricted possibilities for spot checking, i.e., a change of the defender's action space, other (also mixed) equilibria arise.

Analytically, it can be shown that the optimal defense is to "guard" a minimum cardinality cut that is graph-theoretically closest to the asset to protect. Rephrasing this with a folklore metaphor, if you keep your friends close, then Cut-The-Rope's advice is "keep the enemy closer". Essentially, the region that the defender should pay most attention to is computable by walking outwards from v_0 along all attack paths and thereby determining the set of nodes that cuts all attack paths towards v_0, and are possible to inspect and hence safeguard.   ◇

We will let further examples of using the package follow in later sections, and close the discussion at this point.

12.5 Cost of Playing Equilibria

Mixed equilibria prescribe to change action upon every round of the game. This may be optimal, but is not necessarily also feasible; consider the next example:

Example 12.3 (Patrolling in a Building) Imagine a building with 10 floors, and a security guard taking rounds to check each floor. Furthermore, suppose that the room layout is such that more "sensitive" parts of the building are concentrated on some floors, while others may be more or less open to the public. One can then compute an optimal pattern for the security guard to check the floors at random, so as to maximize chances to catch an intruder. Any such mixed equilibrium or security strategy $\mathbf{x}^* = (p_1, \ldots, p_{10})$ could, in a naive attempt, be implemented by approximating the values $p_1, \ldots, p_{10}$ by rationals, and define a roundtrip to check floors in a fixed sequence $(f_1, f_2, \ldots, f_m) \in \{1, \ldots, 10\}^m$ so that the number of times that f_i is checked equals (or at least approximates) the prescribed equilibrium frequency p_i to check that floor. Repeating this patrolling schedule, the security guard eventually ends up doing optimal by playing a mixed equilibrium strategy.

The prescribed schedule can even be further constrained to list the floors in adjacent order, to make patrolling efficient (meaning that we would send the guard from one floor f_k to the next floor f_{k+1} being either directly above or directly below the current floor f_k). While convenient for the security guard, the implementation of the equilibrium makes it easy for the adversary to sneak in, once the schedule and route of the guard is known.

The *proper* way of implementing the equilibrium is thus really doing *random* spot checks on each floor, so as to minimize the foresight that an intruder can gain on where to expect the guard. However, this means that the guard will, from time to time, need to go from floor 1, up to floor 8, then back to floor 2 and then again up to floor 10, and so on. An equilibrium played like this is still optimal and achieves the desired defense, but is apparently inconvenient for the security guard. Eventually, this person will then end up deviating from the equilibrium, simply to ease the daily business. This would already count as bounded rationality, though it roots in reasonable thinking.   ◇

It is a well known effect that cumbersome security can incentivize users to find creative workarounds to it. Examples range from sending an email without encryption because the certificates expired and we cannot get a new certificate in time, up to asking storing sensitive information on privately owned USB sticks because the online cloud storage rejected a login (even though only due to an unnoticed typo in the password).

Overly simple game theoretic models exhibit the same dangers, unless the *efforts* to change the current behavior accordingly is not included in the game model. Intuitively, we can account for such costs by assuming that the defender also plays "against itself", meaning that every defense action needs some investment to change it. That is, to the existing utility structure, we add another loss function $s : AS_1 \times AS_1 \to \mathbb{R}$, whose value $s(a, b)$ measures the efforts (in the sense of losses) incurred when player 1 switches from the current action $a \in AS_1$ to the next action $b \in AS_1$. Typically, we would set $s(a, a) = 0$, since not changing the action between two rounds does not cause any cost.

This function is *always minimized*, and can be included as a further goal in the game. If the strategy spaces are finite, the function s admits a specification as a matrix $\mathbf{S}$. The losses for switching from one to the next strategy are then given by $\mathbf{x}^T \cdot \mathbf{S} \cdot \mathbf{x}$, when $\mathbf{x}$ is a mixed strategy (observe that this is almost the same form as for a normal matrix game, only that player 1 now has become its own opponent in this regard). For a finite two-player matrix game with payoff structure $\mathbf{A}$, the payoff is the bilinear form $\mathbf{x}^T \mathbf{A} \mathbf{y}$, when player 1 chooses $\mathbf{x}$ and player 2 chooses $\mathbf{y}$. Likewise, if player 1 is "its own opponent", the same intuition leads to the quadratic form

$\mathbf{x}^T \mathbf{Sx}$. And this intuition is indeed provably correct, as directly follows from the law of total probability using the stochastic independence of rounds (see [25, 34] for proofs).

Solving the problem with the switching cost as a secondary goal is a humble matter of scalarization and adding the quadratic term to the conditions in (12.1), as a weighted sum with $\alpha \in (0, 1)$ to scalarize the MOG. This gives

$$\left.\begin{array}{rl} & v \rightarrow \min \\ \text{subject to} & v \le \alpha \cdot \mathbf{x}^T \mathbf{Sx} + (1 - \alpha)\mathbf{x}^T \mathbf{Ae}_i \quad \text{for } i = 1, \ldots, m; \\ & \sum_{j=1}^{n} x_j = 1; \\ & x_j \ge 0, \quad \text{for } i = 1, \ldots, n. \end{array}\right\}$$

(12.6)

The choice of α in (12.6) deserves some discussion and practical care: the extreme choices $\alpha = 0$ or $\alpha = 1$ make no sense, since $\alpha = 0$ lets (12.6) become the original linear optimization problem (12.1), so the extension disappears. Conversely, $\alpha = 1$ makes the problem ignore the regular revenues of the game and puts all focus on the switching. Doing so, however, makes all pure strategies into equilibria, if $s(a, a) = 0$ or, equivalently, the matrix $\mathbf{S}$ has all zeroes on the diagonal.

Remark 12.3 The derivation of (12.6) does not depend on the assumption that $s(a, a) = 0$ for all $a \in AS_1$. Thus, one could define a penalty $s(a, a) = s_a$ for a defender who chooses to stay with a strategy for the next round. This could, for example, model an explicit incentive to switch actions.

The choice of α should be made to roughly equalize the changes in the scalarized utility that the terms $\mathbf{x}^T \cdot \mathbf{A} \cdot \mathbf{y}$ and $\mathbf{x}^T \cdot \mathbf{S} \cdot \mathbf{x}$ contribute: if $\mathbf{A}$ is such that a small change in $\mathbf{x}$ or $\mathbf{y}$ causes a huge change of utility, this change may outweigh the gains contributed by the second term. In that case, the resulting equilibrium is not too much different to one without the switching cost term. Likewise, if the quadratic term changes largely relative to the change of the utility $\mathbf{x}^T \mathbf{Ay}$, then the model becomes overly focused on the switching cost, and we suffer a considerable loss of utility. This effect is easy to visualize by examples. Note that the matrix $\mathbf{S}$ is in general not symmetric and hence not definite (Example 12.3 is such a case). This needs to be kept in mind when selecting the proper nonlinear optimization algorithms (as some methods may rely on these properties of the matrix $\mathbf{S}$).

Example 12.4 ([25]) Let the payoff matrix $\mathbf{A}$ and the switching cost matrix $\mathbf{S}$ be given as

$$\mathbf{A} = \begin{pmatrix} 4 & 8 & 6 \\ 9 & 3 & 8 \\ 7 & 6 & 3 \end{pmatrix}, \text{ and } \mathbf{S} = \begin{pmatrix} 0 & 1 & 1 \\ 2 & 0 & 1 \\ 1 & 3 & 0 \end{pmatrix}.$$

Let $u_1(\mathbf{x}, \mathbf{y}) = \mathbf{x}^T \mathbf{Ay}$ be the payoff function, and let $u_{switch} = \mathbf{x}^T \mathbf{Sx}$ be the cost for player 1 to change action when playing a mixed strategy $\mathbf{x}$.

Analyzing the game in the usual way, e.g., by solving (12.6) with $\alpha = 0$ to "deactivate" the switching cost matrix $\mathbf{S}$, yields $v_0 = \text{val}(\mathbf{A}) \approx 6.0889$ at an equilibrium strategy $\mathbf{x}_0^* \approx (0.511, 0.311, 0.178)$.

Instantiating and solving (12.6) with $\alpha = 0.5$, corresponding to equal priority on game payoffs (losses) and switching cost, gives $v \approx 3.36$ at the new equilibrium $\mathbf{x}^* \approx (0.2, 0, 0.8)$. Doubling v to cancel out the equal weights $\alpha = 1/2$ for both goals, gives the total sum cost in the game, coming to $2 \cdot v = u_1(\mathbf{x}^*, \mathbf{y}^*) + u_{switch}(\mathbf{x}^*, \mathbf{y}^*) \approx 6.72$. Naturally, this loss is higher than in the conventional game without switching, so let us look how costly both equilibria are in playing them:

$$(\mathbf{x}_0^*)^T \mathbf{S} \mathbf{x}_0^* \approx 0.88 \quad \text{and} \quad (\mathbf{x}^*)^T \mathbf{S} \mathbf{x}^* \approx 0.32.$$

Not surprisingly, $\mathbf{x}^*$ is cheaper to play than $\mathbf{x}_0^*$. However, the costs of 0.88 have to be added to the loss $v_0 = 6.0889$, making the total payoff $v_0 + 0.88 \approx 6.9689$. So, a disregard of the switching costs comes at additional cost and is suboptimal compared to the "more informed" approach. This makes the account for switching worthwhile.

Now, consider slightly different payoff structure

$$\mathbf{A} = \begin{pmatrix} 4 & 9 & 7 \\ 8 & 3 & 6 \\ 6 & 8 & 3 \end{pmatrix}$$

and the same switching cost matrix $\mathbf{S}$ from before.

Disregarding switching costs, we get $v_0 = \text{val}(\mathbf{A}) \approx 6.08889$ attained at $\mathbf{x}_0^* \approx (0.422, 0.467, 0.111)$ and coming in at cost $(\mathbf{x}_0^*)^T \mathbf{S} \mathbf{x}_0^* \approx 0.892346$. Taking the switching cost into account gives $v \approx 3.69388$, playable at cost $(\mathbf{x}^*)^T \mathbf{S} \mathbf{x}^* \approx 0.816326$ (cheaper, as expected), but being more expensive in total: we have $2 \cdot v \approx 7.38776$, as opposed to $v_0 + 0.986173 \approx 6.98123$. This is a case where the additional cost from the deviation from $\mathbf{x}_0^*$ is higher than the cost reduction from playing the cheaper equilibrium $\mathbf{x}^*$. This is intrinsically attributable to the Pareto-Nash equilibrium tradeoff, and becomes effective due to the low cost values in $\mathbf{S}$. If those costs would increase, e.g., if we redo the calculation with $2 \cdot \mathbf{S}$, then the cost-optimized equilibrium becomes profitable again. This means that the magnitude of the costs modeled by $\mathbf{S}$ should bear a meaningful relation to the magnitude of the other goals in the MOG, when switching cost is an issue. ◇

The fine-tuning of the parameter α is thus not a matter of declaring priorities between the two goals (utility vs. switching cost), but must rather be made by solving (12.6) for several trial values of α, and looking for an equilibrium that achieves a reasonable utility. The savings on the switching costs $\mathbf{S} = (s_{ij})$ should not be all outweighed by the loss of utility that the adapted behavior, which deviates from the original equilibrium without $\mathbf{S}$, gives. For a practical choice, $\alpha := 1/\max_{ij} \{|s_{ij}|\}$ may be an *initial guess* for an (in any case necessary) fine-tuning of the model.

12.6 Making the Most Out of Uncertainty

Perhaps unexpected at first glance, but uncertainty does not only cause difficulties, it can in fact be helpful in practical situations. The following sections consider two ways of turning uncertainty in the modeling into an advantage. The first is a smoothing effect that comes with payoff models constructed from empirical data. One example where this is helpful are disappointment rates from Chap. 7. The probability to "be disappointed" during a game round is a goal function that classical games, i.e., those over $\mathbb{R}$, cannot naturally handle, but distribution-valued games can do this without much effort. We use Sect. 12.6.1 to show how.

The second positive use of uncertainty lies in the intrinsic account for zero-day exploits, which can be seen as a worst case instance of uncertainty in our modeling. *Unexpected attacks*, equivalently inaccurate assumptions on the attacker's action spaces, can be accounted for in the payoff construction by adding uncertainty to the data. We will discuss this in Sect. 12.6.2.

12.6.1 Including Disappointment in Finite Games

Let us now complete the picture of how to handle disappointment rates in general games, which was left open in Sect. 7.4. In the following, we describe a generic procedure to include disappointments in any game whose payoff is modeled with distributions. Assume that we used R [22] and the HyRiM package [23] for the underlying data science, and that we constructed a distribution-valued game that is available as an object G in our R workspace. Then, adding an account for disappointment to this game is a matter of taking the following steps:

1. For every distribution F_{ij} in the payoff matrix $\mathbf{A} = (F_{ij})_{i,j=1}^{n,m}$ of the game, replace this by its expectation, thus recovering a game over real values. This is a matter of invoking the mean function on all entries in the payoff list of the object G, as done in line 3 of Listing 12.3.
2. Likewise, compute the disappointment rate for all distributions F_{ij} in of the game, which is the probability mass assigned by F_{ij} in the region $[E(X), \infty)$. The function disappointmentRate does this upon invocation on a loss-Distribution object; see line 6 of Listing 12.3.
3. Add the so-constructed list of disappointment rates as a second goal to the game (line 9), and solve it using the same techniques as in Sect. 12.4.3.

Remark 12.4 Note that this technique naturally casts a distribution-valued game back into a game with real values, which is an inevitable effect if we want to escape the discontinuity issue of the disappointment rate function. Indeed, the intuitive reason why this works is that the modeling of payoff distributions rather than numbers adds more information to a game than in the other case as if the game would have been set up with real-valued utilities in first place. It is this additional

Listing 12.3 Running an Equilibrium analysis with disappointment rates

```
1   payoffs <- G[,,1]$losses   # get the losses for the first goal
2   # boil down all the loss distributions to their means
3   expectations <- unlist(lapply(payoffs, mean))
4   # compute the disappointment probability for each distribution
5   # using the HyRiM::disappointmentRate function
6   disapp_rate <- unlist(lapply(payoffs, disappointmentRate))
7   # put the two new goals together in a fresh game
8   gameWithDisappointment <- c(expectations, disapp_rate)
9   G2 <- mosg(n=G$nDefenses,
10             m=G$nAttacks,
11             losses=gameWithDisappointment,
12             goals=2, # only the original goal + disappointment
13             goalDescriptions=c("mean_damage", "disappointment_
                  rate"),
14             defensesDescr = G$defensesDescriptions,
15             attacksDescr = G$attacksDescriptions)
16  # our choice of alpha_1 = 0.1, alpha_2 = 0.9
17  # in the scalarization, is *arbitrary* and only
18  # for the example to roughly equalize the goal's
19  # magnitudes (for more meaningful results)
20  eq <- mgss(G2,weights=c(0.1,0.9))
21  print(eq)
```

information collected about the tails of the distribution that Listing 12.3 compiles into the disappointment rate as a second goal. This information would not have been available after a direct definition of the real-valued utilities only.

Example 12.5 (Cut-the-Rope with Disappointments) Let us continue our analysis of the previous instance of Cut-The-Rope, redefined with disappointments in the way as Listing 12.3 instructs. We obtain the following equilibrium as output of the print(eq) statement in the last line:

```
equilibrium for multiobjective security game (MOSG)

optimal defense strategy:
  prob.
1 0.511
2 0.000
3 0.107
4 0.000
5 0.000
6 0.382
7 0.000
8 0.000
9 0.000
```

```
worst case attack strategies per goal:
  mean damage disappointment rate
1    0.0000000           0.583416583
2    0.0000000           0.000000000
3    0.3526474           0.000000000
4    0.0000000           0.002997003
5    0.0000000           0.000000000
6    0.0000000           0.000000000
7    0.6473526           0.413586414
8    0.0000000           0.000000000
```

Since this is a MGSS, the defender knows that by playing the optimal defense, it gets the assured mean losses and disappointments computable by $v_1 = \mathbf{d}^* \cdot \mathbf{A}_1 \cdot \mathbf{a}_1^*$, and $v_2 = \mathbf{d}^* \cdot \mathbf{A}_2 \cdot \mathbf{a}_2^*$, in which $\mathbf{d}^*$ is the optimal defense computed above, and $\mathbf{a}_1, \mathbf{a}_2$ are the worst case attack strategies per goal. The matrices $\mathbf{A}_1, \mathbf{A}_2$ are the variables expectations and disapp_rate in Listing 12.3. We get these matrices easily by issuing the commands

```
A1 <- matrix(expectations,
               nrow=G$nDefenses, ncol=8, byrow=TRUE)
A2 <- matrix(disapp_rate,
               nrow=G$nDefenses, ncol=8, byrow=TRUE)
```

on the R prompt. Given these, we can directly evaluate the above bilinear functions (and coerce the resulting 1×1 matrix to a number by applying as.numeric) to obtain the assured performance in terms of loss and disappointment for the game G2, which are

```
> as.numeric(t(eq$optimalDefense)
        %*% A1 %*% eq$optimalAttacks[,1])
[1] 3.914974
> as.numeric(t(eq$optimalDefense)
        %*% A2 %*% eq$optimalAttacks[,2])
[1] 0.4290704
```

Now, to see the effect of the disappointment to be included, let us construct another game with the same payoffs, but without the disappointment rate. We call this game G3 and compute its equilibrium as follows:

```
G3 <- mosg(n=G$nDefenses,
             m=G$nAttacks,
             losses=expectations,
             goals=1, # only the goal (no disappointment)
             goalDescriptions=c("mean damage"),
             defensesDescr = G$defensesDescriptions,
             attacksDescr = G$attacksDescriptions)
eq_without_disap <- mgss(G3)
```

Note that the exclusion of disappointment here enables the attacker to find another strategy by which it can cause perhaps less damage *on average*, but still gain a larger chance of causing more harm than the defender expects. Letting the adversary still play by the equilibrium from the previous game G2 with equilibrium eq, we can confirm this effect:

```
> as.numeric(
       t(eq_without_disap$optimalDefense)
       %*% A1 %*% eq$optimalAttacks[,1])
[1] 3.892734
> as.numeric(
       t(eq_without_disap$optimalDefense)
       %*% A2 %*% eq$optimalAttacks[,2])
[1] 0.4314893
```

As we see, the average damage disregarding disappointments is 3.892734, which is less than 3.914974 under the previous calculation. However, this seeming improvement comes with the caveat of now having a higher chance to suffer a damage >3.892734, namely with likelihood $\approx 43.15\%$, as opposed to the "safer" defense that has only a chance of $\approx 42.91\%$ to exceed the expectation. $\diamond$

The tradeoff that became visible in the end of Example 12.5 is intrinsic to the scalarization approach when solving multi-objective games, and the magnitudes of how much is gained when we choose between either way of analyzing the game also depends on our choice of α (in this example analysis, we chose $\alpha = 0.1$, but other values would be equally admissible and bring up different results). The bottom line is that including disappointment rates *is possible* and *easy* – Listing 12.3 is in fact not specific for any game, and generically applies to any other game equally well – yet requires manual fine-tuning and some trials to obtain a practically viable (optimal) defense.

Knowing how much loss to expect is important to know for what amount we need to buy insurance. In the same way, knowing how likely a disappointment is means preparing for the case when even the insurance coverage leaves a residue damage to be covered on our own. We can deal with this by setting the insured value to a proportion of the expectation, e.g., we may buy insurance for 120% of how much we expect, to cover a 20% excess on the losses that we actually bear in reality (or any other larger amount).

12.6.2   Risks for Zero-Day Exploits

Zero-day exploits are characterized by them being *unknown* to the security officer by the time when they become effective in a system. From the defender's perspective, counting the date of the exploit becoming known as day zero, the exploit happens on the same day, hence the name "zero-day exploit". Practically, these are relevant for the possibly large damage that they can cause, simply because they are unexpected.

12.6.2.1 A Posteriori Mitigation of Zero-Day Exploits

In a game theoretic model, a zero-day exploit would manifest as the adversary playing a strategy $z \notin AS_2$, when AS_2 is the action set that the defender anticipated for its opponent. In other words, the adversary plays some *unexpected* move, that may well defeat all defense strategies that player 1 could adopt. Let us return to the communication games of Sect. 11.2 to illustrate this.

Example 12.6 (Zero-day attack on a multipath communication game) Let us reconsider the multipath communication game from Sect. 11.2.1. Suppose that the transmission uses a single path from $AS_1 = \{\pi_1, \pi_2, \pi_3\}$ and that the attacker wins if and only if it intercepts exactly that path by eavesdropping on any node from the adversary structure $\{\{v_1\}, \{v_2\}, \{v_3\}\} = \mathscr{A}$. The resulting payoff matrix is actually a plain diagonal game given by

$\mathbf{A}$	v_1	v_2	v_3
π_1	0	1	1
π_2	1	0	1
π_3	1	1	0

A dominating strategy for the adversary may arise if we have mistakenly assumed the threshold as too low or had the wrong adversary structure. In either case, a "zero-day exploit" column $\mathbf{ZD}$ may appears that outperforms all others. This new column corresponds to a strategy that entirely breaks the protocol and makes the whole transmission insecure by Theorem 11.2:

$\mathbf{A}|\mathbf{ZD}$	v_1	v_2	v_3	ZD
π_1	0	1	1	0
π_2	1	0	1	0
π_3	1	1	0	0

has the saddle-point value $\quad v = \mathrm{val}(\mathbf{A}|\mathbf{ZD}) = 0,$

and hence the vulnerability $1 - v = \rho(\mathbf{A}|\mathbf{ZD}) = 1$, so there is no longer a way of secure communication, as Theorem 11.2 tells.   ◇

Upon observing such a dominated strategy, it is useful to know that it *can suffice* to fix the situation only in one circumstance, instead of having to mitigate the attack in all cases. Namely, there is the following fact:

Theorem 12.1 ([27]) *Take a real number $M > 0$. Let $\mathbf{A} \in [0, M]^{n \times m}$ be a game-matrix with value $val(\mathbf{A}) > 0$. Extend $\mathbf{A}$ by one column $\mathbf{c}$ for the adversary, such that $\mathbf{c}$ does not dominate all columns in $\mathbf{A}$, and write $(\mathbf{A}|\mathbf{c})$ for the so-extended matrix. Then $val((\mathbf{A}|\mathbf{c})) > 0$.*

We illustrate the application of Theorem 12.1 by continuing Example 12.6:

Example 12.7 (Mitigating an adversarial winning strategy) Suppose that the defender manages to patch or otherwise avoid the zero-day exploit on the first path only, changing it into the updated strategy π'_1. Then, the new game matrix, including the former zero-day exploit strategy ZD, has become

$$
\begin{array}{c|cccc}
\mathbf{A|ZD} & v_1 & v_2 & v_3 & ZD \\
\hline
\pi'_1 & 0 & 1 & 1 & \mathbf{1} \\
\pi_2 & 1 & 0 & 1 & 0 \\
\pi_3 & 1 & 1 & 0 & 0
\end{array}
\qquad \text{and now has the saddle-point value } v = \mathrm{val}(\mathbf{A|ZD}) = \frac{1}{2},
$$

so that the possibility to do ASMT is fully restored, based on Theorem 11.2. ◇

The lesson to take away from Examples 12.6 and 12.7 is that even upon the discovery of an unexpectedly powerful attack, we *may not need to mitigate the risk in all cases*, but *only need to avoid falling victim to the attack in all cases*. The latter can be a much weaker goal to achieve in practice.

12.6.2.2 Heuristic Modeling of Zero-Day Risks

Consider a situation where we have a hypothesis on the attacker's possible actions, coming as some, in our cases at least compact, action set $AS_2 = \{a_1, a_2, \ldots\}$. Assume that we have assigned losses to each of these actions, in light of all what the defender can do within its action set AS_1. That is, we either have anticipated loss values $u(d_i, a_j)$ for all defense actions $d_i \in AS_1$, or we have assigned losses to the event of action $a_i \in AS_2$ occurring, if our adversary is nature.

Now, suppose that losses have *not* been set to crisp numbers, but rather have been compiled from empirical data, e.g., expert polls (for example, by using a textual questionnaire or a graphical specification like in Fig. 12.1), big data, experience, or other sources. The point is that the valuation of each action $a \in AS_2$ as a loss is *uncertain*, which can mean various things:

- the assessment of the action can be inconsistent and non-consensual between data sources. That is, given action $a \in AS_2$, several experts may have different and diverging opinions about what would happen under event a.
- the assessment comes with some measure of accuracy, e.g., a confidence interval, or another subjective rating of "how sure" the expert is about the loss to expect under event $a \in AS_2$.

As this method of modeling is a kind of data science, we will refer to R in the discussion to follow, and will illustrate the respective steps by using the HyRiM package. This environment conveniently equips us with everything needed in the following.

When constructing payoff distributions from empirical data, a non-parametric loss distribution is constructible by a kernel density estimate, which takes a set of observations $x_1, \dots, x_N$ and constructs an empirical distribution function as the sum

$$f(x) := \sum_{i=1}^{N} \frac{1}{h \cdot N} k \left(\frac{x - x_i}{h} \right), \qquad (12.7)$$

using a *kernel function* k, that needs to satisfy $k(x) \geq 0$ for all x and $\int_{\mathbb{R}} k(t)dt = 1$. The parameter h is the *bandwidth* parameter. For our purposes, it can be interpreted as a measure of *uncertainty* in the data. In R, the construction of f from observations is directly doable by a call to the `density` function from the base installation of R.

Remark 12.5 (Practical handling of the bandwidth parameter h) To assure convergence of the kernel density estimate to the true underlying unknown distribution, the value h needs to depend on n in an explicitly constrained way, and will need to go to zero as n tends to infinity. Choosing h properly is the "art" of constructing a good kernel density estimate, but practically, many systematic and heuristic methods are available to data scientists here. We shall not worry too much about choosing this value, since the way we will practically use (12.7) will internally invoke these heuristics for us. In R, the easiest such rule is accessible via the function `nrd0`, but more sophisticated techniques are available in many packages, such as [20, 35].

A popular choice in practice is taking k as the density of the standard Gaussian distribution (this choice has the additional appeal of putting the resulting kernel density f into $C^{\infty}(\mathbb{R})$, so that Theorem 3.1 would apply to $\preceq$-compare the kernel densities. The `HyRiM` package [23] makes use of this). In the discrete case, a similar smoothing is applicable with a proper discrete distribution (e.g., a discretized version of the Gaussian density).

Since we want to make decisions about security, the next step is to stochastically order the kernel densities to set up and solve decision or game problems.

A neat consequence of this non-parametric modeling of payoff distributions is the fact that their support extends over whole $\mathbb{R}$ (or whole $\mathbb{N}$ in the discrete case), but for the computation of an equilibrium, needs to be truncated at the maximum relevant loss $b > 1$ or to some largest loss category c_1 to $\preceq$-order the distributions, equivalently *prospects*. Table 6.2a on impact categories would, in the description of the largest impact category, provide an example choice for b for a company using this impact scale. Generally, the value b would be the largest loss above which it does not make sense to distinguish any more. As an extreme example, a small start-up company would not bother any more if the liabilities are 1 million or 10 million, since the company would be equally bankrupt in both cases.

If we use Gaussian kernels (continuous or discretized versions thereof), the loss density will assign positive likelihood to losses in the range $Z = (b, \infty)$ of observations $x_1, \dots, x_N$. The value b is what a user can pick freely when computing a multi-goal security strategy, i.e., a zero-sum equilibrium by calling

```
> mgss(G, cutOff = b)
```

The region Z is the range of losses that were never observed. This region is, however, exactly where zero-day exploit events live, if we interpret them as unexpected yet very powerful attacks that cause losses beyond all that was anticipated; formally, losses that are larger than our chosen risk acceptance threshold b. By our construction, however, loss events in Z have a *positive* likelihood to occur in every scenario in $AS_1 \times AS_2$ under the non-parametric loss density model (12.7).

Figure 12.10 illustrates the effect: The lot of mass assigned to the region Z is the heuristic probability estimate for the risk of zero day exploits accumulating in the tails of the loss distribution; specifically, Fig. 12.10a depicts a loss distribution with mass put on the "zero-day area" Z shaded in gray. It has two appealing properties:

1. The heuristically presumed risk for zero-day exploits becomes larger the more data we have that speaks about large losses to be expected. For example, the more experts provide assessments in the large loss range, the larger will the probability for zero-day exploits be assumed in the game; see Fig. 12.10b.
2. The mass put on the region Z also grows if we consider the data to be more uncertain; see Fig. 12.10c.

Practically, one does typically not have to go through the mathematical construction of the nonparametric density given by (12.7), since this formula is a standard tool in statistics. In R, we can plainly use the `lossDistribution` function from the HyRiM package, to construct a loss distribution of the proper form (12.7) in both, a continuous and discrete version, by supplying the data, and the *bandwidth parameter* bw to express the uncertainty h. The call to construct a loss distribution is then a matter of one command on the R prompt:

```
> ld <- lossDistribution(
            dat = observations,
            smoothing = "always",
            bw = h)
```

to construct a loss distribution from the set of observations, and using the bandwidth value h to express the uncertainty. Omitting this parameter makes `lossDistribution` internally invoke a heuristic choice rule to pick a proper value for h. Coming back to the control of zero-day risks, supplying larger values for h and/or adding more observations near the upper end of the loss scale will increase the presumed zero-day risk. The specification `smoothing = "always"` advises the function to do make use of the uncertainty parameter h explicitly, for otherwise, `lossDistribution` leaves the data by default unmodified and "as is".

If the loss distribution for a given scenario is available in our R workspace as the object ld, constructed from a previous call to `lossDistribution` like the above, we can get the probability presumed for a zero-day exploit in this case, equivalently the mass that ld puts on Z, as $\Pr(X > b) = 1 - F_X(b)$ for $X \sim$ ld, by calling

```
> 1 - cdf(ld, x = b)
```

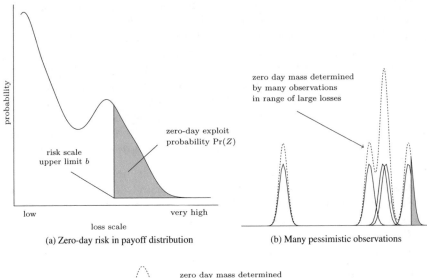

(a) Zero-day risk in payoff distribution   (b) Many pessimistic observations

(c) Many uncertain observations

Fig. 12.10 Implicit heuristic inclusion of zero-day risks in the payoff distribution

where cdf returns the cumulative distribution function's value at x for the supplied loss distribution ld.

Since the construction of loss distributions essentially always works using data and with calls to lossDistribution, we actually have no additional burden to explicitly take zero-day risks into account here, except for a custom choice of the bandwidth and the collection of observations. Besides that, however, the modeling automatically assigns a heuristic lot of risk to zero-day exploits, the exact lot of which is based on the uncertainty in the data. We get this feature "for free".

In connection with the design of our games to yield optimal behavior that minimizes the chances for high losses, the likelihood for zero-day exploits is then automatically kept low as well, since the $\preceq$-optimal decisions are those that have all their masses shifted towards lowest damages as much as possible. Therefore, practically, we can adjust our modeling account for zero-day exploits by adding more pessimistic observations to the data sets from which we construct the loss distributions. But from that point onwards, the construction automatically considers extreme events in the way as we want it without further explicit considerations.

References

1. GNU Octave (2019). https://www.gnu.org/software/octave/
2. Berger U (2005) Fictitious play in $2 \times n$ games. J Econ Theory 120(2):139–154. https://doi.org/10.1016/j.jet.2004.02.003. https://linkinghub.elsevier.com/retrieve/pii/S0022053104000626
3. Blavatskyy PR (2010) Reverse common ratio effect. J Risk Uncertain 40(3):219–241. https://doi.org/10.1007/s11166-010-9093-x
4. Brandt F, Fischer F, Harrenstein P (2010) On the rate of convergence of fictitious play. In: Kontogiannis S, Koutsoupias E, Spirakis P (eds) Algorithmic game theory. Lecture notes in computer science, vol 6386. Springer, Berlin/Heidelberg, pp 102–113. https://doi.org/10.1007/978-3-642-16170-4_10
5. Brenton K, Signorino CS (2014) Estimating extensive form games in R. J Stat Softw 56(8):1–27. http://www.jstatsoft.org/v56/i08/
6. Cano-Berlanga S, Gimenez-Gomez JM, Vilella C (2015) Enjoying cooperative games: the R package gametheory. Working paper no 06; CREIP
7. Carvalho A, Larson KA (2013) Consensual linear opinion pool. In: IJCAI '13. AAAI Press, pp 2518–2524
8. Dietrich F, List C (2017) Probabilistic opinion pooling generalized. Part one: general agendas. Soc Choice Welf 48(4):747–786. https://doi.org/10.1007/s00355-017-1034-z
9. Field AP, Hole G (2003) How to design and report experiments. Sage Publications Ltd, London/Thousand Oaks
10. Goerlandt F, Reniers G (2016) On the assessment of uncertainty in risk diagrams. Saf Sci 84:67–77. https://doi.org/10.1016/j.ssci.2015.12.001. https://linkinghub.elsevier.com/retrieve/pii/S0925753515003215
11. Herath HSB, Herath TC (2011) Copula-based actuarial model for pricing cyber-insurance policies. Insur Markets Companies Anal Actuarial Comput 2(1):14
12. Higgins JJ (2004) An introduction to modern nonparametric statistics. Brooks/Cole, Pacific Grove
13. Hogg RV, Klugman SA (1984) Loss distributions. Wiley series in probability and mathematical statistics applied probability and statistics. Wiley, New York. https://doi.org/10.1002/9780470316634
14. Inc., W.R (2019) Mathematica, Version 12.0. https://www.wolfram.com/mathematica. Champaign
15. Klugman SA, Panjer HH, Willmot GE (1998) Loss models. A Wiley-Interscience publication. Wiley, New York. https://onlinelibrary.wiley.com/doi/book/10.1002/9780470391341
16. Mainik G, Rüschendorf L (2013) Ordering of multivariate risk models with respect to extreme portfolio losses. In: Rüschendorf L (ed) Mathematical risk analysis. Dependence, risk bounds, optimal allocations and portfolios. Springer series in operations research and financial engineering, pp 353–383. Springer, Berlin/Heidelberg. https://doi.org/10.1007/978-3-642-33590-7_14
17. MathWorks (2019) MATLAB – the language of technical Computing. https://www.mathworks.com/products/matlab.html
18. McKelvey RD, McLennan AM, Turocy TL (2007) Gambit: software tools for game theory, version 0.2007.12.04. http://www.gambit-project.org
19. Monderer D (1996) Fictitious play property for games with identical interests. J Econ Theory 68:258–265. http://linkinghub.elsevier.com/retrieve/pii/S0022053196900149

20. Moss J, Tveten M (2019) kdensity: kernel density estimation with parametric starts and asymmetric kernels. https://CRAN.R-project.org/package=kdensity. Pack. ver. 1.0.1
21. Nachbar JH (1990) "Evolutionary" selection dynamics in games: convergence and limit properties. Int J Game Theory 19(1):59–89. http://link.springer.com/10.1007/BF01753708
22. R Core Team (2018) R: a language and environment for statistical computing. R foundation for statistical computing, Vienna. https://www.R-project.org/
23. Rass S, König S, Alshawish A (2020) HyRiM: multicriteria risk management using zero-sum games with vector-valued payoffs that are probability distributions. https://cran.r-project.org/package=HyRiM. Version 1.0.5
24. Rass S, Konig S, Schauer S (2017) Defending against advanced persistent threats using game-theory. PLoS ONE 12(1):e0168675. https://doi.org/10.1371/journal.pone.0168675
25. Rass S, König S, Schauer S (2017) On the cost of game playing: how to control the expenses in mixed strategies. In: Decision and game theory for security. Springer, Cham, pp 494–505
26. Rass S, Rainer B (2014) Numerical computation of multi-goal security strategies. In: Poovendran R, Saad W (eds) Decision and game theory for security. LNCS 8840. Springer, pp 118–133. https://doi.org/10.1007/978-3-319-12601-2_7
27. Rass S, Schartner P (2011) Information-leakage in hybrid randomized protocols. In: Lopez J, Samarati P (eds) Proceedings of the international conference on security and cryptography (SECRYPT). SciTePress – Science and Technology Publications, pp 134–143
28. Rass S, Schauer S (2019) Refining stochastic models of critical infrastructures by observation. In: Proceedings of the 56th ESReDA seminar, European atomic energy community, No JRC118427 in JRC Publications, pp 212–223. Publications Office of the European Union. https://ec.europa.eu/jrc/en/publication/critical-services-continuity-resilience-and-security-proceedings-56th-esreda-seminar
29. Rass S, Schorn A, Skopik F (2019) Trust and distrust: on sense and nonsense in big data. In: Kosta E, Pierson J, Slamanig D, Fischer-Hübner S, Krenn S (eds) Privacy and identity management. Fairness, accountability, and transparency in the age of big data, vol 547. Springer International Publishing, Cham, pp 81–94. http://link.springer.com/10.1007/978-3-030-16744-8_6
30. Robert CP (2001) The Bayesian choice. Springer, New York
31. Robinson J (1951) An iterative method for solving a game. Ann Math 54:296–301
32. Sela A (1999) Fictitious play in 'one-against-all' multi-player games. Econ Theory 14:635–651. https://doi.org/10.1007/s001990050345
33. Wachter J, Grafenauer T, Rass S (2017) Visual risk specification and aggregation. In: SECURWARE 2017: the eleventh international conference on emerging security information, systems and technologies. IARIA, pp 93–98
34. Wachter J, Rass S, König S (2018) Security from the adversary's inertia–controlling convergence speed when playing mixed strategy equilibria. Games 9(3):59. https://doi.org/10.3390/g9030059
35. Wand M (2019) KernSmooth: functions for Kernel smoothing supporting Wand & Jones (1995). https://CRAN.R-project.org/package=KernSmooth. Pack. ver. 2.23-16
36. Washburn A (2001) A new kind of fictitious play. Nav Res Logist 48(4):270–280. http://doi.wiley.com/10.1002/nav.7

Acronyms

APT	Advanced Persistent Threat
ARA	Adversarial Risk Assessment
ASMT	Arbitrarily Secure Message Transmission
BSI	Federal Office for Information Security
CI	Critical Infrastructure
CPS	Cyber-Physical System
CPT	Chinese Postman Tour
CRAN	Comprehensive R Archive Network
CVSS	Common Vulnerability Scoring System
DDoS	Distributed Denial of Service
DH	Diffie-Hellman
DMS	Distribution Management System
DoS	Denial of Service
ENISA	European Anion Agency for Cybersecurity
FP	Fictitious Play
FTP	File Transfer Protocol
GTE	GAME THEORY EXPLORER
ICS	Industrial Control System
ICT	Information and Communication Technology
IDS	Intrusion Detection System
IIM	Input-Output Inoperability Model
IND-CPA	Indistinguishability Against Chosen-Plaintext Attacks
ITM	Interactive Turing Machine
IoT	Internet of Things
LAN	Local Area Network
MAC	Message Authentication Code
MGSS	Multi-Goal Security Strategy
MOG	Multi-Objective Game
NIST	National Institute of Standards and Technology

© Springer Nature Switzerland AG 2020
S. Rass et al., *Cyber-Security in Critical Infrastructures*, Advanced Sciences
and Technologies for Security Applications,
https://doi.org/10.1007/978-3-030-46908-5

NVD	National Vulnerability Database
PBNE	Perfect Bayesian Nash Equilibrium
PDF	Portable Document File
PSMT	Perfectly Secure Message Transmission
RSH	Remote Shell
RTU	Remote Terminal Unit
SCADA	Supervisory Control and Data Acquisition
SMB	Server Message Block
SPE	Subgame Perfect Equilibrium
SSH	Secure Shell
TVA	Topological Vulnerability Analysis
VPN	Virtual Private Network

Glossary

Metric and Metric Space Given a vector space V, a metric is a function $d : V \times V \to \mathbb{R}$ that satisfies three properties:

- positive definiteness: $d(x, y) \geq 0$ for all $x, y \in V$, and $d(x, y) = 0 \iff x = y$.
- symmetry: $d(x, y) = d(y, x)$ for all $x, y \in V$
- triangle inequality: $d(x, y) \leq d(x, z) + d(z, y)$ holds for all $x, y, z \in V$

A *metric space* is a vector space V, together with a metric d as just defined. We stress that the term "metric" introduced on page 46 is to be strictly distinguished from the mathematical concept described above.

Open, Closed, Bounded set In a metric space (V, d), a set $X \subseteq V$ is *open* if for every point $x \in X$, we can find some $\varepsilon > 0$ such that the ball $B_\varepsilon(x) = \{y \in V : d(x, y) < \varepsilon\}$ of radius ε and center at x is entirely enclosed in X, i.e., $B_\varepsilon(x) \subseteq X$. The set of all such inner points in a set X is denoted as X^o. The value of ε may herein depend on x. An example of an open set is the open interval $\{x : 0 < x < 1\} = (0, 1)$, which contains all numbers between 0 and 1, excluding these two. A set is called *closed*, if its complement set is open (note that this admits that there are sets that are neither open nor closed, such as the half-open/half-closed set $\{x : 0 \leq x < 1\}$ in $\mathbb{R}$). The smallest closed set Y that covers a given set as $X \subseteq Y$ is called the *closure* of X and denoted as $Y = \overline{X}$. We call a set X *bounded*, if there is some $\varepsilon > 0$ such that $X \subseteq B_\varepsilon(0)$, i.e., X is itself enclosed in some (sufficiently large) ball centered at the origin.

Compact set A subset $X \subseteq \mathbb{R}^n$ is compact if and only if it is closed and bounded. A typical example in $\mathbb{R}$ is the closed interval $[a, b] = \{x : a \leq x \leq b\}$. In particular, every finite set is compact. The more general topological definition (such as Theorem 3.3 assumes) considers a given (possibly) infinite collection of sets C_i whose union covers X, i.e., $X \subseteq \bigcup_i C_i$, and calls a set X *compact*, if and

© Springer Nature Switzerland AG 2020
S. Rass et al., *Cyber-Security in Critical Infrastructures*, Advanced Sciences
and Technologies for Security Applications,
https://doi.org/10.1007/978-3-030-46908-5

only if every infinite cover of X contains a finite number of sets $C_{i_1}, \ldots, C_{i_k}$ that already cover $X \subseteq C_{i_1} \cup C_{i_2} \cup \ldots \cup C_{i_k}$.

Convexity and Concavity A set X is called *convex*, if the straight line connecting any two distinct points $p, q \in X$ is itself entirely contained in X. For sets in $\mathbb{R}^n$, the smallest convex set covering X is called the *complex hull*, and given by the simplex $\Delta(X)$. A function $f : \mathbb{R}^n \to \mathbb{R}$ with $n \geq 1$ is called *convex*, if the set $\{y : y \geq f(x), x \in \mathbb{R}^n\}$ of points above the function's graph is convex. A function is called *concave*, if $-f$ is convex. Note that this means that there are functions that are neither convex nor concave.

Normal or Gaussian Distribution With $\mu \in \mathbb{R}$ and $\sigma > 0$, a real-valued random variable $X \sim \mathcal{N}(\mu, \sigma)$ has a Gaussian distribution, if its density function is given by

$$f_X(x|\mu, \sigma) = \frac{1}{\sigma\sqrt{2\pi}} \exp\left[-\frac{1}{2}\left(\frac{x-\mu}{\sigma}\right)^2\right].$$

Exponential Distribution Given a parameter $\lambda > 0$, a real-valued random variable $X \sim \mathcal{E}x(\mu, \sigma)$ has an exponential distribution, if its density function is given by

$$f_X(x|\lambda) = \begin{cases} \lambda \cdot e^{-\lambda x}, & \text{for } x \geq 0 \\ 0, & \text{otherwise.} \end{cases}$$

This distribution is useful to model, among others, the time between two independent events (in our case, e.g., independent security incidents, random inspections, etc.). The parameter λ is called the *rate parameter*, and may be understood as the average waiting time between two events.

As an interesting and even characteristic (unique) feature, this distribution is *memoryless*, meaning that: if we know that it has been t time units since the last event, the chances to wait for another x time units is independent of t; formally meaning that $\Pr(X \geq x + t | X \geq t) = \Pr(X \geq x)$.

Poisson Distribution Given a parameter $\lambda > 0$, a random variable $X \sim \mathcal{P}o(\lambda)$ taking values in $\mathbb{N}$ has a Poissonian distribution, if its density function is given by

$$\Pr(X = k) = f_X(k|\lambda) = \frac{\lambda^k}{k!}e^{-\lambda}, \quad \text{for } k = 0, 1, 2, \ldots$$

This distribution describes the number of (stochastically independent) events to occur per time unit, taking λ as the *average* number of such events. It relates to the exponential distribution as follows: if λ events occur on average with a Poissonian distribution, then the pause times between two consecutive occurrences is $\mathcal{E}x(1/\lambda)$-distributed.

Geometric Distribution Given a parameter $0 < p < 1$, a random variable X has a geometric distribution, if its density function is given as

$$\Pr(X = k) = f_X(k|p) = p(1 - p)^{k-1} \quad \text{for } k = 1, 2, 3, \ldots.$$

It describes the number k of trials until the first occurrence of the respective event that p relates to (in an alternative definition, one can also count the number of failures until the first occurrence of the event, in which case we would include $k = 0$ and have the density $p(1 - p)^k$). Like the exponential distribution in the continuous case, this is the only discrete distribution that is memoryless, meaning that even if we know that we have taken t trials until now without the event having occurred, the chances for another x trials until the event happens is independent of the past number t.

Beta Distribution Given two parameters $a, b > 0$, a random variable $X \sim \mathcal{B}e(a, b)$ distribution, if its density function is given by

$$f_X(x|a, b) = \begin{cases} \frac{x^{a-1}(1-x)^{b-1}}{B(a,b)}, & \text{if } 0 \le x \le 1 \\ 0, & \text{otherwise.} \end{cases}$$

with $B(a, b) = \Gamma(a) \cdot \Gamma(b)/\Gamma(a + b)$ and $\Gamma(x) = \int_0^\infty t^{x-1}e^{-t}dt$ being Euler's Gamma-function. This extends the factorial to non-integer values, but for an integer n, we have $\Gamma(n) = (n - 1)!$. The β-distribution is particularly useful as a prior distribution for a Bayesian update, if the likelihood function has a Poissonian, geometric, binomial or negative binomial distribution (see [17, Chp.3.3] for further details). Its primary use for our purpose is for Bayesian learning of probability parameters, such as described in Example 3.2, and Sect. 12.1.2.

Support Given a probability density function f, the *support* of the density, or random variable to which it refers, is the closure of the set $\overline{\{x : f(x) > 0\}}$, where f takes strictly positive values. Intuitively yet with care, this may be thought of the set of values that the random variable can possibly take realizations from.

Continuous function Given a function $f : X \to Y$ between two metric spaces, let for a set V the set $f^{-1}(V) = \{x : f(x) \in V\}$ be the collection of all points that map into V under f. The set $f^{-1}(V)$ is called the *pre-image* of V under f. The function f is called *continuous*, if the pre-image of every open set is again an open set.

In the special case of a function $f : \mathbb{R} \to \mathbb{R}$, one can intuitively think of f to have neither poles or jumps.

List of Symbols

General notation

Typeface	Denotes
Lower-case letters like x, y, v, ...	Scalars (e.g., real values)
Upper case letter like A, B, ...	Sets and random variables
Bold-printed lower-case letter like $\mathbf{x}$, $\mathbf{y}$, ...	Vectors
Bold-printed upper-case letter like $\mathbf{A}$, $\mathbf{B}$, ...	Matrices
Calligraphic letters like $\mathscr{F}$	Probability distributions

Symbol	Explanation	See also
	Special sets and symbols	
$K^{n \times m}$	Set of $(n \times m)$-matrices with entries from the set K	
$A \times B$	Cartesian product of A and B, i.e., set of all pairs (a, b) with $a \in A$, and $b \in B$	
$\prod_i A_i$	Cartesian product of multiple sets or simple product of numbers	
$\mathbf{a}_i$	i-th coordinate of vector $\mathbf{a}$	
$\mathbf{a}_{-i}$	Vector $\mathbf{a}$ excluding the i-th coordinate	
$\mathbb{F}$, $\mathbb{G}$	Finite field or finite group	
$\mathrm{diag}(\mathbf{a})$	Diagonal matrix with elements from $\mathbf{a}$ on the diagonal	
$\mathbf{A}\vert\mathbf{v}$	Block matrix; matrix $\mathbf{A}$ with the additional column $\mathbf{v}$.	
AS	Action space (essentially a set)	pg. 46

© Springer Nature Switzerland AG 2020
S. Rass et al., *Cyber-Security in Critical Infrastructures*, Advanced Sciences and Technologies for Security Applications,
https://doi.org/10.1007/978-3-030-46908-5

Symbol	Explanation	See also
2^V	Power set of V	
$\Delta(X)$	Unit simplex over the set X	pg. 46
$\lvert X \rvert$	Cardinality of a set or absolute value (for scalars like random variables)	
$\mathbb{R}_{\geq 0}$	Nonnegative reals	
$C^\infty(\Omega)$	Set of infinitely often differentiable functions over the set Ω	
$\mathrm{dom}(f)$	Domain of a function f	
$\min_X f$, $\max_X f$	Minimum and maximum values attained by the expression f, with variables varying over the set X	
$\mathrm{argmin}_X f$, $\mathrm{argmax}_X f$	Sets of points in X where a minimum or maximum is attained by the expression f	
$\mathbf{x}^*$	Optimal value (element from argmin or argmax)	
$vec(\mathbf{A})$	Vectorization of the matrix $\mathbf{A}$	pg. 265
$f\vert_{x=t}$	Substitution of t for x in the expression f	
$\lVert \cdot \rVert_p$	General p-norm in $\mathbb{R}^n$, defined as $\lVert (x_1, \ldots, x_n) \rVert_p = \left(\sum_{i=1}^n \lvert x_i \rvert^p \right)^{1/p}$, for $p > 0$	

Probability and statistics

Symbol	Explanation	See also
$X \sim F$ or $X \sim \mathbf{x}$	Random variable X having distribution F or categorical distribution given by the probability mass vector $\mathbf{x}$	
$f_X(x \vert \alpha)$	Conditional density of a random variable X on values (e.g., one or more parameter) α	pg. 286ff
$F_X(t \vert Y)$	Distribution of X conditional on Y, being $F_X(t \vert Y) = \Pr(X \leq t \vert Y)$	
$\mathbb{E}_F$	Expectation w.r.t. distribution F	
$\bar{\eta}$	Empirical mean value for the variable η, i.e., an average over sampled values (data)	
$\hat{p}$	Empirical estimate for the value p or an empirical distribution	pg. 186
$\mu(\theta)$, $\mu(\theta \vert a)$	Belief about adversary type θ, and conditional belief given the information a	pg. 89
$\mathcal{N}(\mu, \sigma)$	Gaussian distribution with mean μ and standard deviation σ	pg. 286
$\mathcal{U}(a, b)$	Uniform distribution over the interval $[a, b] \subset \mathbb{R}$	
$\mathcal{E}x(\lambda)$	Exponential distribution with rate parameter λ, e.g., governing inspection intervals	pg. 286

Symbol	Explanation	See also
$\mathscr{P}o(\lambda)$	Poisson distribution with rate parameter λ	pg. 286
$\mathscr{B}e(a, b)$	β-Distribution	pg. 287

Relations

Symbol	Explanation	See also
$\leq$	General ordering relation	pg. 44
$\leq_1, \geq_1$	Orders on $\mathbb{R}^n$, complement relations to the coordinate-wise $\leq, \geq$ relations (partial orders) on $\mathbb{R}^n$	pg. 47
$\preceq$	Stochastic tail ordering	pg. 52
$\leq_{st}$	Standard stochastic order	pg. 52
$\vartriangleleft$	Preference relation	pg. 44
$\leq_{\text{lex}}, \geq_{\text{lex}}$	Lexicographic orders on $\mathbb{R}^n$	pg. 48
$(R, \leq)$	Ordered set	

Games

Symbol	Explanation	See also
$\Gamma = (N, S, H)$	Game with player set N, strategy sets S and set H of utility functions	pg. 60
BR_i	Best response correspondence for the i-th player in a game	pg. 61
$\text{val}(\Gamma), \text{val}(\mathbf{A})$	Saddle-point value of a game Γ, or of the matrix if Γ is two-player finite zero-sum and representable by a payoff matrix $\mathbf{A}$	pg. 67
$Z(\mathbf{h}), Z_i(\mathbf{h})$	Set of moves of all players, or just the i-th player in an extensive form game, based on the current history $\mathbf{h}$ of moves	pg 71
$p(\mathbf{h})$	Set of players that can move given the current state of an extensive form game, described by the history $\mathbf{h}$	pg. 72
$\ell(\Gamma)$	Length of the extensive form game Γ (depth of the game tree)	pg. 84
$periodic(\lambda)$	Inspection strategy to take exactly λ inspections per time unit with fixed pause times in between	pgs. 192,201
$expon(\lambda)$	Inspection strategy to take a random number of inspections per time unit, with random pause time $\Delta T \sim \mathscr{E}x(\lambda)$ in between	pgs. 196,207

Graphs and networks

Symbol	Explanation	See also
$G = (V, E)$	Graph with node set V and edge set $E \subseteq V \times V$	

Symbol	Explanation	See also
$\mathrm{Pa}(v)$	Parent set of all nodes with edges incoming to v in a directed graph	
$\mathrm{Ch}(v)$	Children of the node v in a directed graph, i.e., the set of all nodes to which edges outgoing from v lead	
Cryptography related symbols		
$E_{pk}(m), D_{sk}(c)$	Encryption of a message m under a public key pk, and decryption of a ciphertext c under a secret key sk	pg. 224
negl	anonymous negligible function in one parameter	pg. 227
$\mathscr{A}^f$	algorithm $\mathscr{A}$ with oracle-access to some function f	pg. 224
$\leq_{asymp}, \geq_{asymp}$	Asymtotic orders	pg. 227
$\leq_{negl}, \geq_{negl}$	Less- or greater than up to a negligible (asymtotic) difference	pg. 227
$\approx_{negl}, >_{noticbl}$	Equal up to negligible error or greater than with noticeable difference	pg. 227

Index

© Springer Nature Switzerland AG 2020
S. Rass et al., *Cyber-Security in Critical Infrastructures*, Advanced Sciences
and Technologies for Security Applications,
https://doi.org/10.1007/978-3-030-46908-5

Printed in the United States
by Baker & Taylor Publisher Services